Calcium
Aluminate
Cements

ORGANIZING COMMITTEE

R.J. Mangabhai, (Chairman) *TH Technology Ltd*

R. Mangabhai graduated in Applied Chemistry from Salford University in 1978 and was a Research Assistant at the University from 1978–1983, carrying out research on cementitious materials. From 1985–1988 he was at King's College London carrying out research on cement grouts, permeability of concrete and he organized a one day seminar on calorimetry of cements in 1986. From 1988–1989 he was at Queen Mary and Westfield College London. He joined TH Technology Ltd, R & D Centre in 1989 where he is a Materials Technologist.

Dr S.A. Jefferis, *Queen Mary and Westfield College*

Dr Jefferis was at King's College London, in the Department of Civil Engineering from 1968 to 1989. On closure of the King's Department he transferred to Queen Mary and Westfield College where he is a Reader. He has been involved in materials research over 20 years. He has a particular interest in grouts for both structural and geotechnical applications. He has been much involved in the use of high alumina cement grouting in offshore structures.

Dr C.H. Fentiman, *Lafarge Special Cements Ltd*

Dr Fentiman graduated from the North East London Polytechnic in 1976 with a degree in Geology. From there he moved to Birkbeck College London where he undertook an MSc in crystallography prior to undertaking cement research leading to a PhD in 1980. He subsequently worked at the Lafarge Special Cement Co. Ltd, firstly as Development Scientist and since 1986 as Technical Manager, throughout this time he has maintained a keen interest in cement research.

Dr P.F.G. Banfill, *University of Liverpool*

Dr Banfill graduated in chemistry from Southampton University in 1972. After two years in the chemical industry he moved to Liverpool University and undertook research in construction materials for a PhD degree. He has been a lecturer there for 11 years in the School of Architecture and Building Engineering. His research interests are mainly in the area of fresh cement and concrete, where he is the author of many papers on rheology and chemistry of admixtures.

Dr J.H. Sharp, *University of Sheffield*

Dr Sharp obtained the degrees of BSc and PhD in Chemistry from the University of Nottingham. He then spent two years in the Materials Research Laboratory at the Pennsylvania State University, and one year as ICI Research Fellow at the University of Aberdeen. He became a lecturer in Ceramics at the University of Sheffield in 1966 and was promoted to senior lecturer in 1973. He served as Dean of the Faculty of Materials from 1985–87. He has carried out research on the chemistry of cements, including calcium aluminate cements, for the last 15 years.

Calcium Aluminate Cements

Proceedings of the International Symposium
held at Queen Mary and Westfield College,
University of London, 9–11 July 1990, and
dedicated to the late Dr H.G. Midgley

Edited by

R.J. MANGABHAI
TH Technology Ltd, R&D Centre

CRC Press
Taylor & Francis Group
Boca Raton London New York

CRC Press is an imprint of the
Taylor & Francis Group, an **informa** business
A TAYLOR & FRANCIS BOOK

CRC Press
Taylor & Francis Group
6000 Broken Sound Parkway NW, Suite 300
Boca Raton, FL 33487-2742

First issued in paperback 2019

© 1990 by Taylor & Francis Group, LLC
CRC Press is an imprint of Taylor & Francis Group, an Informa business

ISBN-13: 978-0-419-15200-2 (hbk)
ISBN-13: 978-0-367-86404-0 (pbk)

British Library Cataloguing in Publication Data
A CIP catalogue record for this book is
available from the British Library.

Visit the Taylor & Francis Web site at
http://www.taylorandfrancis.com

and the CRC Press Web site at
http://www.crcpress.com

Contents

Preface   ix
Dr H.G. Midgley's papers on calcium aluminate cements   xiii
List of abbreviations   xv

1 High alumina cement in construction – a future based on
   experience   1
   H.G. MIDGLEY

PART ONE   CLINKER   15

2 Effect of morphology on the hydration characteristics
   of high alumina cements   17
   I.N. CHAKRABORTY, S. NARAYANAN, D. VENKATESWARAN,
   S.K. BISWAS, A.K. CHATTERJEE

3 High alumina cements based on calcium aluminate clinker
   with different phase compositions and sintering degrees   27
   J. SAWKOW

PART TWO   HYDRATION   39

4 Microstructural development in pastes of a calcium
   aluminate cement   41
   K.L. SCRIVENER, H.F.W. TAYLOR

5 Investigations of the composition of phases formed in low
   cement castables during hydration and after thermal treatment   52
   W. GESSNER, S. MÖHMEL, J. KIESER, M. HÄWECKER

6 Effect of temperature on setting time of calcium aluminate
   cements   65
   A. CAPMAS, D. SORRENTINO, D. DAMIDOT

7 The use of nuclear magnetic resonance (NMR) in the study
   of high alumina cement hydration   81
   D.J. GREENSLADE, D.J. WILLIAMSON

8 On the change of microstructure during the hydration of
monocalcium aluminate at 20° and 50°C   96
W. GESSNER, R. TRETTIN, A. RETTEL, D. MÜLLER

PART THREE   ADMIXTURES   111

9 The influence of superplasticising admixtures on ciment
fondu mortars   113
S.M. GILL, P.F.G. BANFILL, B. EL-JAZAIRI

10 The effect of admixtures on the hydration of refractory
calcium aluminate cements   127
J.H. SHARP, S.M. BUSHNELL-WATSON, D.R. PAYNE, P.A. WARD

11 Properties of fresh mortars made with high alumina cement
and admixtures for marine environment   142
N.C. BAKER, P.F.G. BANFILL

PART FOUR   CALORIMETRY   153

12 Role of foreign cations in solution on the hydration kinetics
of high alumina cement   155
M. MURAT, EL. H. SADOK

13 Calorimetric studies on high alumina cement in the presence
of chloride, sulphate and seawater solutions   167
D.L. GRIFFITHS, A.N.F. AL-QASER, R.J. MANGABHAI

PART FIVE   DURABILITY   179

14 Manufacture and performance of aluminous cement:
a new perspective   181
C.M. GEORGE

15 Long-term performance of high alumina cement concrete
in sulphate-bearing environments   208
N.J. CRAMMOND

16 Behaviour of high alumina cement in chloride solutions   222
W. KURDOWSKI, L. TACZUK, B. TRYBALSKA

17 Acidic corrosion of high alumina cement   230
    J.P. BAYOUX, J.P. LETOURNEUX, S. MARCDARGENT,
    M. VERSCHAEVE

18 The effect of limestone fillers on sulphate resistance of high
    alumina cement composites   241
    W.G. PIASTA

PART SIX   BLENDED SYSTEMS   257

19 Hydration of calcium aluminates in presence of granulated
    blastfurnace slag   259
    A.J. MAJUMDAR, R.N. EDMONDS, B. SINGH

20 The effect of curing conditions on the hydration and strength
    development in fondu : slag   272
    C.H. FENTIMAN, S. RASHID, J.P. BAYOUX, A. BONIN, M. TESTUD

21 The microstructure of blastfurnace slag/high alumina
    cement pastes   282
    I.G. RICHARDSON, G.W. GROVES

22 Effects of microsilica on conversion of high alumina cement   294
    S. BENTSEN, A. SELTVEIT, B. SANDBERG

23 Study of the hydration properties of aluminous cement and
    calcium sulphate mixes   320
    J.P. BAYOUX, A. BONIN, S. MARCDARGENT, A. MATHIEU
    M. VERSCHAEVE

24 Ettringite-based cements   335
    S.A. BROOKS, J.H. SHARP

PART SEVEN   MISCELLANEOUS   351

25 Effects of autoclaving on the strength of hardened calcium
    aluminate cements   353
    A. SARANDILY, R. BAGGOTT

26 Effect of temperature rise on properties of high alumina
   cement grout   363
   S.A. JEFFERIS, R.J. MANGABHAI

27 Activation of hydraulic properties of the compound
   CaO. $2Al_2O_3$ 372
   T.W. SONG, S.H. CHOI, K.S. HAN

   Index   378

Preface

Calcium aluminate cements, also known as high alumina cements (HAC), are a range of cements in which calcium aluminates are the principal constituents.

The first patent was taken out by Bied in 1909, who was working for J. and A. Pavin de Lafarge at Le Teil, France and it was made by fusing together a bauxite or other aluminous and ferruginous material low in silica, with lime in proportions to give a 40 per cent Al_2O_3 clinker rich in calcium aluminate ($CaO.Al_2O_3$).

Calcium aluminate cement was first introduced into the UK in the 1920s and manufactured by Lafarge at West Thurrock on the North bank of the River Thames, 30 miles east of London. A market was rapidly established due to the cement's rapid early strength development and its refractory properties. The latter eventually led to the development of cements with higher purity and even higher alumina contents of up to 80 per cent. During the 1960s, it was discovered that the rapid hardening properties could shorten turnaround time in the manufacture of precast concrete units. However, because of the use of some poor mix designs and higher water/cement ratios, a few widely publicized failures ensued as a result of which, HAC was effectively banned in the 1970s.

However, other major applications have continued to develop. Calcium aluminate cements are used for many special applications in the construction, civil engineering and refractory industries, because of their ability to gain strength rapidly and to withstand aggressive environments and high temperatures. The latter provides some of the most sophisticated uses particularly with the development of low cement castables which, living up to their name, have cement contents well below 10 per cent. Another special application is in mine packing; this uses a two slurry system (one containing calcium aluminate cement and the other containing lime, anhydrite and bentonite) which sets rapidly and contains no less than 91 per cent by volume of water.

Research has been carried out on the properties of calcium aluminate cements in various countries. In 1982, a seminar on the subject was held at the University of Turin. Since then, developments have taken place particularly in mine packing, low cement castables, the use of admixtures and blended systems and more recently, the advent of calcium aluminate cement and blastfurnace slag mixes.

During this period, a considerable contribution to the understanding of the behaviour and properties of calcium aluminate cements was made by Dr H.G. Midgley whilst he was at the Building Research Establishment and subsequently at Hatfield Polytechnic, Thames Polytechnic and finally, but not least, at Ilminster Cement Research. Sadly, Dr Midgley died in late 1988 and this symposium is dedicated to his memory. The volume opens with a review of the development, properties and future potential for HAC which he prepared about three years ago but which has not been published before. It is an honour to include this paper as the first in these Proceedings. The rest of the volume is divided into seven topics:

1. Clinker.
2. Hydration (microstructure, temperature effects and use of NMR).
3. The effect of various admixtures.
4. The use of calorimetry to study the hydration process of calcium aluminate in the presence of various admixtures.
5. Durability of HAC mortars and concretes in various environments.
6. Blended systems particularly the development of calcium aluminate cement/blastfurnace slag cements.
7. Temperature and other miscellaneous effects.

I would like to thank: Lafarge Special Cements, UK for their financial support without which the seminar would not have been possible; Dr Charles Fentiman, for his invaluable help and suggestions in the planning and administration of the event; and the secretarial staff at Lafarge Special Cements, in particular, Wendy Earle, for retyping some of the manuscripts; the organizing committee for their invaluable help in reviewing the abstracts and papers and in planning the programme for the Symposium; the authors of all the papers who have provided an excellent review of the current state of the art on calcium aluminate cements; Nick Clarke of E. & F.N. Spon, for his help and suggestions and for producing the proceedings in such a short time; the Pickup Unit of Queen Mary and Westfield College for taking over some of the responsibilities for the final programme; and finally, TH Technology Ltd, R & D Centre (R.M. Edmeades and M.T. Hutchinson) for allowing me time to organize the seminar.

R.J. Mangabhai
Rickmansworth
March 1990

Publisher's Note

This book has been produced from camera ready copy provided by the individual contributors. This method of production has allowed us to supply finished copies to the delegates at the Symposium.

Henry Midgley, 1918–1988

DR H.G. MIDGLEY'S PAPERS ON CALCIUM ALUMINATE CEMENTS

Midgley, H.G. (1967) The mineralogy of set high alumina cement. **Trans. Brit. Ceram Soc.** 66, 161–187.

Midgley, H.G. (1968) The composition and possible structure of the quaternary phase in high alumina cement and its relation to other phases in the system $CaO-MgO-Al_2O_3$. **Trans. Brit. Ceram. Soc.** 67, 1–14.

Midgley, H.G. and Pettifer, K. (1972) Electron optical study of hydrated high alumina cement pastes. **Trans. Brit. Ceram. Soc.** 71, 55–59.

Wilburn, F.W., Keattch, C.J., Midgley, H.G. and Charsley, E.L. (1975) Recommendations for the testing of high alumina cement concrete samples by Thermoanalytical Techniques. Thermal Methods Group. Analytical Division of the Chemical Society, London. 1–11.

Midgley, H.G. and Midgley, A. (1975). The conversion of high alumina cement. **Mag. Concr. Res.** 27, 59–75.

Midgley, H.G. (1976) Quantitative determination of phases in high alumina cement clinkers by X-Ray diffraction. **Cem. Concr. Res.** 6, 217–223.

Midgley, H.G. (1976) The appraisal of HAC concrete. **Precast Concrete** 7, 23–26.

Midgley, H.G. and Ryder, J.F. (1977) The relationship between mineral composition and strength development of high alumina cement. **Cem. Concr. Res.** 7, 669–672.

Midgley, H.G. and Rao, P.B. (1978) Formation of stratlingite, $2CaO.Sio_2.Al_2O_3.8H_2O$. in relation to the hydration of high alumina cement. **Cem. Concr. Res.** 8, 169–172.

Midgley, H.G. (1978) The use of thermoanalytical techniques for the detection of chemical attack on high alumina cement concrete. **Thermochimica Acta** 27, 281–284.

Midgley, H.G. (1978) The use of thermal analysis methods in assessing the quality of high alumina cement concrete. **J. Thermal Analysis** 13, 515–524.

Midgley, H.G. (1980) The relationship between cement clinker composition and strength recovery of hydrating high alumina cement during conversion. **7th Int. Cong. Chem. Cement,** Paris, Septima, VIII, V68–70.

Midgley, H.G. (1980) The chemical resistance of high alumina cement concrete. **7th Int Cong. Chem. Cement,** Paris, Septima, VIII, V85–87.

Midgley, H.G. (1982) The relationship between hydrate mineral content and compressive strength of set high alumina cement. Eds Murat, M., Bachiorrini, A., Guilhot, B., Negro, A., Regourd, M. and Soustelle, M. Int. Seminar on Calcium Aluminates, held at University of Turin, Turin, Italy, 314–324.

Midgley, H.G. (1984) Measurement of high aluminous cement-calcium carbonate reactions using DTA. **Clay Minerals.** 19, 857–864.

Midgley, H.G. (1985) The use of HAC in Chemical and Civil Engineering in **Corrosion and Chemical Resistant Masonary Materials Handbook,** Ed. W.L. Sheppard, Noyes Publications, New Jersey, 340–363.

Blenkinsop, R.D., Currell, B.R., Midgley, H.G. and Parsonage, J.R. (1985) The carbonation of high alumina cement. Part 1. **Cem. Concr. Res.** 15, 276–284.

Blenkinsop, R.D., Currell, B.R., Midgley, H.G. and Parsonage, J.R. (1985) The carbonation of high alumina cement. Part 2. **Cem. Concr. Res.** 15, 385–390.

Currell, B.R., Grezeskowiak, R., Midgley, H.G. and Parsonage, J.R. (1987) The acceleration and retardation of set high alumina cement by additives. **Cem. Concr. Res.** 17, 420–432.

List of abbreviations

AEM	analytical (transmission) electron microscopy
ASTM	American Society for Testing Materials
AFm	monosulphate
AFt	ettringite
BS	British Standard
Bse	backscattered electron image
CA	calcium aluminate
CH	calcium hydroxide
C-S-H	calcium silicate hydrate gel
C_3A	tricalcium aluminate
C_4AF	tetracalcium aluminoferrite
C_2S	dicalcium silicate
CSF	condensed silica fume
C_3S	tricalcium silicate
DTG	differential thermogravimetry
DSC	differential scanning calorimetry
DC	degree of conversion
DCA	differential calorimetric analysis
DTA	differential thermal analysis
EVA	poly(ethylene-vinyl acetate)
EPMA	electron microprobe analysis
ggbs	ground granulated blastfurnace slag
HVEM	high voltage electron microscope
HAC	high alumina cement
HACC	high alumina cement concrete
IR	insoluble residue
lig	lignosulphate
LCC	low cement castables
LOI	loss of ignition
MLS	modified lignosulphonate
NMR	nuclear magnetic resonance
OPC	ordinary portland cement
pfa	pulverised-fuel ash
Q_{max}	maximum rate of heat evolution for 2nd peak
RH	relative humidity
RHE	rate of heat evolution
SMFC	sulphonated melamine-formaldehyde condensate
SNFC	sulphonated naphthalene-formaldehyde condensate

SSA	specific surface area
STEM	scanning transmission electron microscopy
SEI	secondary electron image
SEM	scanning electron microscope
TEM	transmission electron microscope
TGA	thermogravimetric analysis
t_{max}	time to reach maximum rate of heat evolution for 2nd peak
THH	total heat of hydration
ULCC	ultra low cement castables
UPV	ultrasonic pulse velocity
w/c	water/cement ratio
w/s	water/solid ratio
XRD	X-ray diffraction

1 HIGH ALUMINA CEMENT IN CONSTRUCTION – A FUTURE BASED ON EXPERIENCE

H.G. MIDGLEY
Ilminster Cement Research, Ilminster, UK

Abstract
High alumina cement (HAC) although originally invented as a chemically resistant cement found initial application based on its rapid strength development. Because of its high early strength it was in the 1960's used extensively in the manufacture of precast prestressed beams or joists. Its use enabled the manufacturer to attain mould turnround in less than 24 hours. In 1973-74 there were three failures of buildings which used such beams and there was immediate concern about the many buildings which used these beams or joists. During the period 1974 to 1976 a considerable effort was put into appraising many of these buildings. At about this period the Department of the Environment took the step of no longer recommending high alumina cement as a construction material.

However in 1981 it was reported that on the basis of the appraisals made to that date there were probable 50,000 buildings with structural high alumina cement concrete, of these, thirty-eight needed remedial action; of these in only one could the distress be attributed to the cement and it was said that the distress was a result of the incorrect use of the cement, too much water was used although at that time the importance of low water/cement ratio (w/c) was known.

From these findings it would seem to be now appropriate to use high alumina cement concrete as structural material, especially where its special properties of chemical resistance or high early strength are appropriate. If it is to be used as a structural material then strict control of its use in the making of concrete should be maintained and the design of the members should be on the basis of the "fully converted strength".

Keywords: High alumina cement concrete, Calcium aluminates, Codes of Practice, Conversion, Chemical resistance, Structural properties.

1 Introduction

High alumina cement (HAC) is the name for a range of cements in which calcium aluminates are the principal constituents. Smaller proportions of other materials occur depending upon the composition of the raw materials. Since the raw materials are usually calcium carbonate (limestone) and alumina (bauxite), both of which may be impure and contain iron, silicon and titanium oxides as minor contaminants together with traces of alkalis etc. the phases produced in cement production will include $C_{12}A_7$, CA, C_2AS, C_4AF-C_2F solid solution (fss), CT, C_2S, wustite (FeO), $C_{22}A_{17}F_2S_3$ (pleochroite or fibre) and glass.

Bied (1909) considered that cements based on calcium aluminates would have a greater resistance to sulphates than cements based on calcium silicates, i.e. Portland Cements. The first commercial HAC was produced by Bied (1909) who was working for J and A Pavin de Lafarge at Le Teil, France and it was made by fusing together a bauxite or other aluminous and ferruginous material, low in silica, with lime in proportions to give a mixture of CA and C_2S. This cement proved not only to have the required chemical resistance but showed an additional benefit in the production of high early strength, far in excess of that given by the Portland cements at that time or indeed today.

During the period 1910 to 1913 considerable trouble was experienced in producing a suitable cement on a commercial scale as it was found that while some batches were excellent others of apparent similar chemical composition were of no use. By 1913 a satisfactory method of commercial production had been achieved by the Lafarge company. During the period 1914 to 1918 the cement was supplied to the French government for gun emplacements, although designed originally for corrosion resistance it was first used because of its rapid hardening properties. In 1916 the HAC was used by P.L.M. railway for the reconstruction of the Brass Tunnel (Touche, 1926) which passes through a solid mass of anhydrite, which had a most destructive effect on Portland cement and lime mortars used in its original construction. The HAC proved satisfactory.

HAC was introduced into the UK in the 1920's, one of those structures made with UK cement is the bridge at Fingringhoe in Essex and this structure remains today completely satisfactory.

Thomas and Davey (1929) reporting on laboratory studies on HAC concrete found that under warm wet conditions the concrete could suffer a loss in strength, which they attributed to a change in the state of oxidation of iron compounds in the cement, since they noticed that with the loss of strength there was a change in colour from black to brown. This phenomena they described as "inversion."

By 1935 it had been shown that the loss in strength was due to the change in constitution of the cement hydrates from hexagonal calcium aluminate hydrates to cubic tricalcium aluminate hydrate (Lea and Desch 1935), and by the 1950's the change had become known as "conversion". The term conversion since it refers to a change (conversion) in chemical composition, should be restricted to that connotation and not be used as it is in some quarters to describe the loss in strength since the change in chemical constituents may not be accompanied a the reduction in strength.

The term "converted strength" is meaningless unless accompanied by a description of the conditions under which conversion occurs. Thus rapid conversion, for example caused by curing at too high a temperature, will cause the strength to pass through a lower minimum than one which has suffered a slower conversion.

Research started in 1929 and reported by Watkins and Lea (1960) showed that the residual strength of high alumina cement concrete made with a total w/c ratio of 0.53 and mix proportions by weight of 1:5 after 10 years immersion in concentrated sea water was 36 MPa upwards, in compression.

In 1940 high alumina cements were covered by a British Standard and their use for structural concrete was covered by BS Code of Practice, CP 114, "The Structural Use of Reinforced Concrete in Buildings" published in 1948. This code permitted the use of HAC in reinforced concrete and recommended a mix of 1:2:4 with a total w/c of 0.58. In the explanatory handbook (1965) to the code the manufacturers of the cement were said to recommend that the formwork should be kept wet and struck at eight hours, the concrete being kept wet for at least 24 hours in order to avoid strength from being impaired and objectionable surface dusting. A warning was also given against placing HAC concrete at high ambient temperatures or in large masses which would result in high temperatures being developed due to the evolution of heat by the chemical reactions in the hydration of the cement.

In 1951, BRS Digest No. 27 (1951) was published giving the general characteristics of concrete made with HAC. It recommended the use of mixes from 1:5 to 1:7 with just enough water to enable the concrete to flow easily. At that time a total w/c ratio of 0.50 was recommended and that the ratio should not fall below 0.40. Reference was made to the potential loss in strength of concrete if it was exposed to moist hot conditions at any time in its life but it stated that the residual strength was adequate for structural purposes.

In the early 1950's HAC was being introduced into the manufacture of prestressed concrete members on pretensioning beds. The main reason for the use of HAC was that with the development of high early strength it was possible to achieve a daily production cycle, since it was possible to stress the concrete at an age of 18 hours.

During the next few years the precast concrete industry, on the basis of experiment, was able to conclude that provided the w/c ratio was kept low and the curing temperatures maintained below 30 C the concrete would not lose strength. If however, conversion did occur concretes with low w/c ratios would still retain substantial strength (Newman, 1960).

In 1964 the BS Code of Practice, CP116, "The Structural Use of Precast concrete" dealt in some detail with the use of HAC and concluded that:

1   A w/c ratio of not greater than 0.50 for reinforced concrete and 0.40 for prestressed concrete should be used.
2   The concrete should be kept cool and moist during the first 24 hours.
3   The concrete should be kept reasonably dry and cool after its initial curing.

2 Building failures in the UK during 1973 and 1974

From the early 1960's prestressed beams using HAC went into quantity production by a number of manufacturers with apparent success. However, in 1973/1974 three roofs in which HAC had been used collapsed.

2.1 University of Leicester

The first collapse was in the Bennett Building of the University of Leicester. The building was constructed of precast prestressed HAC concrete beams with precast concrete slab shutter panels. The roof beams were supported by ring beams on columns both made of HAC. The failure was of the support nib on the ring beam, and the prestressed beams fell to the floor. The failure was in the ring beam made with HAC but subsequent examination showed that there was insufficient bearing surface and the concrete had been made to an incorrect specification, i.e. too much water, a probable w/c ratio in excess of 0.5; and the mix design was too lean in cement it had been apparently specified on strength alone and since HAC gives greater strength with the same proportion of cement when compared with Portland cement. No account of the possibility of loss of strength due to conversion was considered relevant.

The conclusion of the investigation into the failure was that there was insufficient bearing surfaces of the ring beams, the loss in strength of the ring beam cement aggravated that failure. **The beams which had fallen were re-used in the reconstruction.**

2.2 Camden School for Girls

The second failure was at Camden, in the roof of the assembly hall of School for Girls. The conclusions arrived by the BRE (Bate, 1973) were:

1   Insufficient bearing of the prestressed concrete roof beams and the edge beams to allow effective reinforcement of the nibs at the end of the roof beams;
2   Termination of the prestressing wires within the span of the roof beams without provision for the continuation of longitudinal reinforcement into the nibs above the bearings;
3   Reduction in strength of the concrete in the prestressed roof beams resulting from the conversion of high alumina cement.

The subsequent collapse of the whole roof was mainly due to the following factors:

4   Insufficient structural cross-tying of the building;
5   Deterioration and corrosion of the continuity reinforcement between the edge beams and the roof.

The report by Bate (1973) concludes that deterioration of the concrete had contributed to the failure of the joint although the strength at the critical section had probably not reduced below that assumed for design. Thus, although the concrete was of the required strength, the margin of safety was insufficient. Had the joint detail, however, been satisfactory, the maximum reduction in strength of the concrete experienced might not have caused failure. In other words the

4

HAC cement had been satisfactory, the failure was due to details of design.

As a result of these failures the Department of the Environment and the Department of Education and Science undertook an appraisal of school halls, gymnasia and swimming pools to see if there were sufficient bearing surfaces for prestressed beams, if insufficient then remedial action was undertaken. A number of buildings of insufficient safety were discovered and in some cases strengthening was used; in a few rebuilding was needed. In all cases the faults were attributed to insufficient safety design.

2.3 Sir John Cass School, Stepney

The third failure was on the 8th February 1974, in the roof of the swimming pool of the Sir John Cass's Foundation and Red Coat School, Stepney, London. The building had been appraised under the original DOE/DES proposals and had been pronounced safe, the appraiser concluded "that there was no evidence of any damage or distress in the joists (beams), nor any undue deflection, to give rise to any cause for anxiety with regard to structural adequacy". The strength of the concrete in the joists (beams) had been checked in his survey with rebound hammer equipment which had indicated a compressive strength of 52 to 83 MPa (Bate, 1974). The rebound hammer does not reflect what is going on inside the concrete and so was of little value to appraise the quality of the beams.

An intensive investigation into this failure was undertaken by the Building Research Establishment, the conclusions were reported on 2nd July 1974; being that the loss in strength was due to the conversion of the HAC followed by sulphate attack leading to the disruption of the concrete.

The loss in strength was greater than expected for the quality of concrete specified and it was inferred that this may have been due to a combination of the following factors:

1   The free w/c ratio may have exceeded the maximum of 0.4 aimed at in production;
2   the temperature during the first day after casting may have been excessive; and
3   the aggregate used, which contained the minerals feldspar and mica, had an accelerating effect on the rate of conversion and loss in strength.

The beams which failed had been in contact with a plaster (gypsum, calcium sulphate hydrate) coating and had been in a wet environment. Calcium sulphoaluminate hydrate, ettringite, was found in large quantities on the surface and was identified to a depth of 50 mm. The concrete was highly converted and the porosity was measured as 22.6%. With the concrete having such a high porosity it was clearly vulnerable to sulphate attack (George, 1979).

It has been reported by George (1974) on the same failure that chemical examination of the HAC concrete revealed the presence of ettringite which generally results from sulphate attack of calcium aluminate hydrates where excessive porosity due to high w/c ratio exists. Calcium sulphate was present in the form of gypsum plaster, and may have been transported to the HAC concrete by condensation or by penetration of rain water. The results of experimental determinations for w/c ratios gave a range from 0.45 to 0.64 with an average of 0.54. An estimate of the permeability was made by measuring the height to which water would rise into concrete by capillary absorption.

Figures ranging from 65 to 120 mm may be compared with values between 25 and 35 mm normally observed for fully converted HAC concrete with a total w/c ratio of 0.4. Porosities were also variable ranging from 7.6 to 13.2%, these variations being linked with the measured differences in cement content. George (1974) concluded that the failures of the beams at Stepney, which in part was due to sulphate attack, was a consequence of to high a w/c ratio in the concrete used for the beam.

3 Chemical resistance of HAC

HAC concrete has been noted for its resistance to many chemicals especially sea water, sulphates, chlorides, weak acids etc. although it may be vulnerable to alkalis.

Because of the intense investigations into HAC concrete some investigations were carried out on the vulnerability of such concrete in aggressive environments.

One such example was the foundations at a school at Edenbridge, Kent Midgley (1985). The foundations were in a soil containing Epsom salts (magnesium sulphate), the most aggressive sulphate to concrete. When samples taken from the outside foundations were examined in the laboratory it was found that only the immediate surface had been attacked and the mineral ettringite (calcium aluminosulphate hydrate) had been formed. The maximum depth of penetration was only 3 mm and at the time of examination the concrete was at least 18 years old. The surface had a measured porosity of 22% whilst the inside had a porosity of less than 15%. Technically this should be classed as an example of chemical attack, but as the attack was only 3 mm in 18 years and since the rate of penetration depends on the square of time the possible time for penetration of 1 m would be in the order of 2 million years, and with these foundations this still would not effect their bearing capacity.

Another example of an investigation into a form of chemical attack was that of a sewer lining by Thistlewayte (1972). HAC concretes have been used extensively as sewer linings (Thistlewayte, 1972) and the failure at Sandwell was unusual. Laboratory examination of the HAC concrete showed that the porosity was high, 19.6%, and it had converted by more than 90% in three years. It had been placed in an environment of "made ground" which contained a high concentration of ferric sulphate which under leaching conditions will hydrolyse to produce sulphuric acid. The sulphuric acid had attacked the HAC concrete to produce ettringite which under the conditions of exposure had proved disruptive. The reason for this failure must be the high porosity and converted state of the HAC concrete. Why this should happen must be conjecture, but the concrete had been made to a Portland cement mix design and with an artificial lightweight aggregate of high porosity, this high porosity would need a higher than usual amount of water to obtain workable mixes. The possible use of too high a w/c ratio together with a possible too high temperature of initial cure would produce a highly converted, porous concrete which would be highly vulnerable to chemical attack. If the cement is misused it will misbehave, like any other product.

4 Reaction to failure of buildings with HAC beams

Following the collapse of the roof of the school swimming pool at Stepney, the Department of the Environment issued a document on the 20th July 1974, Ref. BRA. 1068/2, entitled "High Alumina Cement Concrete in Buildings". The report concluded that:

1   All existing buildings incorporating high alumina cement concrete must now be regarded as suspect, at least in the longer term;

2   A further programme of action is required starting immediately with those buildings with the greatest risk;

3   High alumina cement concrete should not be used for structural work in buildings until further notice. An amendment to the building regulations will be proposed accordingly.

The report continued:- "The action began by being related to :

Stage (a) precast prestressed isolated roof beams of high alumina cement concrete, particularly in conditions of heat and humidity.

Stage (b) precast prestressed isolated roof beams of high alumina cement concrete regardless of environmental conditions.

Stage (c) all remaining precast prestressed non-composite roof or floor members (also columns) of high alumina cement concrete, dealing in the first place with those buildings with roof of floor spans exceeding 5 metres.

Within this class, buildings with structural members which are of sensitive cross section or which are isolated, those subject to the more extreme or variable environmental conditions, and those where failure would constitute the greatest risk to public safety, should be appraised first."

The document BRA/1068/2 also in very general terms gave guidance on suitable methods of appraisal.

Following this warning document many hundreds of appraisals were undertaken, a number of buildings were found to be possible hazard. Some were replaced as it was considered to be cheaper than strengthening.

The Building Regulations Advisory Committee set up in April 1975 a sub-committee under the chairmanship of Mr. C.B. Stone to consider the use of high alumina cement concrete. They reported their findings in 1976 as the Report by sub-committee P (high alumina cement concrete) (1975).

The report recommended that "in buildings consisting solely of houses, maisonettes and flats, floors and roofs of standard factory-produced joists up to 10 inches in depth, which is the maximum size used in such construction, may be considered safe and exempt from appraisal provided:

a   the buildings do not consist of more than 4 stories,
b   in the case of roofs of HAC construction:
  1   they are of joist and block or composite construction;
  2   they are not used for access, except for maintenance;
  3   the spans do not exceed 6.5 metres clear in the case of an X7 or less than 8.5 metres in the case of an X9 or X10 joist roof.
c   there is no persistent leakage or sustained heavy condensation.

If these recommendations are followed it is clear that there are many situations in which high alumina cement concrete would be satisfactory."

5 Post appraisal reports

Following the failures of the three buildings and the considerable activity in appraising structures containing precast pretensioned high alumina cement concrete beams there was a considerable effort into research into HAC.

During 1975-1976 a number of important reports were published; Midgley and Midgley (1975) on the basic chemical and mineralogical changes involved in the conversion reactions; Teychenne (1975) on the long-term changes in high alumina cement concrete; George (1974) on the structural use of high alumina cement concrete; Bate (1976) on high alumina cement concrete in buildings; Neville (1975) on various aspects of high alumina cement failures; Bungey (1974) and Mayfield and Bettison (1974) on ultrasonic testing of concrete; the Chemical Society (1975) on methods of testing high alumina cement concrete using thermoanalytical techniques; Roberts and Jaffrey (1974) on a rapid chemical test for the detection of high alumina cement concrete; the Institute of Structural Engineers on guidelines for the appraisal of structural components in high alumina cement concrete (1974).

In 1980-1981 although there had been no more failures of buildings since 1974 there was still considerable interest in the subject; the Institute of Structural Engineers called a special meeting on 15th January 1981, to discuss the subject on the basis of two papers by Safier (1980) and Bate (1980); this meeting proved successful so a further meeting was called for about a month later. **The consensus of the meeting was that the banning of high alumina cement from structural concrete was unnecessary and that a new revision of the code of** practice on the use of high alumina cement should be undertaken. However, to date this has not yet been issued.

In 1980 Bate (1980) issued from the BRE a summary of the position of high alumina cement concrete in existing building superstructures.

He reported that "in all, between 30,000 and 50,000 buildings were estimated to contain high alumina cement concrete. Minor uses of high alumina cement were in foundations where adverse levels of sulphate existed, in cast in-situ concrete, in mortar joints in precast construction and for grouting tendons and earth anchors. These applications were not included in the investigations, since it was considered that any structural defects due to loss in strength in concrete would either be insignificant or would give ample warning of structural inadequacy; where the use was below ground, temperatures would be lower and therefore conversion and loss of strength would progress at a very much lower rate than in building superstructures."

Bate reports that "1022 appraisals were undertaken, of these thirty eight buildings, mainly in educational use, were identified as being distressed. Of these the owners decided to replace the units in fifteen and to strengthen in seventeen cases. The reasons for strengthening or replacing were excessive deflections in nine instances, poor bearings at ends of beams in three buildings, a case of chemical attack in one where the original design provided an inadequate margin of safety."

"Of the failures or near failures experienced in service, only one was the result of loss in strength due solely to conversion, and it was built before the importance of keeping the w/c ratio low was appreciated and before the advent of the X-joist or similar section. In nearly all cases of failure or near failure, the possible exception being the swimming pool roof at Stepney, prior inspection would have shown signs of distress."

The general conclusion from the Bate report is that only 1 in 30,000 buildings have suffered failure due to limitations of the concrete, and this could have been avoided if the proper mix design, especially a low w/c ratio had been used.

High alumina cement concrete was frequently used as foundation material and this material was used in the extensive trial by the Building Research Establishment into the sulphate resistance of various cements in an aggressive soil environment at Northwick Park, UK and in the laboratory. The report by Harrison and Teychenne (1979) stated that after 5 years "there has been no significant corrosion of the HAC concretes in the sulphate solutions although there has been a loss in strength due to the conversion. It should be noted that the test programme used a w/c ratio of 0.5 whereas 0.4 or less is normally recommended for corrosive environments. On the site the HAC concretes showed little penetration of sulphates but the cores taken from the basement walls and floors and in the precast cylinders showed there had been loss in strength due to conversion."

Midgley (1980) reported on the chemical resistance of high alumina cement concrete and concluded that the concrete was resistant to sulphates, chlorides and sea water if correctly made and used in the correct manner. If however, the high alumina cement concrete is highly porous due to a high w/c ratio, or cracked (for example prestressing cracks) then it may be disruptively attacked by sulphates. Sea water and chlorides appear to have little disruptive effect. Alkaline hydrolysis, carbonation where the cement part of the HACC by alkaline hydoxides can be completely disrupted, would further assist in chemical attack of the concrete.

6 Conclusions

It must therefore be concluded that there is no valid reason for not using high alumina cement as a construction material.

If it is to be used the recommendations of George (1974) must be adhered to. These recommendations are as follows:

"For the manufacture of good quality durable high alumina cement concrete the fundamental requirement is that the total w/c ratio shall not exceed 0.4. Although at low w/c ratios high alumina cement concrete exhibits better workability than an equivalent mix based on Portland cement, the adoption of a minimum cement content of 400 kg/m^3 is a sound practical safeguard against errors in the control of w/c ratio in practice. As an added precaution against the possibility of alkaline hydrolysis aggregates and sands containing releasable alkalies are proscribed."

"Spraying or soaking in water is necessary during the initial curing period to prevent drying out of the concrete as a result of heat evolution, whenever conditions of placing lead to substantial increase in temperatures."

"When these requirements are met, experience shows that high alumina cement concrete even when rapidly and fully converted is reliable structural material. While exceptionally high strengths can be obtained under special circumstances prudence dictates that structural design should be based on minimum converted strength."

The final conclusions are that, possibly with the exception of prestressed beams, there is every reason to use high alumina cement as a construction material, especially where its special properties, such as resistance to chemical attack, or high early strength can be used to advantage.

In the author's considered view, high alumina cement, the properties of which are now well documented, is a valuable construction material, not as a substitute for Portland cement, but as an alternative where requirements of rapid hardening and specific corrosion resistance merit its use. Like any other product, certain conditions of use must be respected if its specific properties are to be successfully exploited, these have been outlined in the present text.

Like any other material, high alumina cement will perform well when properly used. When problems arise with Portland cement the reaction is to correct misuse, not to incriminate the cement. There is no reason to adopt a different attitude to high alumina cement.

7 References

Bate, S.C.C., (1973) Report on the collapse of the roof of the assembly hall of the Camden School for Girls. Report of Special Investigation by BRE, HMSO.

Bate, S.C.C. (1974) Report on the failure of roof beams at Sir Johns Cass's Foundation and Red Coat Church of England School, Stepney. Building Research Establishment Current Paper, CP 58/74.

Bate, S.C.C. (1976) High alumina cement in buildings, Building Research Establishment Current Paper CP 34/76.

Bate, S.C.C. (1980) High alumina cement concrete- assessment from laboratory and field studies. **The Structural Engineer**, 58A, 12, 388-394.

Bate, S.C.C. (1980) High alumina cement concrete in existing building structures, BRE Special Report, PD 101/80.

Bied, J. (1909) British Patent 8193.1909.

British Standard 915 (1940), British Standard Institute, London.

British Standards Institution (1948), Code of Practice for the Structural Use of Reinforced Concrete in Buildings.

British Standards Institution (1964), Code of Practice for The Structural Use of Precast Concrete.

British Standard Institution (1972) The Structural Use of Concrete, CP110, London.

Building Research Station (1951) High Alumina Cement, Digest No. 27, HMSO.

Bungey, J.H. (1974) Ultrasonic testing of high alumina cement concrete on site, **Concrete** 8, (9), 39-41.

George, C.M. (1974a) **Revue des Materiaux de Construction** 4, 201-208

George, C.M. (1974b) The structural use of high alumina cement, Lafarge Fondu International, France.

George, C.M., (1979) High Alumina Cement with special reference to Ciment Fondu Lafarge, Seminar on Cement and Concrete Science at Thames Polytechnic, April.

Harrison, W.H. and Teychenne, D.C. (1979) Second report on long-term investigation of sulphate resistance on concrete buried at Northwick Park. Building Research Station Special Report.

Lea, F.M. and Desch, C.H. (1935). **The Chemistry of Cement and Concrete**, 1st Edition, Edward Arnold, London.

Lea, F.M, and Watkins, C.M. (1960) The durability of reinforced concrete in sea water, Building Research National Building Studies, Research Paper No. 30, HMSO.

Mayfield, B. and Betison, M. (1974) Ultrasonic pulse testing of high alumina cement concrete in laboratory, **Concrete** 8, (9), 36-38.

Midgley, H.G. and Midgley, A., (1975) The conversion of high alumina cement, **Mag. Concr. Res.** 27, 59-77.

Midgley, H.G. (1980) The chemical resistance of high alumina cement 5th **Int. Symp. Chem of Cement.**, Paris, Septima, VIII, V85-V87.

Midgley, H.G. (1985), The use of high alumina cement in chemical and civil engineering in **Corrosion and Chemical Resistant Masonry Materials Handbook**. Editor. W.L. Sheppard, Noyes Publications, New Jersey, 340-363.

Neville, A.M., (1975) **High Alumina Cement Concrete**, The Construction Press, Lancaster.

Newman, K., (1960) Design of concrete mixes with high alumina cement **Reinforced Concrete Rev.**, 5, 269-294.

Roberts, M.H. and Jaffrey, S.A.M.T. (1974) Rapid chemical test for the detection of high alumina cement concrete, BRE Information Sheet IS 15/74.

Safier, A.S. (1980) High alumina cement concrete- apprisals, the problems and some findings. **The Structural Engineer**, 58A, 12, 381-387.

Scott, W.L., Glanville, W.H., and Thomas, F.G. (1965), Explanatory handbook on the Code of Practice for Reinforced Concrete, 1950 revised 1957 and 1965, Concrete Publications Ltd.

Stone Committee (1975) High alumina cement concrete. Report by sub-committee P: Building Regulations Advisory Committee, Department of the Environment and Welsh Office.

Teychenne, D.C. (1975) Long-term research into the characteristics of high alumina cement **Mag. Concr. Res.** 27, 78-102.

Thistlewayte, D.K.B. (1972) Control of Sulphides in Sewerage, Butterworths, London.

Thomas, W.H, and Davey, N. (1929) The effect of temperature on the setting time of cements and on mortars and concretes, Building Research Report No. 13, HMSO.

12

Touche, Le Cement. (1926), 31, 240.

Wilburn, F.W., Keattch, C.J., Midgley, H.G. and Charsley, E.L. (1975) Recommendations for the testing of high alumina cement concrete samples by Thermoanalytical Techniques, Thermal Methods Group. Analytical Division of the Chemical Society, London. p. 1-11.

PART ONE
CLINKER

2 EFFECT OF MORPHOLOGY ON THE HYDRATION CHARACTERISTICS OF HIGH ALUMINA CEMENTS

I.N. CHAKRABORTY, S. NARAYANAN,
D. VENKATESWARAN, S.K. BISWAS, A.K. CHATTERJEE
The Associated Cement Cos Ltd, Thane, India

Abstract
The role of various calcium aluminate phases on the hydration characteristics of high alumina cements are well documented in the literature. This work investigates the effect of grain growth and morphology of these phases on hydration behaviour of high purity high alumina cements as a function of alumina content. The paper attempts to correlate grain growth, morphological changes and hydration characteristics. Based on these results, pyroprocessing parameters for various high purity high alumina cements have been optimised.
Keywords: Calcium Aluminate, Hydration, Microstructure, Grain growth, Setting time, Cold Crushing Strength.

1 Introduction

The main phases in high purity industrial alumina cements are CA, CA_2 and $C_{12}A_7$ depending on the exact CaO to Al_2O_3 ratio. Opinion differs in the literature regarding the hydration characteristics of CA_2 phase. CA_2 is normally believed to be a non-reactive phase by itself and reacts with water only in the presence of CA. The hydration products of the calcium aluminate phases are identical, only the relative concentrations differ which is a function of CaO to Al_2O_3 ratio of the starting calcium aluminate phase. The most predominant hydration products are CAH_{10}, C_2AH_{10}, C_2AH_8, C_3AH_6 and AH_3. However, the only stable hydrates at ordinary temperature are gibbsite and cubic C_3AH_6. The metastable hexagonal calcium aluminate hydrates, CAH_{10} and C_2AH_8, are known to form only at low temperature.

The reactivity of calcium aluminate increases with lime content. The presence of CA, as a major phase in calcium aluminate cement, is desirable since this results in higher initial strength development. Cements containing appreciable quantities of CA_2 are penalised artificially by standard test. The standard evaluation procedure involves curing for 24 hours only at 20°C which is too short a duration for a lesser reactive CA_2 phase.

When anhydrous calcium aluminates are placed in water, lime and alumina dissolve. As the CaO/Al_2O_3 ratio of the solution is increased above 1.06, in the presence of $C_{12}A_7$ for example, the normally long induction period for the formation of calcium aluminate hydrate, is progressively diminished until the immediate precipitation occurs at C/A ratio $\gg$ 1.2. It is also observed that the induction period decreases with the degree of concentration of the solution phase even though the C/A ratio is held constant. The stage at which the measurable mechanical resistance begin to develop, i.e. setting as measured by Vicat apparatus, correspond to the onset of bulk hydrate precipitation from the solution phase in contact with the cement.

This paper investigates the effect of morphological changes of CA/CA_2 phase on the setting and strength development characteristics for high purity (CaO and Al_2O_3 make up 98.5% of total) calcium aluminate cements with varied concentration of CA and CA_2 phases. For this investigation, cements with nominal Al_2O_3 concentrations of 65 (Cal-Al-65), 70 (Cal-Al-70) and 75 (Cal-Al-75) percents were selected. Since two extreme compositions are predominantly CA and CA_2, respectively, an intermediate composition, viz. Cal-Al-70 which is almost equimolar in CA and CA_2 content, was also included in the investigation.

2 Experimental Procedure

2.1 Preparation of Sample

The calcium aluminate cements, viz. Cal-Al-65, Cal-Al-70 and Cal-Al-75 were prepared by mixing commercial grade hydrated lime (CaO:71%, SiO_2 :1% and loss on ignition (LOI):28%) and Bayers alumina (Al_2O_3:65% ,SiO_2 and Fe_2O_3 : 0.05% each, Na_2O:0.5% and LOI:34.5%) in right proportion. The mixes were blended and ball milled for one hour with alumina grinding media for appropriate particle intimacy. The milled materials were made into briquettes (9 x $4\frac{1}{2}$ x1") in double acting hydraulic press at a pressure of 2000 kg/cm^2. The briquettes were dried at 150°C and then fired at 1450°C, 1500°C and 1550°C for 1 hr , 2 hrs and 4 hrs in LPG – air fired furnace. The samples were analysed for their free lime contents which did not exceed 0.2% for any cement. The sintered products were crushed and ground in ball mills using alumina grinding media. The Blaine surface area of the cements was maintained in the range of 4000 - 4200 cm^2/ gm.

2.2 Quantification of Phases Present

The phase concentrations of the cements were determined by X-ray diffraction (XRD) reference intensity ratio (RIR) technique. The technique was modified to account for the phases which are not identified by the diffraction pattern. This was necessary, since low concentration of unreacted α-Al_2O_3 and certain intermediate phases formed due to non-equilibrium in calcium aluminate cements are not easily identified by XRD and thus, cannot be quantified by conventional RIR technique.

2.3 Determination of Setting Time and Cold Crushing Strength (CCS) of the Cements

The setting time and cold crushing strength of these cements were carried out as per BS 915 Part 2. Six cubes for each cement have been tested for CCS measurement and their averages are reported.

2.4 Microscopic Analysis of Cements

The samples in pellet forms were mounted in epoxy resin and impregnated under vacuum for $\frac{1}{2}$ hour. After curing, the samples were ground on silicon carbide discs of grit 120 through 600 then polished stepwise with standard diamond compounds. The samples were etched with 5% aqueous solution of NaOH for phase contrast and examined under incident light at 100 x to 1000 x magnification. The crystal size was measured using semi-automatic image analysis system.

3 Results and Data Interpretation

The chemical and phase composition of the cements investigated are reported in Table I. The setting time and cold crushing strength (CCS) of the cements are presented in Table II. From the data available, the effect of the three independent variables, viz. the sintering temperature, soaking time and composition on the setting time and CCS of the cements can be summarised as under.

Table I. The Chemical and Phase Compositions of High Alumina Cements

%	Cal-Al-65		Cal-Al-70		Cal-Al-75	
	1450°C/ 1 hr	1550°C/ 4 hrs	1450°C/ 1 hr	1550°C/ 4 hrs	1450°C/ 1 hr	1550°C 4 hrs
Al_2O_3	64.36		69.33		74.30	
CaO	34.66		23.70		24.80	
SiO_2	0.53		0.47		0.40	
Fe_2O_3	0.04		0.05		0.06	
Na_2O	0.40		0.43		0.46	
CA	86.0	80.7	64.4	57.5	25.7	21.9
CA_2	13.5	16.4	35.1	42.0	72.3	76.3
Others	0.5	2.9	0.5	0.5	2.0	1.8

Table II.   Setting Time and Cold Crushing Strength (CCS) of High Alumina Cements

Temp./ Soaking Time	Composition	Water requirement (%)	Setting Time (minutes) Initial	Final	CCS (kg/cm²*) 1 D	1+1 D	Remarks
1550°C/ 4 hrs	Cal-Al-65	22	7 hrs	–	515	630	
	Cal-Al-70	22	7 hrs	–	550	665	
	Cal-Al-75	22	7 hrs	–	385	775	
1550°C/ 2 hrs	Cal-Al-65	22	160	360	585	775	
	Cal-Al-70	22	60	170	265	470	
	Cal-Al-75	22	150	415	475	705	
1550°C/ 1 hr	Cal-Al-65	22	280	7 hrs	590	760	
	Cal-Al-70	22	145	280	550	715	
	Cal-Al-75	22	105	280	390	725	
1500°C/ 4 hrs	Cal-Al-65	22	290	–	570	765	
	Cal-Al-70	22	250	–	550	800	
	Cal-Al-75	22	7 hrs	–	260	595	
1500°C/ 2 hrs	Cal-Al-70	22	140	310	500	765	
	Cal-Al-75	22	155	305	385	820	
1500°C 1 hr	Cal-Al-65	22	160	400	595	710	
	Cal-Al-75	22	120	320	425	800	
1450°C/ 2 hrs	Cal-Al-65	22	130	350	570	670	
	Cal-Al-75	22	70	140	335	820	
1450°C/ 1 hr	Cal-Al-65	22	40	95	505	610	
	Cal-Al-75	29	230	270	510	820	(Flash-set tendency observed)

* After curing at 110°C for 1 day

3.1 Setting Time

The setting time increases with increasing firing temperature and soaking time for all cements.

In case of Cal-Al-65 cements, the firing temperature shows better correlation with setting time, whereas in Cal-Al-75, soaking time rather than sintering temperature, correlates better with the setting time. With the limited data available on Cal-Al-70, it is not possible to say which one of these two factors has greater influence on the setting time of the cement of this composition.

At extreme conditions of firing, i.e.at 1550°C 4 hrs all the three cements do not show initial setting even after 7 hours curing.

From the data available, no definite trend could be established between the setting time and cement composition.

3.2 Cold Crushing Strength (CCS)

In general, the higher firing temperature and longer soaking time results in higher strength development.

As in the case of setting time, this increasing trend in CCS 'is not evident in cements prepared at extreme firing condition,i.e.at 1550°C, 4 hours.

Unlike in the case of setting time, the composition does seem to have a significant influence on the CCS of the cements. The mean one day CCS for Cal-Al-65, 70 and 75 compositions are 561, 531 and 396 kg/cm^2, respectively. While the one day strength is higher for Cal-Al-65 and 70, the percentage gain in strength after treating at 110°C for 24 hours,(1+1 day) is appreciably more for Cal-Al-75 cements.

3.3 Inter-relation between CCS and Setting Time

A comparison of these two dependent variables suggest that increasing setting time results in higher CCS for a given type of cement. However, this is not true for cements having very delayed initial setting time (i.e. more than 7 hours), as for example, in case of cements sintered at 1550°C for 4 hours.

The above observations are discussed in the following section.

4 Discussion

4.1 Influence of Phase Composition on Setting Time

As the sintering temperature and soaking time are increased, the setting time of the cement also increases irrespective of the chemical composition. The phase composition of the cements determined by XRD analysis reveal that with increasing firing temperature and soaking period, CA_2 forms at the expense of CA . Quantitatively, the phase concentration of CA differs by about 6% for all cements fired at two extreme conditions, viz. 1450°C for 1 hour and 1550°C for 4 hours. The concentration of unidentified phases(i.e.phases other than CA and CA_2) remains almost constant except in case of Cal-Al-65 where it changes to the extent of about 2.5%. It is,thus evident that the variation in setting time and CCS for a given cement with change in firing conditions cannot be explained by this small differences in phase concentrations.

Silica and Na$_2$O are the major impurities in these cements. The first formation of liquid phase is likely to take place at 1500°C, the invariant point of CA-CA$_2$-C$_2$AS system, if only the presence of SiO$_2$ is considered. The liquid content will increase with temperature, which in turn, aids the grain growth of the phase formed. A similar effect will be expected if the materials are soaked at a temperature lower than 1500°C for longer duration. In this event, however, the grain growth will take place in the absence of liquid phase.

4.2 Microstructure of Unhydrated Cement and its Effect on Setting Time

The polished sections of the high alumina cement clinkers, synthesised at two extreme conditions were examined under microscope to ascertain the grain growth of CA and CA$_2$ phases. The microstructure of these samples reveal distinct differences which are brought out in photomicrographs 1-4.

Fig.1 and 2 corresponds to the samples sintered at 1550°C for 4 hours. While Cal-Al-75 displays granoblastic texture with xenomorphic crystals of CA$_2$, the Cal-Al-65 consists predominantly of panidiomorphic CA crystals. In both the cases, the crystals are coarse and interlocked with little inter or intragranular porosity. Gehelenite is present in both the samples in minor amounts. The mean crystal size of CA, CA$_2$ is of the order of 25-35 microns as against 10-15 microns usually observed in other samples sintered at lower temperature with lesser retention time. The excessive grain growths of CA, CA$_2$ and the granoblastic texture devoid of any porosity do not favour the normal hydration and hence these samples, irrespective of their phase composition, show extremely slow setting characteristics.

The sample of Cal-Al-65 composition, sintered at relatively low temperatures (1450°C and 1500°C for 1 hour) exhibit a different microstructure as is revealed in Fig.3 and 4. At 1450°C, the texture shows phenocrysts of equiaxial CA, CA$_2$ crystals in an aphanetic matrix of the same composition. The crystals have a mean size of about 10 microns. In such a microstructure, the aphanetic matrix, probably reacts first followed by the phenocrysts of CA and CA$_2$. As the sintering temperature is raised to 1500°C, the phenocrysts grow at the expense of the aphanetic matrix. The crystals also have a tendency to develop interlocking texture. The hydration of these coarser phenocrysts are likely to be slow owing to their improved crystallinity and hence the comparatively longer setting time for the cements produced at higher temperatures. Thus, it is evident that the crystal growth and texture of unhydrated phases greatly influence the setting time characteristics of the cement regardless of its chemical compossition.

4.3 Strength Development and Setting Time

The 1 day CCS is higher for both Cal-Al-65 and 70 as the more reactive CA phase is the predominant constituent in these cements. Their respective CA contents are 83% and 60% approximately, as against only 24% for Cal-Al-75 composition. Despite a difference of about 25% in their CA contents, the mean 1 day CCS is more or less of the same order i.e.561 and 531 kg/cm^2, respectively for Cal-Al-65

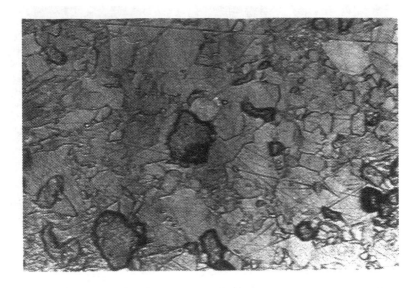

Fig. 1. Cal-Al-75/1550°C - 4 hrs - Granoblastic Texture
with interlocked crystals of CA_2 x 400

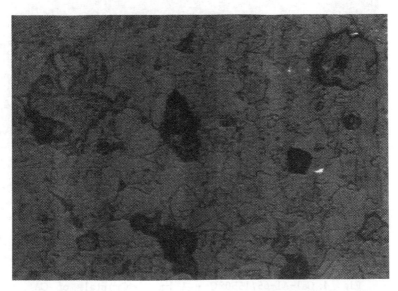

Fig. 2. Cal-Al-65/1550°C - 4 hrs - Interlocked crystals
of CA in Cal-Al-65. The dark patches are voids
created during sample preparation x 250

Fig. 3. Cal-Al-65/1450°C - 1 hr - Equi-axial,
    porphyritic crystals of CA in an aphanetic
    (fine grained) matrix x 600

Fig. 4. Cal-Al-65/1550°C - 1 hr - Crystals of CA
    grown at the expense of aphanetic matrix.
    Note the tendency towards interlocking texture.
    Aphanetic matrix is less abundant compared to
    Fig.3 x 600

and 70 compositions. Thus it appears that beyond certain threshold value, increase in 1 day CCS of the cement is not proportional to the CA content of the cements.

The rate of gain in CCS after treating at 110°C is higher for Cal-Al-75 cements, since at elevated temperature, the usually docile CA_2 phase is activated and the acceleration of the rate of reaction between this phase and water results in much higher strength as compared to the first 24 hours of hydration which is carried out at 18°±2°C.

Both the physical properties, i.e. strengths development and setting time, are the manifestations of the hydration reactions. While setting time is influenced by the rate of hydration and the hydration products, CCS is determined by the resultant porosity due to the above two factors. At temperatures below 20°C, both CA and CA_2 on hydration, give rise to CAH_{10}. Since this phase is metastable, it converts to stable C_3AH_6 through C_2AH_8. During this conversion, approximately 0.57 cm^3 of water and 0.49 cm^3 of solid hydrate are formed for every cm^3 of CAH_{10} hydrate and an equivalent amount of porosity is developed in the system. This increase in porosity causes loss of strength of the cement paste structure. If the conversion takes place sufficiently slowly, the porosity that may develop because of this change may be filled with newly formed hydration products from the secondary hydration. Hence the rate of conversion, in addition to the extent of conversion, determine the change of strength. Since setting time is a function of rate of hydration, which in turn determines the rate and extent of conversion, the cements with slower setting, generally result in higher CCS.

5   Conclusion

The influence of the three independent variables (viz. sintering temperature, soaking time and chemical composition) on the setting time and CCS of the high purity, high alumina cements has been discussed. While the first two variables have shown significant effect on both the physical properties, the third variable, viz. the composition, appears to influence only the CCS of the high alumina cements being investigated. The increase in firing temperature and sintering time aids crystal growth of the principal phases CA and CA_2, which in turn affects the setting characteristics of the cements, regardless of their chemical composition.

Fortunately, the crystal growth is sensitive to variations in temperature and soaking time and, therefore, optimisation of this microstructural feature is possible through a set of process conditions Once this standardisation between the microstructure (crystal size in particular) and process parameters is achieved, the setting characteristics of HAC can be monitored through simple measurements of grain growth of CA and CA_2 phases. The phase composition, however, will remain the most important criterion as far as the strength development characteristics are concerned. In conclusion, it can be said that by considering a fourth quantifiable variable

25

(i.e. crystal size) it is possible to formulate a probablistic multivariate modelling for the simplification of behavioural prediction of high alumina cement phases on hydration.

6 References

Bauret, P. and Menetner, D. (1974) Contribution to the study of the kinetic mechanism of aluminous setting cement. **Cement and Concrete Res.**, 4, 723-733.

Chou, K.S. and Burnet, G. (1981) Formation of calcium aluminates in the lime sinter process, **Cement and Concrete Res.**, 11,167-174.

Chung, F.H. (1974) Quantitative interpretation of X-ray diffraction patterns of mixture II. Adiabatic principle of X-ray diffraction analysis of mixtures. **J. Appl. Cryst.**, 7, 526-531.

Chung, F.H. (1975) Quantitative interpretation of X-ray diffraction patterns of mixtures III. Simultaneous determination of a set of reference intensities. **J. Appl. Cryst.**, 8, 17-19.

George, C.M. (1980) The hydration kinetics of refractory aluminous cements and their influence on concrete properties. **Trans. Brit. Cer. Soc.**, 79, 82-90.

Midgley, H.G. (1967) The mineralogy of set high alumina cement. **Trans. Brit. Cer. Soc.**, 66, 161-187.

Osborn, E.F. and Mann, A. (1964) System $CaO-Al_2O_3-SiO_2$, **In Phase Diagrams for Ceramists** (Eds. E.M. Levin, Carl R. Robbins and Howand F. McMurdie). The American Ceramic Society Inc., PP 219.

Parkar, K.M. and Sharp J.H. (1982) Refractory calcium aluminate cements - Review. **Trans. Brit. Cer. Soc.**, 81, 35-42.

3 HIGH ALUMINA CEMENTS BASED ON CALCIUM ALUMINATE CLINKER WITH DIFFERENT PHASE COMPOSITIONS AND SINTERING DEGREES

J. SAWKOW
Institute of Refractory Materials, Gliwice, Poland

Abstract

The phase composition and service properties of high alumina cements based on sintered clinkers, containing mainly CA and admixtures of $C_{12}A_7$ or CA_2 were investigated. On the basis of these investigations an initial clinker was chosen for obtaining cements with an Al_2O_3 content above 66%. Also properties of cement produced from clinkers with a different degree of sintering in a rotary kiln were studied.

The phase composition of cements was analysed by x-ray diffraction. The quantitative determination was carried out by the x-ray diffraction method with an β quartz as internal standard. In some cement slurries the hydration degree of $C_{12}A_7$, CA and CA_2 phases were also determined by the x-ray. In these investigations the forming progress of calcium aluminates in the range up to 1500°C was determined and the role of the $C_{12}A_7$ phase was explained because - as detected - in the presence of this phase the rate and degree of hydration of monocalcium aluminate increases considerably. On the base of investigation results of the phase composition, supplemented with cement service properties, a limit was set to which may vary the calcium aluminate clinker composition, assigned for manufacture of cements with an alumina content near 70% and 80%. From these clinkers and from α-Al_2O_3, cements with an alumina content near 70% and 80% were produced. Such properties of these cements such as: refractoriness, specific surface, strength of hardened cement slurries and of alumina concretes after 12 and 24 hours and after firing at 1500°C to 1800°C were investigated.

Also linear changes in high temperatures of the concrete were determined. Taking into account the achieved results an optimal addition of α Al_2O_3 to the cement with a 70% alumina content was defined. It has been found that corundum as a high alumina cement component allows controlling the cement shrinkage in high temperatures. It also has been ascertained that the alumina cement "70" with an optimal amount of α Al_2O_3

protects the alumina concrete against excessive, inadmissible expansion, caused by forming hexacalcium aluminate during concrete firing. However after testing the composition and properties of concretes from clinkers, with a different sintering degree, a limited sintering rate of the clinker in a rotary kiln was defined.

1 Introduction

In the last years we observe a constant development of application of refractory concretes, mixes and various monolithic materials based on refractory alumina cements. These cements contain usually a considerable amount of CA, a small amount of CA_2 and in some of them are admixtures of $C_{12}A_7$ and C_2AS, also exceptionally of CA_6.

The hydraulic activity of refractory alumina cements depends mainly on the CA/CA_2 ratio as well as on the form and amount of α-Al_2O_3 and also on the remaining admixtures (Kurdowski 1965, Sawkow 1972 and Kurdowski 1981). $C_{12}A_7$ with strongly marked hydraulic properties is an important addition. In the unit cell of $C_{12}A_7$ occur 6 combinations of calcium pyroaluminate with an unsymmetrical oxygen co-ordination i.e. $6[ICa_2Al_2O_5]$ and one part of Al_2O_3 (Grzymek 1988). The unsymmetrical co-ordination of cations I Ca++ with oxygen ions is the reason of a high hydraulic activity of $C_{12}A_7$. Therefore the hydraulic activity of alumina cement can considerably change the effect of the $C_{12}A_7$ phase. The service properties of alumina cements depend not only on the phase composition, but also on the thermal conditions in which the aluimina clinker is produced. Therefore in this paper the formation of calcium aluminate in a range of temperature up to 1500°C is given, also the effect of the $C_{12}A_7$ phase on the hydration degree of CA is explained as well as the limited rate of the clinker sintering degree in a rotary kiln is defined. It also has been shown that α -Al_2O_3, as a cement constituent, can protect against the inadmissible expansion during firing of high alumina concrete.

2 The composition of refractory alumina cements

The chemical composition of alumina cements comply with changes of the Al_2O_3 content in the range of 51-81%, CaO 18-37%, SiO_2 0.1-70%, Fe_2O_3 0.15-2% and Na_2O 0.1-0.5%. The change of the phase composition in refractory alumina cements depends mainly on the Al_2O_3/CaO ratio and on the occurring admixtures occurring (Sulikowski et al 1978 and Sawkow 1978). Phases occurring in alumina cements, produced in different countries, are shown in table 1 (Bushnel-Watson et al 1986). Because alumina cements may contain $C_{12}A_7$, CA, CA_2 and also corundum as well as gehlenite (C_2AS) therefore for the clinkering process the knowledge of an advantageous phase composition and of the maximum firing temperature is necessary. For that reason, developing the Polish technology of manufacture, a quantitative x-ray analysis has also been carried out to determine the content of CA and CA_2 in alumina cements containing up to 2% admixtures.

Table 1.   Mineralogical composition of refractory calcium aluminate cements

Phases	Secar 51	Secar 71	Alcoa CA 14	Alcoa CA 25	Secar 80
CA	+++	+++	+++	+++	+++
CA_2	nd	++	+++	+	++
$C_{12}A_7$	+	tr/+	tr	+	tr
$\alpha-A$	+	+	+	++	+++
C_2S	tr	nd	nd	nd	nd
C_2AS	nd	nd	nd	nd	nd
C_4AF	nd	nd	nd	nd	nd

+++ major amount present   tr trace amount detected
 ++ substantial amount present
  + minor amount present   nd not detected

The interval of measurement precision of the intensity ratio of adequate diffraction lines has been calculated by using the variance formulae for the complex measurement, assuming that the examined samples of cements were sufficiently ground (under 7 μm). As an internal standard in the x-ray analysis of the quantity of CA, CA_2 phases β -quartz was used. The achieved interval of precision determination is $\pm 3\sigma$, reaching up to $\pm 4\%$ (Sawkow). The main analytic line of quartz was the 3.35 Å line, which is sufficiently isolated from the CA 2.97 Å and CA_2 3.49 Å lines. Estimating the accuracy of this method it has been shown that the influence of all admixtures in cement will not be larger than the influence of 2% Fe_2O_3; i.e., the component with the largest mass absorption coefficient.

2.1   Calcium aluminate forming
In order to examine the formation of calcium aluminate the mixture of aluminum hydroxide and limestone with a grain size under 60 μm was fired. The aluminum hydroxide

contains 64.41% Al_2O_3 and 0.94% of all admixtures;
limestone 55% CaO and 1.07% of admixtures (SiO_2 + Fe_2O_3 + MgO).

Two batches have been prepared which meet the requirements of alumina cements Gorkal 60 and Gorkal 70. In one of the batches the cement contained 67% Al_2O_3 and 33% CaO and in the other one 72% Al_2O_3 and 28% CaO. During ¼ hour the mixtures were fired in alumina (corundum) crucibles in a temperature range of 900° to 1500° C.

In table 2 phases occurring in the two fired batches are given. All existing aluminates form almost simultaneously. With the temperature rise the sinter composition approaches to the state of equilibrium, defined by the initial composition. In both raw material batches, fired at 1100°C, the amount of $C_{12}A_7$, CA and CA_2 are getting close in quantity.

Table 2.   Phases in raw material mixtures after firing at 1000°C - 1500°C

Batch firing temperature	Phases detected by x-ray analysis	
	A batch containing 67% Al_2O_3 /Gorkal 60/	A batch containing 72% Al_2O_3 /Gorkal70/
1000° C	Ca(OH)$_2$, CaCO$_3$, α-A CA, C$_{12}$A$_7$	CaCO$_3$, Ca(OH)$_2$, α-A CA, C$_{12}$A$_7$
1100° C	CA, CA$_2$, C$_{12}$A$_7$, CaO	CA, CA$_2$, C$_{12}$A$_7$, CaO
1200° C	CA, CA$_2$, C$_{12}$A$_7$, CaO	CaO, C$_{12}$A$_7$, CA, CA$_2$ α-A
1300° C	CA, CA$_2$, C$_{12}$A$_7$ CaO	CA, CA$_2$, C$_{12}$A$_7$
1400° C	CA, CA$_2$, C$_{12}$A$_7$	CA, CA$_2$
1500° C	CA, CA$_2$	CA, CA$_2$

These aluminates form almost simultaneously. With the temperature rise the amount of CA increases, it also follows up to 1300°C a growth of CA$_2$ (Fig. 1).

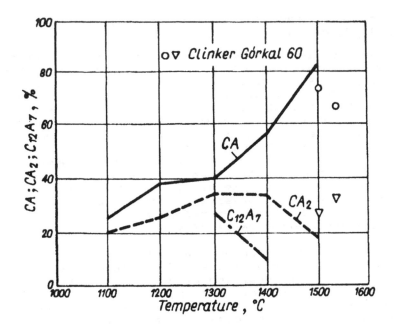

Fig. 1   Cement composition depending on firing
           temperature. Initial batch: 67% Al_2O_3 +
           33% CaO. Holding 15 min at firing
           temperature.

Then the amount of CA_2 increases and of $C_{12}A_7$ decays, this
can be explained with the reaction in a solid state :

$$C_{12}A_7 + 5CA_2 = 17CA$$

On Fig. 1 has been marked with wheels and triangles the
content of CA and CA_2 in an industrial clinker Gorkal 60.
Otherwise is the course of aluminates forming in a batch
with 72% Al_2O_3 content. The real CA + CA_2 content, also
quantitative x-ray analysed in sinters fired at 1300°,
1400° and 1500°C, amounts respectively: 20 + 48%, 42 +
50%, 42 + 57%. Therefore already at 1300°C 48% of CA_2 is
forming which amount increases slightly at 1500°C (up to
57%). Also a good agreement with the stoichiometrical
content of CA_2 (53.2%) and with the real amount (57.0%) in
this batch was achieved. But such an alumina cement with
majority of CA_2 has worse service properties than a cement
with a considerable CA amount. Therefore for further
investigations initially was accepted a clinker with a
greater amount of CA (67% Al_2O_3).

31

3 Properties of synthetic alumina cements with χ -Al$_2$O$_3$ content

The refractory alumina cements were manufactured from two clinkers, containing 63.51% Al$_2$O$_3$ and 66.90% Al$_2$O$_3$. The α -Al$_2$O$_3$ phase was introduced in the form of sintered aluminium oxide. A co-grinding of sintered clinker and of corundum with 98% Al$_2$O$_3$ content was used. The admixture level in each of the clinker did not exceed 2%. The amount of sintered aluminium oxide in synthetic cements reached 10 and 15%. From the data given in table 3 we can see, that α -Al$_2$O$_3$ increases refractoriness of cement and effects advantageously the strength of cement slurry and of standard alumina concrete after 12 and 24 hours. The corundum addition also causes the increase of refractoriness under load, therefore extends the temperature application range of cement in refractory concrete.

It should also be stressed, that the strength of synthetic cements is higher than of the industrial cement Gorkal 60 and even exceeds the strength of the French cement Secar 71 after 1 day setting (Gouda et al 1975).

Of special interest are the linear changes of the alumina concrete based on cement with 10% of χ -Al$_2$O$_3$ content (Fig. 2).

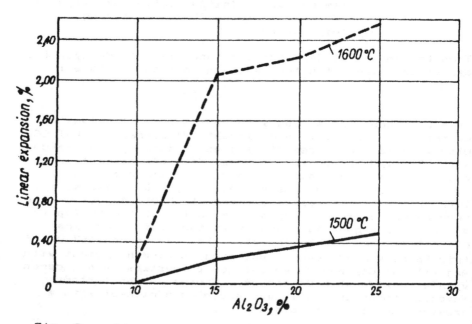

Fig. 2   Linear changes of alumina concretes based on cements with 10-25% of α -Al$_2$O$_3$ after firing at 1500°, 1600° C

Table 3. Properties of refractory alumina cement with corundum addition

Properties	Components of the cement					
	100% of clinker 63.51% Al₂O₃	90% of clinker 63.51% Al₂O₃ + 10% corundum	85% of clinker 63.51% Al₂O₃ + 15% corundum	100% of clinker 66.9% Al₂O₃	90% of clinker 66.9% Al₂O₃ + 10% corundum	85% of clinker 66.9% Al₂O₃ + 15% corundum
Al_2O_3 content in cement, %	63.51	67.33	69.51	66.90	70.22	71.66
Refractoriness of cement °C	1540	1610	1650	1580	1670	1710
Specific surface area blaine, cm²/g	4.180	4.480	4.260	4.260	4.320	4.140
Cold crushing strength of hardened cement slurry after 12h/24h MPa	22.3/30.7	29.8/35.9	28.6/30.8	24.8/25.1	33.5/42.1	27.7/35.6
Cold crushing strength of standard alumina concrete after 24h, W/C = 0.4 MPa	39.0	61.1	50.4	34.9	65.8	46.3
Cold crushing strength of standard alumina concrete after firing at 1500°C 2h MPa	31.4	1600°C 38.6	35.4	28.3	37.9	34.2
Refractoriness under load of standard alumina concrete $t_{0.0}$ °C / $t_{0.4}$ °C	1430 / 1560	1590 / 1660	1620 / >1700	1440 / 1580	1680 / 1700	1650 / >1700

Notes:
1. The cement slurry produced with a proper water amount is determined with the Vicat apparatus;
2. The composition of standard alumina concrete used in strength estimations (beams 40x40x160mm), 35% grains - 0.15mm diameter, 30% grains - 0.55mm diameter, 35% grains - 2.20mm diameter, cement content - 25% fused alumina content - 75%;
3. Refractoriness under load was determined on standard alumina concrete samples after drying at 110°C (cement content - 20% fused alumina content - 80% W/C = 0.4).

The expansion of such concrete after firing at 1600°C amounts to 0.19% and the cold crushing strength to 38.6 MPa. With an α-Al_2O_3 content above 10% during firing of the concrete can form at the corundum grain boundary a considerable amount of CA_6 causing cracking of alumina concrete (Sawkow).

Because a high hydraulic activity of cement with the content of $C_{12}A_7$ phase has been stated therefore investigations of the phase composition of a hydrated cement were carried out using the method accepted in the Institute of Refractory Materials.

Cement samples were treated with water in the water-cement ratio of 0.50. Then they were stored at a temperatures of 22° to 24°C for 24 hours. After this period the phase composition was x-ray analysed and the hydration degree determined.

The hydration degree was determined according the following law :

$$SH = 1 - \frac{I_t}{I_o}$$

where I_t - intensity of diffraction line of the phase in hydrated cement
I_o - intensity of diffraction line in a dry cement

The results of carried out investigations are given in table 4.

In spite of a low W/C coefficient in hydrated cements occur only CAH_{10} and C_2AH_8 phases. The hydrate C_3AH_6 does not occur and it can be formed with CA by a much lower W/C of 0.46 than the CAH_{10} hydrate. If the transformation of hexagonal hydrates into C_3AH_6 and AH_3 takes place, then at the ratio W/C = 0.4 a decrease of the cold crushing strength of about 50% will follow (Taylor 1966). But at the temperature 20° to 24°C such a transformation in cement slurries and alumina concretes has not been found (Sawkow 1988).

One can see that then - when $C_{12}A_7$ occurs in cement - the hydration degree of CA is much higher. Also the hydration degree of CA_2 is higher in a cement with CA than in that one containing only CA_2. On account of the highest hydration degree in a cement containing 63.51% Al_2O_3 an high refractory alumina cement was produced with 50% of α-Al_2O_3. The refractoriness of this cement amounts to 1770°C, 1 day cold crushing strength: of a slurry - 20,6 MPa, of a standard concrete - 38,4 MPa, the cold crushing strength of the standard alumina concrete after firing at 1800°C (2 h) - 58,8 MPa. This cement is characterised by a moderate annealing time (beginning - 54 min., ending - 92 min. at H_2O = 26.6%) and a specific surface area close to refractory cements - 4580 cm²/g

Table 4. The phase composition of hydration degree of hydrated cements

Al_2O_3 content in cement	Phases of cement slurries after 24h	Hydration degree (SH) of the phase after 24 h		
		$C_{12}A_7$	CA	CA_2
63,51%	CAH_{10}, C_2AH_8	100%	76,0%	-
66,90%	CAH_{10}, CA, CA_2	-	18,4%	25,5%
62,88%	CAH_{10}, C_2AH_8	100%	68,8%	-
65,86%	CAH_{10}, CA_2	-	29,7%	20,4%
76.90%	CA_2, C_2AH_n, CAH_{10}	-	-	13.7% 3 days after

Notes: 1. The hydration degree of x-ray analysed phases in dry cement
2. The specific surface of examined cements oscillated between 4180 - 4520 cm^2/g

Blaine. Other properties of the cement containing nearly 80% of Al_2O_3 were not examined because a similar alumina cement was produced in Poland before (Sawkow 1978).

On the base of investigation results it was acknowledged that to produce such a cement a calcium aluminate clinker with with $C_{12}A_7$ content is required, whereby the development of the surface, both of cement and its component corundum is essential.

4   Properties of refractory alumina cements based on clinker with a different sintering degree

Investigations were carried out on cements produced of industrial clinkers, after determining the litre weight, which were ground in a steel vibrating mill. Those clinkers with a sintering degree above 1400 g/l were ground 2-4 hours longer than the other ones. The most important properties and the chemical composition of cements are given in table 5.

Table 5. Properties of cements based on different calcium aluminate clinkers

Investigated Properties	Litre weight of clinker g/l							
	Clinker CA				Clinker CA$_2$			
	1260	1310	1460	1540	1250	1300	1350	1400
Al$_2$O$_3$ content in cement, %	63.46	62.24	62.14	62.31	76.66	77.25	75.70	76.20
X-ray analysed phased composition main phase	CA	CA	CA	CA	CA$_2$	CA$_2$	CA$_2$	CA$_2$
remaining phase (traces)	-	C$_{12}$A$_7$	C$_{12}$A$_7$	C$_{12}$A$_7$	C$_6$AS	C$_6$AS	C$_6$AS	C$_6$AS
Refractoriness °C	1540	1520/1540	1540	1540	1750	1770	1750	1750
Specific surface according to Blaine cm²/g	4320	4480	5020	4210	5040	4520	4870	4450
Cement density g/cm³	2.89	2.90	2.90	2.92	2.90	2.91	2.91	2.92
Crushing strength of standard alumina concrete after 24 hours (W/C = 0.4), MPa	66.0	76.8	69.1	83.4	12.2	11.7	12.4	7.9
Crushing strength of standard alumina concrete after firing at 1500°C (2 hours), MPa	41.2	30.9	25.2	31.9	34.1	37.8	59.5	23.8
Linear changes after firing at 1500°C (2 hours) of standard concrete, %	+0.47	+0.68	+0.74	+1.00	-0.45	-0.75	-0.92	-0.33

Note : The composition of standard alumina concrete used in strength estimations
(beams 40x40x160mm), 35% grains 0.15mm diameter
30% " 0.55mm "
35% " 2.20mm "
cement content - 25%
fused alumina content - 75%

Cements based on clinkers weighing (litre) 1310, 1460, 1540 - except CA - contained also $C_{12}A_7$. In cements containing 76-77% Al_2O_3 CA_2 and C_2AS (traces) were x-ray detected. Among cements containing 62-63% Al_2O_3 a similar 1 day cold crushing strength (66-83 MPa) have also those based on clinkers 1310, 1460, 1540 g/l. The higher strength of these cements may be explained with a complete hydration of the $C_{12}A_7$ phase and - as already shown - with an increase of the hydration degree of CA in the presence of $C_{12}A_7$. The strength of CA_2 cements is lower than of cements containing CA and changes slightly with the sintering degree of the clinker. Alumina concretes based on cements containing CA_2, in comparison with concretes based on cements containing mainly CA, have a firing shrinkage (2 h at 1500°C). This is due to a low hydraulic activity of CA_2 and to the presence of the non-hydraulic gehlenite /C_2AS/. During CA_2 hydration a decrease of CAH_{10} forming rate takes place, because of an amorphous AH_3 forming. That also effects the strength (Leers).

The achieved results entitles to state that the sintering degree of alumina clinker, containing mainly CA or CA_2 should not exceed 1300 g/l.

5   Conclusion

The real CA and CA_2 content in pure alumina cements containing 67% and 72% Al_2O_3 and fired up to 1500°C has been analysed by the quantitative method with an internal standard. As an internal standard β-quartz was used, which main diffraction line 3,35 Å was an analytical line. Beginning from 1400°C changes of the phase composition are proceeding in the direction defined by CaO and Al_2O_3 content and they end at 1500°C. After firing at 1500°C a cement containing less Al_2O_3 contains 81% CA and 17% CA_2, the other one 42% CA and 57% CA_2.

Alumina cements having a considerable CA content and a small amount of $C_{12}A_7$ are hydrating earlier than cements with CA_2 admixtures. This shows the hydration degree of CA in alumina cements containing below 64% of Al_2O_3. The hydration degree of CA determined by x-ray after 24 hours in hardened cement slurry containing 63.51% of Al_2O_3 amounts 76% whereas the hydration degree of CA in a cement slurry containing 65.86% Al_2O_3 amounts 29.7%. Therefore cements based on clinkers containing mainly CA and a proper quantity of $C_{12}A_7$ are characterised by good service properties.

These properties are additionally improved by corundum which is a component of cement containing Al_2O_3 near to 70% and also to 80%.

Investigating the properties of cements based on alumina clinkers with different sintering degree there could not be found out for the standard alumina concrete a proportional strength dependence on clinker sintering degree.

6 References

Bushnel-Watson, S.M. and Sharp, J.H. (1986) The effect of temperature upon the setting bahaviour of refractory calcium alumina cements. Cement and Concrete Research, 16, 6, 875-884.

Grzymek, J. (1988) Wykorzystanie mineralnych odpadow wtornyck. Wydawnictwo PAN, Warszawa, 184-185.

Gouda, G.R. and Roy, D.M. (1975) Properties of hot pressed calcium aluminate cements. Cement and Concrete Research, 5, 6, 551-563.

Kurdowski, W. (1965) Rodzaje cementow produkowanych we Francji. Cement-Wapno-Gips, 20, 7, 209-210.

Kurdowski, W. (1981) Poradnik technologa przemyslu cementowego. Wydawnictwo Arkady, Warszawa, 128-131.

Leers, K.J. Reaktionen bei der Erhartung von Tonerdezementen. Doktor-Dissertation. Bergakademie Clausthal.

Sawkow, J. (1972) Mozliwosci wytwarzania ogniotrwalych cementow glinowych. Cement-Wapno-Gips, 27, 12, 370-374.

Sawkow J. Wplyw stosunku CA/CA_2 na wlasnosoci uzytkowe ogniotrawalych cementow glinowych. Praca doktorska AGH, Promotor: Prof. Dr. J. Sulikowski.

Sawkow, J. (1988) Polish high-alumina refractory cement grade. Polish Technical Review, No.6 5-7.

Sawkow J. (1978) Wysokoogniotrwale cementy glinowe. Cement-Wapno-Gips, 31, 12, 348-356.

Sulikowski, J., Sawkow J. (1978) Einfluss des Verhaltnisses von CA/CA_2 auf die Eigensochaften von Tonerdezement und korundhaltigem Feuerbeton. Silikattechnik, 29, 6, 165-169.

Taylor H.F.W. (1966) The chemistry of cements. Royal Inst. of Chemistry, Lecture Series, No.2, 18-20.

PART TWO
HYDRATION

4 MICROSTRUCTURAL DEVELOPMENT IN PASTES OF A CALCIUM ALUMINATE CEMENT

KAREN L. SCRIVENER, H.F.W. TAYLOR
Department of Materials, Imperial College, London, UK

Abstract
The hydration of white calcium aluminate cement pastes at 5° and 40°C was investigated. At both temperatures CAH_{10} was the earliest product detected. At 5°C CAH_{10} persisted up to at least 7 days hydration. At 40°C C_2AH_8 and AH_3 formed from the CAH_{10} after a few hours, C_2AH_8 subsequently reacting to give C_3AH_6 and more AH_3. A distinct 'inner product' was noted, the nature of which is discussed.
Keywords: Calcium aluminate cement, Hydration, Microstructure, Bse images, CAH_{10}, C_2AH_8, C_3AH_6, AH_3.

1 Introduction

The hydration chemistry of calcium aluminate cements is relatively well understood, but few studies have been made on the development of micro-structure. Cottin (1971) examined replicas by TEM and described a number of microstructures, in some of which prismatic crystals of CAH_{10} or platey ones of C_2AH_8 could be seen. Halse and Pratt (1986) reported SEM studies, mainly on fracture surfaces. Backscattered electron (bse) images of polished sections, coupled with image analysis and X-ray microanalysis (Scrivener 1989), have provided much information on the development of microstructure in Portland cement pastes and concretes. In the present work, these techniques are applied to pastes of a calcium aluminate cement. Because of its relatively simple chemical composition and microstructure, Secar 51 was selected. This cement is characterised by intermediate aluminium, low iron and relatively high silica content.

2 Experimental

The pastes were mixed at w/c = 0.38 in CO_2-free air at room temperature and cast in plastic tubes 10 mm in diameter. They became noticeably warm during mixing. The tubes were sealed and, within 15 min from commencement of mixing, were placed in water at 5° or 40°C. After 6 h, one end of the tube was opened to allow access to a large volume of water. Except where otherwise stated, the results relate to material within 1-2 mm of the surfaces of the cylinders, which sometimes differed from that further in, perhaps due to temperatures gradients

Table 1. Backscattering coefficients

Perovskite	0.18-0.19	C_3AH_6 *	0.140	
Belite	0.166	C_2AH_5 *	0.135	
Gehlenite	0.156	C_2ASH_5 *	0.134	* Compositions following
Glass	0.151	$C\bar{A}H_4$ *	0.125	partial dehydration in
CA	0.146	AH_3	0.110	the high vacuum

caused by the heat evolved by the hydration reactions.

The principal technique used was examination of flat, polished sections in the SEM, operated at an accelerating voltage of 15 kV. The sections were imaged with bse using a detector designed to maximize contrast from compositional differences. In general phases were ident- ified in the bse images from a combination of grey level (brightness), morphology, composition determined by X-ray microanalysis and infor- mation from X-ray powder diffractometry (XRD) of the bulk specimen.

The grey level of a given phase depends approximately on its average atomic number, or, more exactly, on its backscattering coefficient. This quantity may be calculated from the composition (Goldstein et al 1981). Table 1 gives values for the phases encountered. They take into account observed ionic replacement found by X-ray microanalysis and, for the hydrated phases, the probable effects of partial dehydration in the high vacuum.

X-ray microanalyses were made on the polished sections using an energy dispersive detector. ZAF corrections were applied. Analysis totals include oxygen calculated by stoichiometry from the cations determined, but exclude H_2O.

For some samples, the volume percentages of phases were determined by image analysis of the bse images. It was possible to distinguish between anhydrous phases, hydration products and pores larger than about 500 nm (smaller pores being included with the hydration products). Individual anhydrous phases could be distinguished from each other in most cases, and in a few specimens some of the hydration products could also be differentiated. However, the values thus obtained were less reliable than those for the three broader categories. All the image analysis results are averages for at least 10 fields about 150 μm square, and for up to 20 if the heterogeneity of the sample was particularly coarse. It is difficult to estimate the accuracy of the image analysis results, but the error is probably less than 10 % of the average.

3 Results

3.1 Starting Material
The bulk composition of the Secar used is given in Table 2.

Table 2. Chemical analysis of the cement (weight %)

Al_2O_3	52.4	FeO	0.8	SO_3	0.05	loss	0.2
CaO	38.0	Fe_2O_3	0.4	S	0.01	IR	2.0
SiO_2	5.5	Na_2O	0.09	MgO	0.0		
TiO_2	2.6	K_2O	0.1	CO_2	0.0		

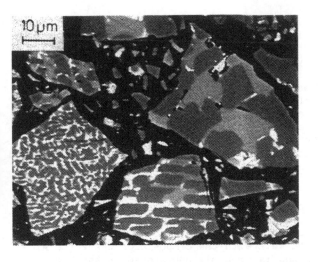

Fig. 1.
Bse image of cement.
The two major phases
are CA (darker grey)
and gehlenite
(lighter grey).
The bright minor
phase is perovskite

Bse images of the clinker and cement (Fig. 1) with X-ray microanalysis showed the main phases to be CA and gehlenite, with smaller
proportions of perovskite, glass and belite. XRD results were in
agreement. The CA and gehlenite were intergrown on a scale varying
from < 1 μm to about 20 μm. Most grains of the cement were poly-
mineralic. X-ray microanalysis showed that the CA contained < 1 % of
substituents. The gehlenite composition (wt %) was: MgO, 0.1; Al_2O_3,
39.4; SiO_2, 15.7; CaO, 41.1; TiO_2, 2.8; (FeO+Fe_2O_3), 0.7. This gives
the formula $(Ca_{4.13}Fe^{II}_{0.03})(Mg_{0.01}Al_{4.36}Si_{1.44}Ti_{0.19}Fe^{III}_{0.02})O_{14}$,
referred to 14 oxygens atoms and assuming $Fe^{2+}/Fe^{3+} \approx 2.0$ as in the
bulk analysis. The deficiency in tetrapositive cations compared with
the ideal formula $Ca_4(Al_4Si_2)O_{14}$ thus appears to be balanced by
incorporation of extra Ca^{2+} ions. This situation may alternatively be
described as solid solution with C_5A_3, which is structurally closely
related to gehlenite. The perovskite regions were too small for
satisfactory analysis, but there were indications of partial
replacement of Ti^{4+} by Al^{3+} balanced by O vacancies.

The results of image analysis (Table 3) showed the volume
composition to be approximately 59 % CA, 3 % perovskite and 38 %
gehlenite and other phases. This 38 % probably consists of about 28 %
gehlenite, 5 % glass and 5 % belite. Calculations from the bulk
analysis and the compositions and densities of the individual phases
were in broad agreement, but suggested smaller proportions of the more
siliceous phases. The bse images and laser diffractometry indicated
that the particle size distribution of the Secar was similar to that of
a typical OPC but with fewer large particles (> 50 μm).

3.2 Individual Hydration products
The term 'inner product' is used to denote products formed in space
clearly seen to have been occupied previously by anhydrous phases,
usually CA. All other hydration products are called 'undesignated
product'. This term comprises products formed in space previously
occupied by water ('outer product') or by anhydrous grains of which the

outlines cannot be seen, either because they were too small or because of microstructural reorganisation. These forms of undesignated product could not be distinguished from each other.

CAH_{10}, when sufficiently well crystallised, formed networks of interlocking prisms, individually about 1 μm long and 0.1 μm wide. These prisms are barely resolved in bse images (Fig. 2(a)). X-ray microanalyses of regions of these networks gave Ca/Al ratios usually above the theoretical value of 0.5 (typically around 0.7) and analysis totals below that of 69 % calculated for CAH_4, to which CAH_{10} is probably dehydrated in the high vacuum. Observations with different counting times, or analysing 1-2 μm rasters instead of spots, indicated that Al was probably lost under the electron beam. A little Si was usually present (Si/Al = 0.05-0.1), perhaps due to intermixed small crystals of gehlenite.

C_2AH_8 was often recognizable as well formed plates, seen on edge in polished sections, and typically about 1 μm thick and 20-30 μm across (Fig. 2(b)). They were usually clustered into aggregates. X-ray microanalyses usually agreed well with the composition C_2AH_5 (Ca/Al = 1, total 70 %), to which C_2AH_8 is probably dehydrated in the high vacuum. Some analyses showed up to 3 % of SiO_2, perhaps due to intergrown strätlingite (C_2ASH_8). The latter, when present in larger amounts, formed plates similar in size to those of C_2AH_8, but of lower grey level and higher in SiO_2. The Si/Ca ratios were typically about 0.4, suggesting the presence of intergrown C_2AH_8.

C_3AH_6 was characteristically observed as aggregates of equi-dimensional nodules, individually about 0.2 μm in diameter (Fig. 2(c)). X-ray microanalyses of such aggregates gave Ca/Al ratios of 1.2-1.3, with totals of 60-70 % and up to 4 % of SiO_2. The composition and morphology suggest that the nodules were of C_3AH_6 (Ca/Al = 1.5; total, 71 %), probably admixed with some combination of AH_3, gehlenite and strätlingite.

AH_3 was generally observed as apparently structureless regions of low and uniform grey level (areas of matrix in Fig 2(c)), which could be readily identified by X-ray microanalysis. Rarely, it formed pseudomorphs after C_2AH_8.

An inner product replacing CA was observed in many paste samples. It was apparently structureless and gave Ca/Al ratios of 1.2-1.3 and analysis totals around 70 %. It was also observed in a clinker grain that had partially hydrated during storage (Fig 2(d)); on the surface of this grain, adjacent to the inner product, was a deposit of AH_3. This observation shows that the inner product retains all the CaO that was present in the CA, and part of the Al_2O_3, the rest of which is expelled. The nature of this inner product is uncertain. The Ca/Al ratio and analysis total could be explained by assuming it to be C_3AH_6 plus as much of the accompanying AH_3 as the space allows; however, neither C_3AH_6 nor AH_3 could be detected by XRD in some of the pastes for which bse images showed substantial amounts of this inner product to be present, so it may be poorly crystalline or amorphous.

The glass particles reacted relatively quickly and formed reaction rims often with a complex structure of up to four layers. These were too narrow for satisfactory analysis.

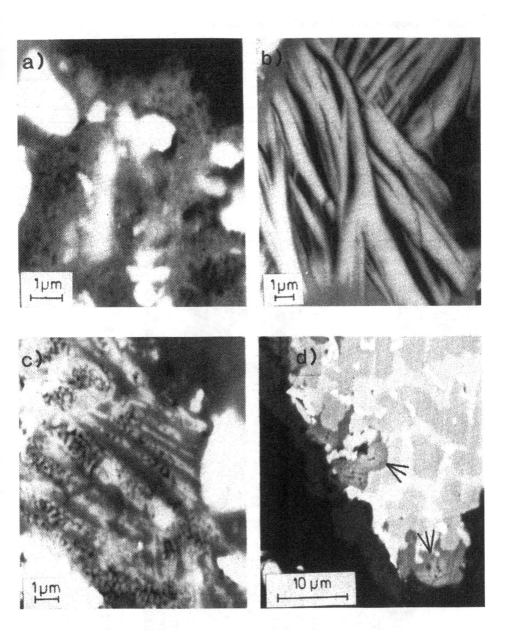

Fig. 2. Microstructures of hydration products.
a) Network of prisms of CAH_{10} (grey) and unreacted cement (bright).
b) Plates of C_2AH_8.
c) Array of fine nodules of C_3AH_6 in a matrix of AH_3 with some unrected cement.
d) Areas of inner product (arrowed) which has replaced CA in a partially reacted clinker grain; along the right edge of the grain is a deposit of AH_3.

Table 3. Image analysis results (volume %)

	Per	G+ G+B	CA	Total Anhyd	Products	Pores
cement	3	38	59	100	-	-
paste on mixing (theoretical)	1	18	28	47	-	53
5°C						
6 hours				29	50 (CAH_{10},UD)	21
7 days (edge)	1	11	14	26	71 (8 IP, 63 UD)	3
7 days (int)	1	8	7	16	61 (16 IP, 45 UD)	23
40°C						
2 hours				36	38 (CAH_{10}, UD)	26
4 hours	2	8	6	16	74 (11 IP, 22 AH_3, 41 other UD)	10
7 days	2	5	1	8	76 (31 AH_3, 31 C_3AH_6, 14 strät & other)	16

Per - perovskite; G+G+B - gehlenite, glass and belite
IP - inner product; UD - undesignated product; strät - strätlingite

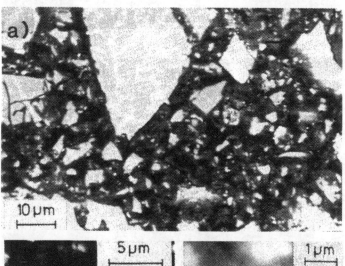

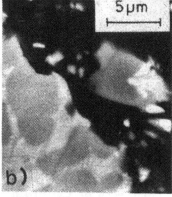

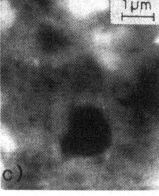

Fig. 3. Microstructure of paste hydrated at 5°C for 6 hrs.
a) Overview: partly reacted anhydrous grains (bright) in a matrix of CAH_{10} with large open pores (black).
b) Preferential reaction of CA has left protruding edges of gehlenite and perovskite.
c) Hollow shell of CAH_{10} surrounding a void probably left by the complete hydration of a small grain.

46

3.3 Development of microstructure at 5°C

The bse images showed that, after 6 hours, the microstructure consists of partly reacted grains in a matrix of CAH_{10} prisms, together with large open pores (Fig. 3(a)). Examination of the larger grains showed that reaction was mainly of the CA, which in places had dissolved away to leave protruding gehlenite and perovskite (Fig. 3(b)). CAH_{10} was also observed as hollow shells surrounding voids a few micrometres in size which may have been formed by the complete hydration of small grains (Fig. 3(c)).

With longer hydration times the amount of CAH_{10} increased and it could be seen that more of the cement had reacted. Reaction was still predominantly of the CA and small amounts of inner product, as described previously, could be seen. The 7-day old paste showed a particularly marked change in microstructure with distance from the edge of the specimen. In the outermost millimetre, very little porosity was detectable in the bse images (Fig. 4(a)). Further in, the microstructure of the undesignated product was much more open, and its motley texture suggested that conversion to C_3AH_6 and AH_3 could have occurred; much more inner product had formed from the CA (Fig. 4(b)). Image analysis confirmed that the ratio of inner to undesignated product was higher in the interior than in the surface region and also showed the percentage reaction to be higher.

XRD showed CAH_{10} as the only detectable product in any of the specimens examined. Comparisons of peak heights of CA and gehlenite with those of the perovskite, which was assumed to be inert, indicated substantial reaction of the CA, but, except possibly in the interior of the 7-day-old paste, no significant reaction of the gehlenite. Image analysis showed considerable reaction of the fraction that included gehlenite (Table 3); this was presumably largely or wholly of glass or belite. In agreement with the image analysis results, XRD showed that

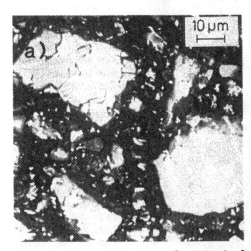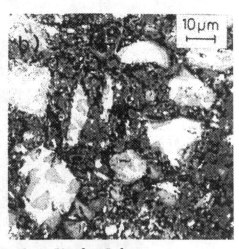

Fig. 4. Microstructure of paste hydrated at 5°C for 7 days;
a) near surface - partially reacted cement in a matrix of CAH_{10};
b) interior - motley undesignated product with high porosity and significant amounts of inner product

more CA had reacted in the interior of the 7-day-old sample than near the edge. The SEM evidence is compatible with the hypothesis that temperature increases within this specimen caused increased reaction and partial or complete conversion, but the XRD evidence suggests that any products other than CAH_{10} are of low crystallinity or small in amount or both.

3.4 Development of microstructure at 40°C

The samples hydrated for 1 or 2 hours at 40°C were microstructurally similar to the younger 5°C samples, with CAH_{10} as the only identifiable product. This was confirmed by XRD. Image analysis indicated that 23% of the cement had hydrated at 2 hours

By 4 hours, the microstructure had changed markedly. It contained highly porous regions some 100-200 μm in size separated by regions that were relatively dense (Fig. 5(a)). The undesignated product consisted mainly of fine nodules, probably of C_3AH_6 but too small for analysis, embedded in a matrix of AH_3. Occasionally large plates of C_2AH_8 were observed. The fine nodules formed arrays, sometimes mixed with remnants of C_2AH_8 plates (Fig. 5(b)); this suggested that C_3AH_6 had

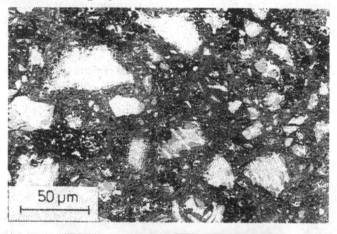

Fig. 5. Microstructure of paste hydrated at 40°C for 4 hours. a) Overview: distribution of products is very uneven with large porous areas (dark), a fair amount of inner product can be seen in the partially reacted grains. b) Detail: fine nodules of C_3AH_6 and remnants of C_2AH_8 plates (both mid grey) can be seen in a matrix of AH_3 (v. dark grey), in the partially reacted grains (bright) some of the CA has been replaced by inner product (mid grey).

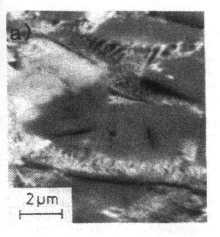 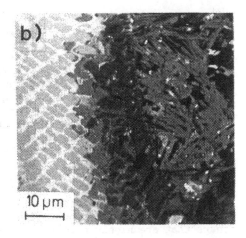

Fig. 6. Microstructure of paste hydrated at 40°C for 7 days.
a) Motley areas were originally finely mixed gehlenite and perovskite;
   the gehlenite has reacted to leave voids in the perovskite network.
b) From the left: a mixture of unrected CA (l.grey), gehlenite and
   perovskite (bright); inner product (m.grey); plates of stätlingite
   (d.grey) spanning original grain boundary; plates of C_2AH_8 (m.grey).

formed locally from reaction of the C_2AH_8. A considerable amount of
inner product was observed, mainly replacing the CA. Image analysis
indicated that some gehlenite had also reacted. XRD indicated that
over 50% of the CA had reacted and that the main product was C_2AH_8 with
a little residual CAH_{10} but no detectable C_3AH_6.

By 7 days, the bse images showed that almost all the CA had reacted.
In some grains in which the gehlenite and perovskite were finely mixed,
the gehlenite had clearly reacted, leaving voids (Fig. 6(a)). Much of
the undesignated product consisted of AH_3 interspersed with regions of
brighter material which was sometimes nodular, sometimes platey and
sometimes intermediate between the two. X-ray microanalyses suggested
that most of these bright areas were of C_3AH_6, though some analyses
indicated that some C_2AH_8 remained. The usual inner product of CA was
present. Occasionally, plates of strätlingite were observed in the
undesignated product and also extending from the latter into the partly
reacted cement grains (Fig. 6(b)). Some plates consisted partly of
C_2AH_8 and partly of strätlingite, which had possibly formed from the
C_2AH_8 by reaction with silicate ions. XRD confirmed that nearly all
the CA and a substantial proportion of the gehlenite had reacted and
showed the major products to be C_3AH_6 and AH_3, with some strätlingite.

4 Discussion and conclusions

The microstructures of some of the pastes examined varied markedly with
distance from the surface on a millimetre scale, possibly resulting
from increase in temperature within the specimen due to heat evolved in
the hydration reactions. It is therefore essential to examine a number
of suitably chosen fields in any given specimen.

At 5°C the earliest hydration process of CA detected in the present work was dissolution followed by precipitation of CAH_{10} in the undesignated product, leaving empty space where the CA previously existed. As reaction proceeds, this process is accompanied by one in which an inner product is formed. This product has a Ca/Al ratio of 1.2-1.3. It was not detected by XRD and is either amorphous or poorly crystalline, in the latter case probably a mixture of C_3AH_6 and AH_3. The factors favouring its formation are not clear. Possibilities include local temperature increase and high concentrations resulting from the constricted nature of the regions in which it is formed.

At 40°C CAH_{10} and inner product were again the earliest products to be detected. C_2AH_8 and AH_3 appear to form by dissolution of CAH_{10} followed by precipitation in the undesignated product, and C_3AH_6 and additional AH_3 by a similar mechanism starting from C_2AH_8. These products give rise to microstructures that are very uneven on a scale of around 100 μm, with some regions that are relatively compact and others that are highly porous. Major reorganisation of the structure evidently occurs. Apart from the formation of the inner product which may contain C_3AH_6, there was no evidence of the direct formation of C_3AH_6 and AH_3 from CA.

Evidence on the reaction of gehlenite at 5°C or in times less than a few days at 40°C was conflicting, but both SEM and XRD indicated that substantial reaction had occurred by 7 days at 40°C. The reactivity of this phase probably increases with the extent to which Si is replaced by Al plus extra Ca, since this is equivalent to solid solution with C_5A_3, which is highly reactive (Brisi et al 1986). The formation of strätlingite, detected both by SEM and XRD, provides further evidence for reaction of the gehlenite. Strätlingite seems to form partly by alteration of C_2AH_8, but crystals of it appeared also to displace C_3AH_6 and to extend into the dissolving cement grains forming inner product.

5 Acknowledgements

We would like to thank Lafarge Coppée Recherche for the Secar 51 cement and the chemical analysis. The XRD work was carried out at the University of Aberdeen. KLS would like to thank the Warren Research Fund of the Royal Society for financial support. HFWT would like to thank the London Centre for Marine Technology for a Visiting Professorship, which allowed him to take part in this work.

6 References

Cottin, B. (1971) Etude au microscope electronique de pâtes de ciment alumineux hydratées en C_2AH_8 et en CAH_{10}. **Cem. Concr. Res.** 1(2), 177-186.

Brisi, C., Bolera, M.L., Montanero, L. and Negro, A. (1986) Hydration of $5CaO.3Al_2O_3$. **Cem. Concr. Res.** 16(2) 156-160.

Goldstein, J.L., Newbury, D.E., Echlin, P., Joy, D.C., Fiori, C. and Lifshin, E. (1981) page 75 in **Scanning Electron Microscopy and X-ray Microanalysis**, Plenum Press, New York & London.

Halse, Y. and Pratt, P.L. (1986) The development of microstructure in high alumina cement, in **8th International Congress on the Chemistry of Cement**, Abla Graphica e Editora, Rio de Janeiro, vol. 4, pp. 317-321.

Scrivener, K.L. (1989) The microstructure of concrete, in **Materials Science of Concrete I** (ed J. Skalny), Am. Ceram. Soc., pp 127-161.

5 INVESTIGATIONS OF THE COMPOSITION OF PHASES FORMED IN LOW CEMENT CASTABLES DURING HYDRATION AND AFTER THERMAL TREATMENT

W. GESSNER, S. MÖHMEL
Central Institute of Inorganic Chemistry of the Academy of
Sciences of the GDR, Berlin, GDR
J. KIESER
Institute of VEB SKET, Weimar, GDR
M. HÄWECKER
College of Architecture and Building Materials, Weimar, GDR

Abstract
As a contribution to the discussion on the chemical
reactions during the hydration and the thermal treatment
of castables low in HAC - content (LCC), investigations on
the pure binder-system of LCC without aggregates are
demonstrated.

Samples containing HAC, microsilica, alumina and
deflocculant were hydrated on different conditions and
investigate by x-ray methods, ^{27}Al NMR spectroscopy and
the so called molybdate method. Information were expected
from the ^{27}Al NMR spectroscopy regarding the possible
occurrence of AlOx[Si] units in the hydration products,
and from the molybdate method concerning the reaction rate
of the microsilica.

The results show the formation of CASH phases in the
LCC binder-systems especially in those, which were treated
under hydrothermal conditions. It seems that the CASH
phases are of zeolitic nature with a range of stability up
to 600°C. The existence of gehlenite hydrate discussed in
literature cannot be established with certainty in the
investigated samples. A whole explanation of the relation
between the development of strength in the LCC - matrices
and the observed low concentration of CASH phases cannot
be given at this time. Nevertheless, the results show
that such relation exists.
Keywords: High Alumina Cement, Low Cement Castable, NMR
Spectroscopy, X-ray Diffraction, Gehlenite hydrate,
Thermal Treatment.

1   Introduction

It is generally known that conventional refractory
concretes containing High Alumina Cement (HAC) in an
amount of approximately 20% show a loss of strength during
thermal treatment and dehydration leading to negative
effects regarding the possibilities of application
especially at operating temperatures between 800°C and
1000°C. In recent years as alternative materials the so

called low cement castables (LCC) and ultralow cement
castables (ULCC) have been established which do not
exhibit this disadvantage (Fisher 1985 and Eguchi et al
1989). LCC and ULCC are characterised by a drastic
reduction of the cement content and an addition of highly
dispersed compounds such as silica fume (microsilica) and
alumina. The high reactivity of these materials leads to
modified reactions already during the hydration process in
respect to commercial castables. Also the dehydration
reactions and the formation of the ceramic bond are
influenced by this special reactivity. Whereas the
processes of hydration and dehydration of conventional
hydraulically bound refractory castables may be regarded
as largely explored, the adequate reaction in LCC are far
more complex and therefore still under discussion.

Relevant literature does not yet offer any
comprehensive interpretation of that subject. The results
presented here should be considered as a contribution to
the question of a possible relationship between the
formation of special phases and the strength behaviour.

2   Research program and methods

To improve the detectibility of the hydrate phases and
their dehydration products the investigations were carried
out on the pure binder system of LCC ("LCC binder matrix")
without aggregates. Therefore samples were prepared from
HAC, silica fume and a deflocculant respectively in
different portions as given in Table 1.

The samples were than mixed with water in an amount
also shown in Table 1 and treated at normal conditions
(20°C), at 105°C and in an autoclave at 180°C,
respectively.

For characterising the phase formation and conversion
qualitative x-ray diffraction methods were used.
Information concerning the degree of reaction of the
silica fume and the degree of condensation of silicate
anions in possibly occurring silicate hydrates and alumino
silicate hydrates, respectively, were expected by applying
the so called molybdate method (Thilo et al 1965).
Furthermore the solid-state ^{27}Al NMR spectra of the
samples were recorded which indicate to the coordination
number of aluminium and the eventual existence of AlOx[Si]
units in the reaction products (Müller et al 1984).

3   Results and discussion

During the hydration of samples containing HAC at normal
hardening conditions CAH_{10} and gibbsite (γ-AH_3) occur as
was to be expected; C_2AH_8 could not be observed with
certainty. By a treatment at 105°C the conversion of the

Table 1.   LCC binder matrix samples under investigation

Series		A	B	C	D	1	2	3
Mixture :								
HAC	%	60	33	33	100	37.5	23	
Alumina L	%			45			38.5	50
Alumina A	%		45					
Microsilica I	%	40	22	22				
Microsilica II	%					62.5	38.5	50
Hydration :								
Deflocculant	%	0.4	0.4	0.4	0.4	0.4	0.4	0.4
Water content	%	17.5	13.1	21.7	24.0	25.0	27.7	27.5
Hardening :								
Normal hardening (7 days)		x	x	x	x	x	x	
Drying at 105°C (24 hours)		x	x	x	x	x	x	x
Autoclaving at 180°C (24 hours)		x	x	x	x			
(8 hours)						x	x	x

Thermal Curing :

	Normal hardening	105°C	Autoclaving
200°C/24 hours	x	x	x
300°C/24 hours	x	x	x
400°C/24 hours	x	x	x
500°C/24 hours	x	x	x
600°C/24 hours	x	x	x

x = samples tested

calcium aluminate hydrates low in lime content into C_3AH_6 takes place. Hydrothermally treated samples show an increase in C_3AH_6 content. Gibbsite is dehydrated partly or completely (series D) into boehmite (γ-AH; see Fig. 1 Table 2). The presence of calcium hydroxide cannot be proven. However, the significant occurrence of calcite in the autoclave samples indicates primary portlandite formation followed by carbonation (Table 2). Reflexes with low intensity in the x-ray diffractograms in the range of d = 2.5 up to 13 Å are difficult to assign. But a statistical frequency distribution derived from the investigation of 32 samples suggests the existence of

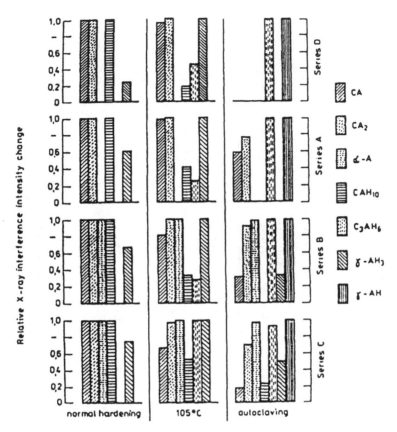

Figure 1. Relative concentration change in crystalline main components as a function of hardening conditions. LCC binder matrix / test series A-D.

calcium alumino silicate phases (CASH) having a zeolitic structure (Fig. 2). The formation of gehlenite hydrates C_2ASH_8 (stratlingite) discussed in literature (Seltveit et al 1986 and Monsen et al 1985) may be present, but cannot be established with certainty in the samples under consideration.

The low intensities of the reflexes characterising the CASH phases have to be attributed to small concentrations and/or to a high degree of disorder in the crystal lattices. The proven peak locations cannot be related to any unhydrated phases and CAH or ASH phases. Reflexes typical for CASH phases can be recorded under any hardening conditions. However, a dependence of number and intensity of the reflexes on the composition of the samples and the temperature conditions is not given. The difficulties as how to interpret these phenomena have to be regarded in connection with further process variables

Table 2

Distribution of hydrate phases detectable in
LCC binder matrix samples by x-ray diffractometry.

| Mixture — Binder Matrix | | | Hardening | Gibbsite AH_3 | Boehmite AH | Portlandite $CH^{x)}$ | CAH_{10} | Hydrogros-sularite C_3AH_6 | Straetlingite C_2ASH_8 | Zeolite CAS_mH_η |
HAC	Micro-silica	Alumina								
x			normal	xx			xx			
x			105°C	xxx			x			
x			autocl.		xxxx	x		xxx		
x	x		normal	x			xx			
x	x		105°C	xx			x		(x)	x
x	x		autocl.	x	x			xxxx	(x)	x
x		x	normal	x			xx			
x	x		105°C	xx			x	x		
x	x		autocl.	x	x		(x)	xx		
x	x	x	normal	} no hydrate phases detectable		x				xx
x	x	x	105°C	} no hydrate phases detectable	not investigated			x	(x)	x
x	x	x	autocl.	} no hydrate phases detectable				xx	(x)	x

x) indirectly classified by calcite identification

xxxxx maximum reflex intensity
x   minimum reflex intensity
(x)   no reliable proof

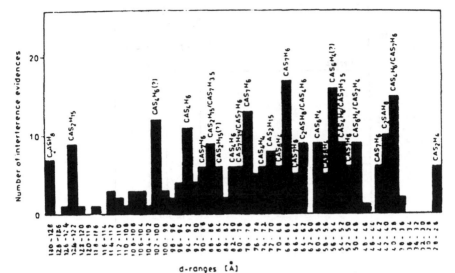

Figure 2. Statistical frequency distribution of
low-intensity x-ray peak locations based on 32 LCC
binder matrix samples and their possible classification
into CASH phases.

affecting the preparation of samples e.g. water content,
densification, grain-size effects and pH conditions.
Contrary to what might be expected, x-ray diffractometric
testing does not show any increase of CASH content at
hydrothermal treatment.

A somewhat clearer picture is obtained by proven
changes occurring in the content of the x-ray amorphous
silica fume under varying conditions of sample composition
and hardening. The distinctly marked maximum of the rise
of the base line in the x-ray diffractograms at d = 4.23 Å
of samples containing silica fume is suitable for a
semiquantative analysis. The results regarding the
difference in height Δ h (mm) between the continuously
running base line and its maximum rise at 4.23 Å show
(Fig. 3,4) a decrease of the silica fume content in the
hydrothermally treated samples and that the concentration
of silica fume correlates with Δh.

Whereas the differences between normal hardening and a
treatment at 105°C, respectively, and autoclave conditions
are significant in this respect a decrease of the portion
of silica fume cannot be verified unambiguously for the
change from normal hardening to hardening at 105°C (Table
3). The diminishing of concentration of the silica fume
does not lead to the formation of modifications of free
crystalline silica. Obviously the presence of calcium
aluminate phases is a premise for the reaction of the
silica fume as it is shown by comparison of the results of
investigation of the samples from series 1 and 3 (Fig. 3).

For getting information about the degree of reaction of

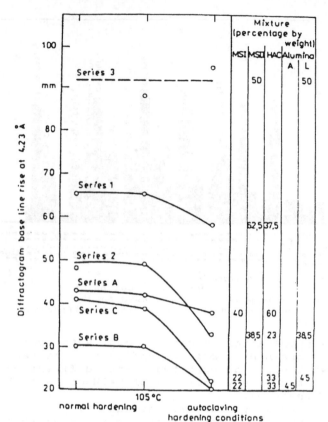

Figure 3. Degradation of amorphous microsilica components as a function of hardening and mixture conditions (LCC binder matrix samples).

the silica fume and the degree of condensation of the silicate anions formed, the samples were investigated by the molybdate method. This method is based on the reaction of the silicic acids being released from the silicates in the acid medium of the reagent solution, with molybdenic acid into the silico-molybdenum complex; the lower the degree of condensation of the silicate was, the faster this complex is formed (Thilo et al 1965).

Therefore, from the course of increase of the concentration of this complex in dependence of time (recorded spectrophotometrically as an increase of extinction) information is available about the condensation degree of the occurring silicate anions. From the final extinction value (E_∞) the total content of molybdate active species can be determined. In the case of the LCC binder system samples the portion of molybdate active species reflects the degree of reaction of the SiO_2 since the silica fume itself is molybdate inactive, whereas the formed CASH phases are molybdate active due to the fact that (Al-O-Si)-bonds will quickly be hydrolized in an acid medium. The condensation degrees determined from the time dependence of the increase of extinction are those of the remaining (Si-O-Si)-fragments. Therewith also a hint is given to the Al/Si-ratio of the CASH phases. High condensation degrees of the remaining silicate species are due to a small Al/Si-ratio ($\ll$ 1), whereas low degrees correspond to a high Al/Si-ratio ($\leq$ 1). For instance at Al/Si = 1 only monomeric silicate anions have to be expected.

In the case of the LCC binder system samples treated under normal hardening conditions only very small degrees of reaction of the silica fume are observed. After drying at 105°C the degree of reaction actually increases but still remains at small maximum values of 2.9% (Table 3). The condensation degree of the silicate anions is in the range of monomeric and dimeric species, respectively, as

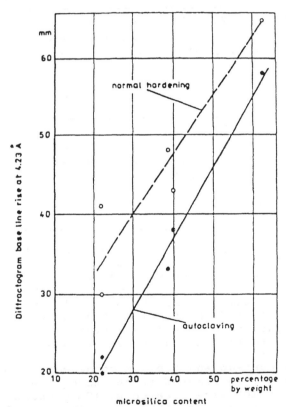

Figure 4. Relationship between the x-ray diffractogram base line rise and the microsilica content in LCC binder matrix samples containing HAC tested after normal hardening and autoclaving.

Table 3
Results characterizing the hydraulic bond in
LCC binder matrix samples

Series	A	B	C	D	1	2	3	Hydration
Composition %								
HAC	60	33	33	100	37,5	23		
microsilica	40	22	22		62,5	38,5	50	
alumina		45	45			38,5	50	
SiO_2 reacted %	0,8	1,2	1,1	–	–	–	–	
^{27}Al NMR	AlO_4	AlO_4	AlO_4	AlO_4				normal
	AlO_6	AlO_6	AlO_6	AlO_6				hardening
base line rise								
Δh (mm)	43	30	41	–	66	48	–	
XRD								
SiO_2 reacted %	2,9	2,0	1,4	–	–	–	–	
^{27}Al NMR	AlO_4	AlO_4	AlO_4	AlO_4	AlO_4	AlO_4		
	AlO_6	AlO_6	AlO_6	AlO_6	AlO_6	AlO_6		drying at
base line rise								105°C
Δh (mm)	42	30	39	–	64	49	88	
XRD								
SiO_2 reacted %	5,4	7,3	8,3	–	2,7	5,3	–	
^{27}Al NMR	AlO_4	AlO_4	AlO_4		AlO_4	AlO_4		
	AlO_6	AlO_6	AlO_6	AlO_6	AlO_6	AlO_6		auto-
	$AlO_4(Si)$	$AlO_4(Si)$	$AlO_4(Si)$		$AlO_4(Si)$	$AlO_4(Si)$		claving
base line rise								
Δh (mm)	38	20	22	–	57	33	94	
XRD								

it is shown in Figure 5 where the molybdate reaction curves are plotted for selected samples. Figured are the portions of SiO_2 still non reacted at a time t_i versus the time on a logarithmic scale. These portions were calculated from the photospectrometrically determined extinction at t_i and the final extinction value E_f. Decreasing gradient of the curves plotted in Figure 5 corresponds with increasing condensation degree of the silicate anions.

The small content and the low condensation degree of the SiO_2 which has been reacted in the normal hardened samples and in those which were treated at 105°C could be interpreted in such a way that small amounts of CASH phases featuring a high Al/Si-ratio have been formed under these conditions. However, concerning the samples dried at 105°C one would equally have to discuss the possibility

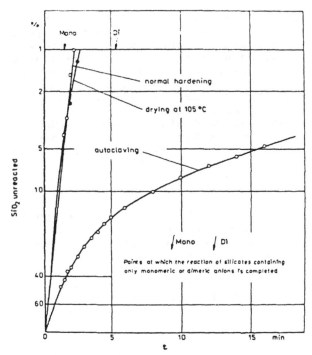

Figure 5. Molybdate reaction curves of selected LCC binder pastes.

that in addition monomeric SiO₄ units might have been inserted into the lattice of the hydrogarnet phase C₃AH₆.

Autoclaving, on the other hand, results in a marked increase of the degree of reaction of the silica fume, as compared with samples hardened under normal conditions and those dried at 105°C. This observation is supported by the diminished rise of the base line in the x-ray diffractograms of these samples (Table 3). Also the condensation degree of the silicate anions is higher in this case (Figure 5), which indicates a decreasing Al/Si-ratio in the occurring CASH phases.

Further solid state ²⁷Al NMR spectroscopic measurements were carried out. The position and shape of the signals in the spectra permit conclusions in regard to the coordination of aluminium. So, the parameter "chemical shift" (δ) derived from the spectra depends as well on the first coordination sphere (Al in tedrahedral or octahedral coordination - AlO₄ or AlO₆ groups) as on the occurring of other nuclei in the second coordination sphere (e.g. AlO₄[Si] groups). The chemical shift of the signals typcially for aluminium in AlO₄ tebrahedra (as occurring e.g. in the calcium aluminates C₁₂A₇, CA, CA₂) has been determined to be in the range of 60-80 ppm (versus a standard of an aqueous solution of AlCl₃.6H₂O). On the other hand, signals typically for aluminium in AlO_6 octahedra (as occurring in calcium aluminate hydrates like CAH_{10}, C_2AH_8, C_3AH_6 and in aluminium hydroxide and α-Al_2O_3) are recorded in the range of chemical shifts of 0-20 ppm. However, AlO_4 tetrahedra with Si in the second coordination sphere, as to be expected for CASH phases, are characterised by NMR signals with a chemical shift of 50-60 ppm (Müller et al 1984).

By using these characteristic data the AlOx building units occurring in the different LCC binder system samples were determined and listed in Table 3. As mentioned above the AlO_4 tetrahedra have to be attributed to unhydrated calcium aluminates, AlO_6 octahedra to CASH phases, aluminium hydroxide and α-Al_2O_3 and AlO_4[Si] groups to CASH phases formed.

As it is shown in Table 3 and illustrated by Figure 6 only in the case of hydrothermally treated LCC binder system samples an unambiguous evidence is given by NMR for the existence of AlO_4[Si] groups and, thus, the formation of (Al-O-Si) bonds in CASH phases.

The fact that in the samples treated at 105°C AlO_4[Si] groups could not be observed by NMR may be in connection with the above mentioned possibility of insertion of SiO_4 units into C_3AH_6.

At a thermal curing in the range of 200 up to 600°C , investigated in series C, the following changes in the primary and secondary composition of phases occur in dependence of the hardening conditions. Whereas in the

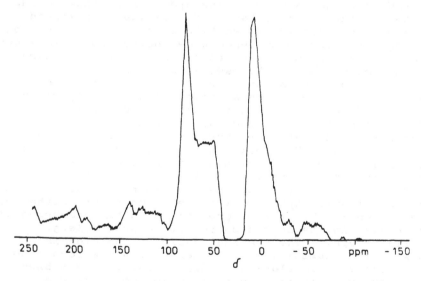

Figure 6. ^{27}Al NMR spectrum of a LCC binder matrix sample (autoclaving, series 1).

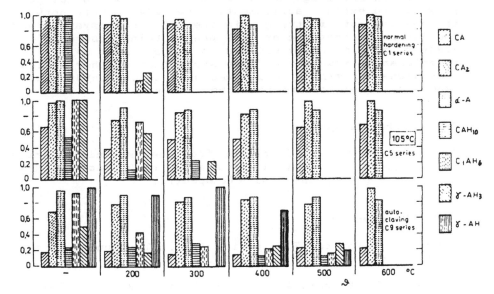

Figure 7. Relative concentration change in crystalline main components as a function of hardening conditions and thermal curing LCC binder matrix / test series C.

samples treated under normal conditions and at 105°C, respectively, CAH phases and gibbsite are detectable up to 300°C only, those which were hydrothermally hardened contain small concentrations of CAH_{10}, C_3AH_6, gibbsite and boehmite up to 600°C (Figure 7).

A progressive degradation of the silica fume can be established at increasing temperatures by following up the decrease of the maximum rise of the base line at 4.23 Å in the x-ray diffractograms. The increase of degree of reaction of the silica fume is also reflected by the molybdate method.

4   Conclusion remarks

The results demonstrated here show unambiguously the formation of CASH phases in the LCC binder systems at hydrothermal treatment. Regarding the existence of these phases in samples hardened at normal conditions and dried at 105°C, respectively, the results of x-ray investigation partly disagree with those obtained from the molybdate method and the ^{27}Al NMR at this stage.

It seems that the CASH phases are of zeolitic nature with a range of stability up to 600°C. The existence of gehlenite hydrate cannot be established with certainty in the investigated samples.

With the knowledge currently available, it is not yet possible to explain sufficiently the relationship between the development of strength in the LCC binder systems and the low concentrations of CASH phases being observed. However, complementary studies do suggest such a relationship because they have shown that neither silica fume alone nor silica fume mixed with calcium hydroxide generates strength when reacting with water.

5 Acknowledgement

We wish to thank Dr. D. Müller from the Central Institute of Inorganic Chemistry of the Academy of Sciences of the GDR for recording the ^{27}Al NMR spectra.

4 References

Eguchi, T.; Takita, I.; Yoshitomi, J.; Taikabutsu Overseas Kiritani, S.; Sato M. $\underline{9}$ (1989) 10

Monsen, B.; Seltweit A.; Adv. in Ceram. $\underline{13}$ (1985) 201

Müller, D.; GeBner, W.; Scheler, G. Mittbl.Chem.Ges.DDR $\underline{31}$ (1984) 107.

New developments in monolithic refractories Advances in Ceramics, vol. 13 Edited by R.E. Fisher Columbus, American Ceramic Society, $\underline{1985}$

Seltveit, A.: Dhupia, G.S.; Krönert, W. Ceram.Proc. $\underline{1986}$ 243

Thilo, E.; Wieker, W.; Stade, H., Z.anorg.allg.Chem. $\underline{340}$ (1965) 261

6 EFFECT OF TEMPERATURE ON SETTING TIME OF CALCIUM ALUMINATE CEMENTS

ALAIN CAPMAS
Lafarge Fondu International, Neuilly, France
DANIÈLE MÉNÉTRIER-SORRENTINO
Lafarge Coppée Recherche, Viviers, France
DENIS DAMIDOT
Faculté des Sciences Mirande, Dijon, France

Abstract
In the range 18°C-30°C, the setting time of Calcium Aluminate Cements becomes progressively slower as the temperature rises, and only above 30°C does the rate of set accelerates again. It has been said that this effect was due to difficult nucleation of C_2AH_8. Experiments on $C_{12}A_7$ show that C_2AH_8 does not present difficult nucleation, and precipitates almost immediately after supersaturation is achieved, even at 30°C. The longer setting time seems to be due to difficult nucleation of CAH_{10}, partly to massive gel formation, and partly because the supersaturation level becomes very low at this temperature.

1 Introduction

It is a well known fact that Calcium Aluminate Cements present an anomalous behaviour with the setting time at about 28°C. This longer setting time appears with most commercial Calcium Aluminate Cements[1], and with pure CA[2], thus proving that it is linked with the hydration of this phase. However, the retarding effect seems to be more or less important depending on the presence of other phases like C_2S[3] or admixtures[4].

The hydration of Monocalcium Aluminate has be studied extensively: it is generally observed that the hydrates formed depend on temperature:
- at low temperature (below 15°C), CAH_{10} is the main hydrate found
- at temperatures between 15 and 30°C, CAH_{10} is precipitated together with C_2AH_8 and AH_3 gel.
- at temperature above 30°C, C_2AH_8 and AH_3 gel are formed together and convert rapidly - if not simultaneously above 45°C - to C_3AH_6 and gibbsite. The water / cement ratio has no influence on the behaviour of Calcium Aluminate Cements.

A.Capmas and D.Ménétrier-Sorrentino[5] indicated that the thermodynamical study of the solubility curves, proposed by P.Barret & D.Bertrandie[6] could provide a good explanation as to the effect of temperature on the hydrates formed.

It has been shown by A.Rettel et al[7] that amorphous phases were appearing before or jointly with crystallised phases, depending on temperature, and RN.Edmonds and AJ.Majumdar[8] added that the amount of the amorphous (or gel) phases was increased with temperature. DR.Payne and JH.Sharp[9] analysed the gel phase and concluded that under 29°C this phase was a proto-form of CAH_{10} with a C / A ratio close to 1.1, thus reinforcing the hypothesis of the formation of crystallised CAH_{10} through a gel CAH_{10} proposed by Currel et al[10].

K.Fujii and W.Kondo[11] interpreted the dormant period as the formation of a protective barrier which critical thickness varies upon temperature from 3 nm at 10-20°C to 12 nm at 30°C thus explaining a retarding effect, but the authors did not study higher temperatures than 30°C to understand why this hypothetic barrier was suddenly quicker to break at these temperatures.

The retardation at 28°C has been associated with difficult nucleation of C_2AH_8 by SM.Bushnell-Watson and JH.Sharp[1]. RN.Edmonds and A.Majumdar[8] added later that it could also be due to a difficulty of formation of both CAH_{10} and C_2AH_8.

These hypothesis could be verified by hydrating CA and $C_{12}A_7$ at different temperatures, as it is predictable by the study of theoretical solubility curves that $C_{12}A_7$ precipitates more C_2AH_8 than CA at a given temperature. This result has also been found experimentally by RN.Edmonds and AJ.Majumdar[12].

2   Experimental

2.1   Materials
CA and $C_{12}A_7$ have been synthesised from reagent grade limestone and aluminium hydroxide; the later however contains 0.18% Na_2O, thus explaining that the analyses of CA and $C_{12}A_7$ show approximately 0.11% Na_2O.

The raw materials were heated twice in an electric furnace and air atmosphere. In between heatings, the solid is ground and rehomogeneised. The sintering process is followed by a slow cooling stage.

The purity of the calcium aluminates was checked by XRD and, more accurately, by SEM associated with EPMA: their composition, homogeneity, crystal shapes are good. $C_{12}A_7$ contains a few percent of CA. The crushed solids were stored in dessicator, to avoid ageing and contamination. The weight of fine powder needed for one experiment, was freshly ground at 300 m^2/kg SSA, just before the beginning of each hydration study.

2. 2   Hydration experiments and experimental techniques
The hydration studies were conducted under nitrogen on stirred suspensions: pure water is used and unless otherwise stated, W/S = 10.

<u>Electrochemical measurements: qualitative approach</u>

In a preliminary stage, the main course of hydration was followed by means of a special device, the so called chemical reactor. It basically consists of a thermoregulated Pyrex vessel, in which the suspension is stirred at a constant speed. A pH and electrical conductivity electrodes and a temperature captor are installed in the vessel. The numerical values are continuously recorded and processed on a microcomputer (H.P 85 model), using our own programme.

The pH and electrical conductivity electrodes were calibrated before each experiment with adequate standard solutions (solutions of pH 7 and 12.45 and a 0.1N KCl solution are used). The electrical conductivity, which is representative of the whole negatively and positively charged ions in the liquid phase, is expressed in milli-Siemens /cm (mS/cm) and abbreviated as K. Such a technique has already been applied to follow cement phases hydration, mainly on CA^2,[13] and on C_3S[14],[15]

Liquid and solid phase analysis: quantitative approach

Following the initial qualitative experiments, more precise determinations were carried out in order to detail and possibly quantify the different steps constituting the hydration process.

Taking into account the shapes of the pH, K and temperature curves (see for examples figures 1, 2 or 7) especially their irregularities and their time of occurrence, a second experiment was conducted the following day using the same experimental conditions. At selected times, small volumes of the suspension were taken from the reactor and quickly filtered on millipore filters (0.45 μm).

The liquid phase was stabilised with HCl, before being quantitatively analysed by I.C.P (A.R.L 3520): We use our multistandard hydration programme: in these studies only Ca^{2+}, $Al(OH)^-_4$ and Na^+ were determined.

Further hydration of solid phases was prevented by washing with acetone which was in it's turn removed by ether. The samples were then stored in plastic boxes in dessicators untill the different analysis were completed: the determination of the bound water by loss on ignition gives us the overall quantity of hydrates formed: XRD has been used to give a rough idea on the rate of the anhydrous solid disappearance and of the nature of the crystalline hydrates formed. DTA studies and SEM observations have also been extensively carried out but with the exception of two micrographs these results are not included in the present paper.

Before entering the sections dealing with experimental results, we should mention that privilege will be given to results on C12A7 due to the facts that hydration of CA has been more extensively studied and that our techniques, allowing to seize the very first development of hydration, have revealed new aspects of the hydration of C12A7.

3 Electrochemical aspects of CA and C12A7 hydration at 20°C

Because we have not yet made external publications of the typical curves obtained with the chemical reactor, it is necessary to give in a preliminary experimental section their characteristic features. Their significance (which will only be briefly explained since it is outside the scope of this paper) are however required knowledge to discuss the hydration results as a function of temperature.

3.1 Example of curves obtained with CA

From a series of curves in figure 1, representing the course of CA hydration at 20°C, 4 main periods are observed, thus confirming previous results.

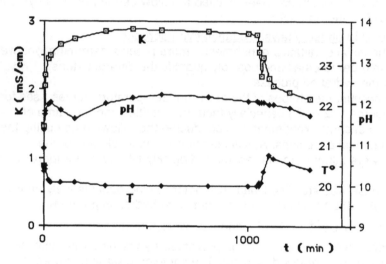

Fig 1: pH, K and T°C evolution during hydration of CA at 20°C

During the first period, the dissolution of CA, bringing Ca^{2+} and $Al(OH)_4^-$ ions into solution is the prevailing reaction: K and pH increase accompanied by temperature. The second period where very slight changes of K, pH and temperature are detectable is often called in the literature the dormant period: it is indeed mainly the nucleation period of the hydrates: the dissolution reaction as well as the precipitation reaction are slow. The third part of the curves evidences a strong solid / liquid reactivity: K decreases promptly due to a massive precipitation of the hydrates which consumes ions from the solution: the dissolution of the anhydrous CA also accelerates but somewhat later on as evidenced by the temperature and conductivity raise. In the fourth period both reactions continue to compete but with a decreasing rate, one reason being that the remaining quantity of the anhydrous solid becomes low.

3. 2 Example of curves obtained with C12A7

Studied at the same temperature (20°C) C12A7 behaves quite differently as shown in figure 2. It's hydration develops very quickly: a maximum of conductivity occurs within the first 10 minutes, accompanied by a sharp temperature raise. The pseudo plateau of conductivity is very short (approximately 180 min) compared to that of CA (15 hours with this sample). Afterwards the conductivity drops promptly, showing an accelerated stage of hydrates precipitation.

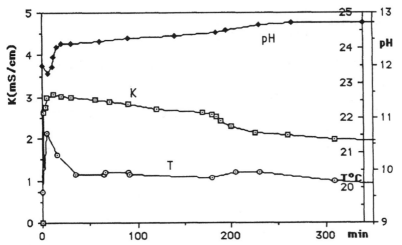

Fig 2: pH, K and T°C evolution during hydration of C12A7 at 20°C

4 Hydration of CA as a function of the temperature

Only the qualitative results of the experiments conducted with the reactor will be given here, the quantitative programme being still in progress.
The network of conductivity curves established in a small domain of temperature, 20°C to 35°C, in which the lengthening of the setting time is supposed to occur, is shown in figure 3.

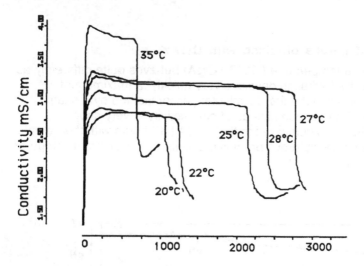

Fig 3: Conductivity curves for CA at different temperatures

The differences are drastic, particularly the length of the plateau. If we plot on a graph, these values as a function of the temperature, a sharp maximum is evidenced at 28°C: figure 4.

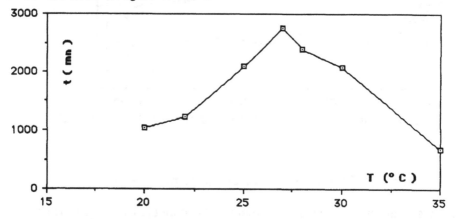

Fig 4: Length of the conductivity plateau for CA as a function of T°C

This has several meanings and implications: it confirms previously found results on CA by different techniques such as conduction calorimetry[2] or very recently by the embeded thermocouple method. Most of the work which has been done with Fondu on the topic is the study of the setting time of cement pastes[1]. This also shows that the lengthening of the plateau can be adequately studied by means of electrochemical techniques.

5 Hydration of C12A7 as a function of the temperature

5.1 Qualitative results

Figure 5 assembles a series of conductivity curves obtained in the same narrow range of temperature: the pseudo plateau of conductivity shortens progressively and finally disappears (at approximately 30°C), when the temperature increases.

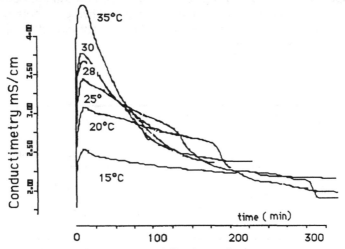

Fig 5: Conductivity curves of C12A7 at different temperatures

It is important to note that there is no delay of precipitation at 27-28°C: the higher the temperature, the faster the precipitation. This is true indeed in a much larger domain of temperature. In order to have a more complete view of the modifications of the hydration kinetics versus the temperature, experiments have been conducted from 5°C to 80°C.

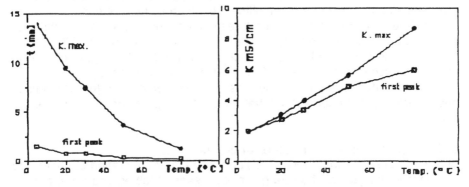

Fig 6: Conductivity peaks: Left: time of occurence
Right: conductivity values at maximum

The characteristic features gathered in figures 6, show interesting points concerning the initial period of hydration: the maximum of conductivity, whose position changes slightly (6 - 9 min) in figure 5, shifts from 14 min at 5°C to 1 min at 80°C (figure 6 left): the related conductivity values also increase with increasing temperature: 2 mS/cm to 8.75 mS/cm (figure 6 right). The real effect on the solubility cannot however be determined in this manner, because the electrical conductivity depends on the concentration of the ionic species, at a given temperature but also for a given concentration on the temperature. Therefore what we observe is a cumulative effect of two parameters and it is only with quantitative analysis of the liquid phases, that the effect of temperature will be precisely known.

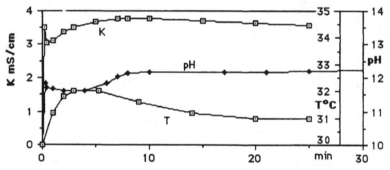

Fig 7 : Short term hydration of C12A7 at 30°C

If the very first part of the curves is observed on an extended scale of time, new features appear (an example is given in figure 7 presenting the very short term hydration at 30°C): a very early and sharp peak of conductivity, a less pronounced one of pH and an significant exothermic heat release.
This is true for all the other temperatures but the time of occurrence shifts to even shorter values when temperature increases: 90 seconds at 5°C, 24 seconds at 80°C (curve labelled first peak in figure 6, left): higher conductivity values are also observed: 2.03 mS to 5.98 mS (figure 6, right).
The modification of the parameters as a function of temperature is continuous: once more, it can be pointed out that no lengthening occurs around 28°C or what so ever.

5. 2 Liquid and solid phases analysis
In order to describe with more precision the course of hydration of C12A7, the quantitative analysis will be now given, by means of tables and / or graphs. Experiments have been conducted at 3 temperatures of particular interest: 20°C, 25°C and 30°C.

Table 1 summarises results obtained at 20°C, during the main hydration (corresponding figure 8);

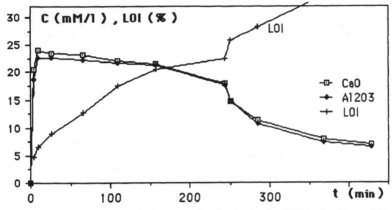

Fig . 8 . Liquid and solid phases analysis during C12A7 hydration , 20° C

Similar results have been collected at 25°C and 30°C but will just be graphically expressed in the series of figures 9. Part A displays the comparative evolution of the calcium ions in the liquid phase at the three considered temperatures. Results with aluminate ions will not appear. As the curves show similar trends we can rather present the modifications of the C/A molar ratio of the liquid phaseon Fig9B. Part C reports the comparative evolution of bound water as determined by loss on ignition on the related solid phases.

Table 1:   Liquid & solid phase analysis of C12A7 at 20°C

Time mn	CaO mM/l	Al2O3 mM/l	C/A	Loss on ignition %	Ki mS/cm	pH
0	0,00	0,00			0	
3	20,48	18,74	1,09	4,80	2,82	11,7
9	24,04	22,65	1,06	6,65	3,00	11,75
26	23,55	22,71	1,04	8,85	2,92	12,2
65	23,16	22,28	1,04	12,75	2,83	12,2
109	22,13	21,70	1,02	17,45	2,74	12,3
157	21,43	21,34	1,00	20,55	2,68	12,4
243	17,82	17,75	1,00	22,40	2,40	12,5
250	14,77	14,69	1,01	25,90	2,25	12,5
284	11,34	10,79	1,05	28,25	2,11	12,7
368	7,91	7,36	1,07	33,90	1,95	12,75
428	6,89	6,52	1,06	35,25	1,94	12,75

First of all we can get the real effect of temperature on the maxima of concentrations reached in approximately 10 minutes: 24 mM/l of lime at 20°C,

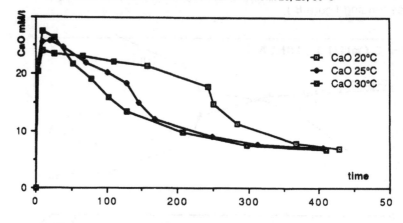

Fig 9 A : C12A7: CaO concentration versus time at 20, 25, 30°C

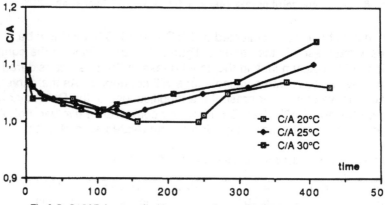

Fig 9 B: C12A7: C/A ratio versus time at 20, 25, 30°C

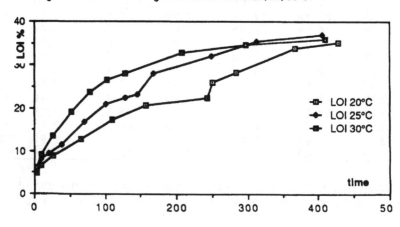

Fig 9 C: C12A7: Loss on Ignition versus time at 20, 25, 30°C

27 mM/l at 25°C and 27.6 mM/l at 30°C: 22.5 mM/l of alumina at 20°C, 25 mM/l at 25°C and 26.5mM/l at 30°C.

From Fig 9 A and B it is clear that the higher the temperature, the faster the hydration kinetics. The Calcium evolution ascertains the qualitative results of conductivity, in particular the shortening and disappearance of the pseudo plateau about 30°C. The quantity of bound water is also higher for a given time when temperature is higher. It can be seen however that at 20°C and at 25°C the formation of hydrates does not develop at a constant rate as it seems to be the case at 30°C (the curves of lime and LOI develop smoothly). Somewhat before the end of the pseudo plateau the hydrates formation is lowered: in the following period on the contrary, an acceleration process is observed.

When analysed by XRD and as expected the main hydrate formed is C_2AH_8 whatever the temperature. It is early and easily detected: it does not have problems of crystallisation. CAH_{10} has also been found but lately and in a much less degree. It's time of occurrence is well correlated with the beginning of the increase of the C/A ratio, around 1.01 (figure 9 B). No other crystallised hydrates have appeared within the period under study, which is not dealing with the long term hydration.

In a complementary approach concerning the related liquid phase analysis, a nice and useful way to present the results consists in placing the successive couple of points (lime, alumina), on the hydrates solubility diagram of the $CaO- Al_2O_3- H_2O$ system. We have chosen to show such results at 20°C(figure 10). We should mention that the presence of approximately 1 mM/l of Na2O in solution has been taken into account to calculate the solubility curves. The same construction can be done at any other temperature, but the appropriate solubility diagram must be used, as the position of the solubility curves varies with the temperature. In a recent publication[5], we presented the calculations of such diagrams: using computerised programmes the solubility curves can be easily drawn. This is a real progress compared to the limited possibilities offered by the use of experimentally determined diagrams for a given temperature.

One of the interests of placing the kinetic path described by the liquid phase on this network of curves at equilibrium is the possibility to see in which part of the diagram the solution is located especially the supersaturations areas. From this, we can conjecture which hydrates may or may not be formed.

It can be seen that the point corresponding to the earliest time (30 seconds) is already located in the area of supersaturation towards C_2AH_8. We will detail this point in the following paragraph. The path moves up to higher lime and alumina concentrations, reaches the maximum within 9 minutes and goes down following a different path much closer of the C_2AH_8 solubility curve. This does not mean however that during the ascending part of the curve the only reaction is the dissolution of $C_{12}A_7$: it does mean neither that the precipitation process starts after the liquid phase has gone through the maximum. It is is clear from the high bound water values obtained at this maximum (6.65 % at 20°C, 8.2% at 25°C and 9.20% at 30°C) that the precipitation process starts

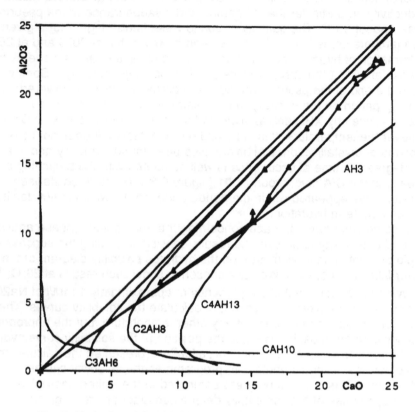

Fig 10: Hydration of C12A7. Kinetic path on the CAH solubility diagram

very soon. We will detail these points now, focusing our observations on the very first minutes of hydration.

Table 2:   Liquid phase analysis of C12A7 at 20°C, 25°C and 30°C

20°C				25°C				30°C			
time mn.ss	CaO mM/l	Al2O3 mM/l	C/A	time mn.ss	CaO mM/l	Al2O3 mM/l	C/A	time mn.ss	CaO mM/l	Al2O3 mM/l	C/A
0,00	0,00	0,00		0,00	0,00	0,00		0,00	0,00	0,00	
0,26	15,16	10,91	1,39	0,22	14,96	10,66	1,40	0,24	15,61	11,47	1,36
0,27	15,07	11,66	1,29	0,26	14,52	10,47	1,39	0,45	16,82	13,44	1,25
0,31	15,00	10,91	1,37	0,34	15,09	11,40	1,32	1,01	17,57	14,23	1,24
1,12	16,13	12,52	1,29	1,16	16,54	13,33	1,24	1,47	18,68	15,86	1,18
1,20	16,21	12,70	1,28	1,57	16,98	13,84	1,23	2,49	22,18	19,68	1,13
3,24	17,61	14,88	1,18	3,11	19,23	16,79	1,15	4,51	24,45	23,22	1,05
4,28	19,79	17,46	1,13	5,30	23,21	21,97	1,06	8,57	26,05	24,84	1,05
4,55	21,02	19,25	1,09	10,22	27,20	25,10	1,08				
6,33	22,30	20,94	1,07								
9,27	24,11	22,46	1,07								

Table 2 contains the quantitative analysis of calcium and aluminate ions into the liquid phase for the short term hydration experiments at the three temperatures. In the corresponding figures 11 are expressed the comparative evolution of the Calcium and of the C/A ratio.

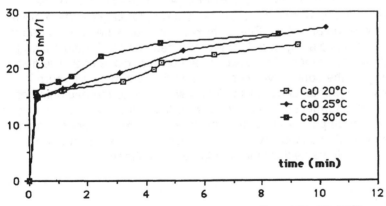

Fig 11 A : C12A7 : CaO concentration versus time at 20, 25, 30°C

Within the first 30 seconds, the dissolution is extremely fast: the concentrations reached are in the range of 15 mM/l of lime and 11mM/l of alumina. The C/A ratio is close of 1.40 and tends to decrease as a function of the time: figure 11 B

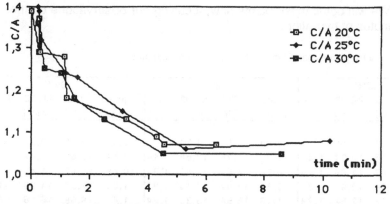

Fig 11 B : C12A7 : C/A ratio versus time at 20, 25, 30°C

Two hypotheses can be proposed to explain this result: either the dissolution of C12A7 is not congruent (C/A = 1.71), or the precipitation process is already initiated, the composition of the formed hydrates being richer in lime than 1.71. To answer this, we have developed a complementary work: the idea is to try to isolate a period of pure dissolution by increasing the dilution of the suspension. Experiments have been conducted with a W/S ranging from 10 to 200, the time of contact of the solid with water being as short as possible. The ratio 1.70 has been obtained for 3 values of dilution (50, 100,150). These results privilege the hypothesis of a congruent dissolution and of the almost instantaneous precipitation of a lime rich hydrate (C2AH8 or C4AHn): the position of the figurative point in figure 10 and it's displacement on the diagram are more in favour of the formation of C2AH8. Scarce hexagonal plates have been detected by SEM on C12A7 samples hydrated for few minutes: their C/A ratio although not quantitatively determined (fine powder) is however close of 2. The first peak appearing on the conductivity curves within the first minute (figure 7) is now explained: the prompt conductivity decrease is due to the beginning of hexagonal hydrates formation: the pH lowers for the same reason. This shows also that the period of pure dissolution with C12A7 is extremely brief, ranging from 90 seconds to 24 seconds, when the temperature varies from 5 to 80°C and for a W/S = 10. In paste this time should be even shorter !

6 Discussion

The results presented in this paper give valuable information about the hydration behaviour of the two main components of Calcium Aluminate Cements. These results also lead us to consider that the hydrates formed play an important role on the effect of temperature on the hydration of these cements

6.1 Role of C_2AH_8

- Hydration of pure CA presents an anomalous setting behaviour at 28-29°C, which seems to coincide with the disappearance of CAH_{10} above this temperature. The setting time actually lengthens above 20°C to reach a maximum at 28°C, then shortens very quickly above this maximum.
- Hydration of pure $C_{12}A_7$ is accelerated with temperature, the retardation effect being completely avoided in this case, and the hydrates formed being mainly C_2AH_8 at any temperature under 35-40°C, with a very short nucleation period, if any.

These results prove that C_2AH_8 presents neither difficult nucleation nor retarded setting time at any temperature, and cannot be taken as an explanation to the retardation at 28°C. The difficult nucleation has been attributed to C_2AH_8, because this hydrate does not seem to provoke the massive precipitation.

To understand what happens with C_2AH_8 in the case of $C_{12}A_7$, it is interesting to rely on P.Barret's work on solubility curves[6]: the dissolution of $C_{12}A_7$ is congruent, thus bringing very quickly the solution with a C / A ratio of 1,71 in supersaturation toward C_2AH_8. A precipitation is immediate to lower the ionic concentration, with a C / A ratio of 2. The competition between the dissolution of $C_{12}A_7$ and the precipitation of C_2AH_8 can be seen on the evolution of the loss on ignition, linked to the bound water and the associated path of the ionic concentration. It is interesting to note that this path joins the minimum instability curve described by P. Barret. When this happens, there is a competition between AH_3 gel and C_2AH_8, slowing drastically the formation of C_2AH_8. Afterward it seems that the precipitation of C_2AH_8 is only done on existing nuclei, and no new nuclei can be traced by SEM.

6.2 Role of CAH_{10}

The long setting time can be explained as the combination of 3 factors:
- Amorphous CAH_{10}: CAH_{10} precipitates partly as an amorphous compound which proportion is increased with temperature. Above 30°C, no evidence of crystallised CAH_{10} has been found by RN.Edmonds and A.Majumdar[8] even after 48h.
- Chemical affinities: Using again P.Barret's thermodynamic interpretation of hydration, it is possible to calculate the chemical affinity for each hydrate when the maximum ionic concentration is obtained. These computations are not very accurate, but allow a description of hydration: for C_2AH_8 and AH_3, the chemical affinity is about the same (-1500) between 10 and 35°C. For CAH_{10}, the chemical affinity varies from -3300 at 10°C down to -2200 at 35°C. In other words, at 10°C, the ratio of the chemical affinities between CAH_{10} and C_2AH_8 is 2.2, while at 35°C, the ratio is down to 1.5. This alone explains why CAH_{10} intends to decrease with the rising temperature to the benefit of C_2AH_8 and AH_3.
- Protecting layer: SEM observation of CAH_{10} shows that this hydrate is nearly always present as a shell around unhydrated cement, while C_2AH_8 and AH_3

are obviously formed separately to the cement. It is not possible to discuss here the possibility of a topochemical reaction, but this observation means that a possible protective layer can explain a retardation when the CAH_{10} gel formed does not crystallise and stays around the anhydrous.

7 Conclusion

The mechanism of retardation between 20 and 29°C can be described as a more and more difficult nucleation of CAH_{10} with temperature until it becomes nearly impossible above 29°C, and subsequently a formation of gel phase which acts as a slowing barrier to further dissolution / precipitation. The combination of C_2AH_8 and AH_3 presents a difficult nucleation and slows the growing of nuclei, while temperature acts as an accelerator. When this reaction becomes the predominant one, the setting time accelerates very quickly.

These results also help to understand why commercial cements behave differently, some containing more $C_{12}A_7$ than others thus having a higher proportion of C_2AH_8 during the massive precipitation.

Acknowledgements: the authors wish to thank Professor P. Barret and Professor HFW. Taylor for fruitful discussions.

[1] S.M. Bushnell-Watson & J.H. Sharp, CCR [16] 875-884 (1986)

[2] B.Guilhot & M.Soustelle,Sem.Int.Torino,97-111 (1982)

[3] CM.George,B.Cottin & R.Ranc, Sem.Int.Torino, 138-149 (1982)

[4] S.M. Bushnell-Watson & J.H. Sharp, CCR in the press

[5] A.Capmas & D.Ménétrier-Sorrentino, UNITECR'89,vol. 2, 1157-1170 (1989)

[6] P.Barret & D.Bertrandie,7th Int.Congr.Cement Paris, 3V,134-139 (1980)

[7] A.Rettel, W.Gessner & D.Müller, Br.Ceram.Trans.J. 84,25-28 (1985)

[8] RN.Edmonds & AJ.Majumdar, CCR [18] 311-320 (1988)

[9] DR.Payne & JH.Sharp CCR in the press

[10] BR.Currell, R.Grzeskowlak, HG.Midgley & JR.Parsonage, CCR [7] 420-432 (1987)

[11] K.Fujii & W.Kondo,J.Am.Ceram.Soc., 69 [4] 361-364 (1986)

[12] RN.Edmonds & AJ.Majumdar, CCR[19] 848-856 (1989)

[13] P.Galtier &B.Guilhot, CCR[14] 679-685 (1984)

[14] A.Zeiwer, Rev. des Mat. 681-701 (1973)

[15] FA.Shebl,AB.Summan & D.Helmut, CCR[10] 121-130 (1980)

7 THE USE OF NUCLEAR MAGNETIC RESONANCE (NMR) IN THE STUDY OF HIGH ALUMINA CEMENT HYDRATION

D.J. GREENSLADE, D.J. WILLIAMSON*
University of Essex, Colchester, UK

Abstract
HAC hydration has been investigated by proton (^{1}H) and aluminium (^{27}Al) NMR. Proton measurements have been performed by a simple method which allows an inexpensive bench top spectrometer to be utilised. NMR was used to give a measure of "free" water with time in order to produce a hydration profile rather than determine NMR magnetisation or relaxation parameters which may also be used to study hydration. The suitability of ^{27}Al NMR to study HAC hydration has also been demonstrated again by a hydration profile although this requires more specialised equipment.
Keywords: Nuclear Magnetic Resonance, Spin Echo, Hydration.

^{1}H NMR
1.1 Introduction

The use of proton NMR to follow cement hydration was investigated some time ago by a number of workers especially Blaine (1962) and Seligman (1960 and 1968) whilst others (Rettel et al. (1983), Blinc et al. (1978), Murphy et al. (1980) and Barbic et al. (1982)) continued this work. More recently extensive studies have been pursued by MacTavish et al. (1985a and b) and Milkjovic (1986 and 1988). Hydration is monitored by evaluation of proton relaxation times or a measurement of the total "free" water proton magnetisation.

Using a spin-echo (90°-t-180°) pulse sequence a measure of the amount of "free" proportioning water at any time after mixing is given by the amplitude of the refocussed NMR signal, or echo, which occurs after the 180° radio frequency pulse. Monitoring hydration by proton NMR gives a direct measurement of the behaviour of the proportioning water. The use of the spin-echo technique allows low

---

+ Present Address: School of Chemical Sciences, University of East Anglia, Norwich, England.

power inexpensive bench top instruments to be used, for certain classes of cements, as the signal does not have to be observed immediately after the initial radio frequency pulse. Spin-Echo formation attributed to E.L. Hahn (1950) is described in Appendix 1 for a 90_x-τ-180_y pulse sequence.

1.2. Experimental

The instrument primarily used in these measurements was a Newport N20 Spectrometer made by Newport Instruments (now Oxford Analytical Ltd., Oxford). This spectrometer is built around a large gap permanent magnet with a field strength of 0.05T operating at approximately 2.5 MHz for proton NMR. The large gap allows a big sample to be used, standard 0.5 inch test tubes are acceptable, and the entire instrument measures 17.0 x 17.5 x 27.5 cm and weighs under 6 kgs. The use of a Spin-Echo pulse sequence, see Appendix 1, gives several advantages. The echo is unaffected both by any moderate field inhomogeneity, as such effects are refocussed, and by any signal due to bound water as the echo occurs well after any such signal will have decayed. Further the construction of the apparatus is simplified, no complicated electronic techniques are required to obtain rapid recovery of the receiver from saturation by the radio frequency pulse, which does occur, as the spin-echo extends the timescale of the experiment so that relatively simple construction gives fast enough recovery. Although the Newport N20 is no longer commercially available successful "in house" construction of a similar instrument has been carried out at UEA Norwich (Strike (1988)).

Samples were prepared by mixing cement either SECAR 71 (6g) or ORDINARY PORTLAND CEMENT (6g) or a mixture of the two with distilled water (2.4g) at 22°C and transferring to a sample tube within one minute when the initial measurement of the echo height was made. Subsequent measurements were made at hourly intervals when practicable. Echo height could either be measured directly from the oscilloscope or by scanning through the memory of the [++] Versatile Laboratory Aid (VELA) which was used in fast transient mode to digitise the signal. All measurements were made using a 1ms pulse interval between the 90° and 180° radio frequency pulses. Further studies were performed at UEA Norwich, on a home built high power spectrometer operating at 60 MHz for proton NMR known as the "Magnion". The higher sensitivity of this instrument allowed measurements to be made on high iron HAC and also possessed the added advantage of a pulse programmer and signal

---

[++] Available from R S. Components, Corby. UK Stock No. 611-975, a member of the Electrocomponents plc group.

averager/data recorder. The latter facilities permitted automatic operation of the experiments.

Samples for the "Magnion" were prepared by mixing cement (1g) with distilled water (0.4 g).

1.3. Results

Figure 1 shows plots of spin echo height against time measured using the Newport N20 spectrometer for (a) a sample of hydrating Secar 71, upper profile, and (b) OPC lower profile. Secar 71 exhibits a sharp step at 6-7 hours whereas the OPC curve shows a slight decrease after a few hours with the main decreases in echo height occurring between 10 and 40 hours. The spin-echo height is a direct measure of the mobile water remaining in the hydrating sample. This signal decreases as water which becomes bound on hydration is not detectable by the experiment. The signal of the bound fraction does not give a normal spin-echo and hence does not interfere with the mobile fraction signal. Hence the experiment records bulk take up of the proportioning water. Figure 1 shows that this occurs as an avalanche process for Secar 71 whereas in OPC the process occurs gradually at a much later time.

Figure 2 shows a hydration profile for a mixture of Secar 71: OPC of 70:30. The avalanche process occurs some two hours earlier than for the same Secar 71 without the addition of OPC, Figure 1(a).

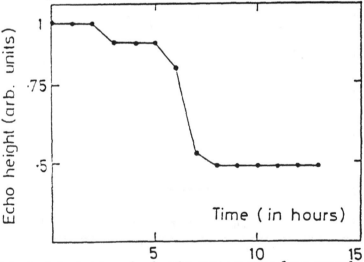

Fig. 1(a) A plot of spin-echo height against time for a sample of hydrating Secar 71

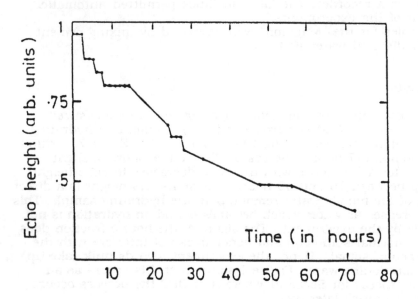

Fig. 1(b). A plot of spin-echo height against time for a sample of hydrating OPC.

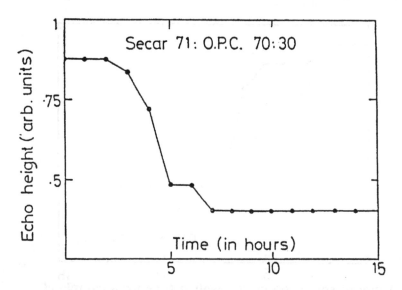

Fig. 2. A plot of spin echo height against time for a hydrating sample of a mixture of Secar 71 and OPC in the ratio of 70:30.

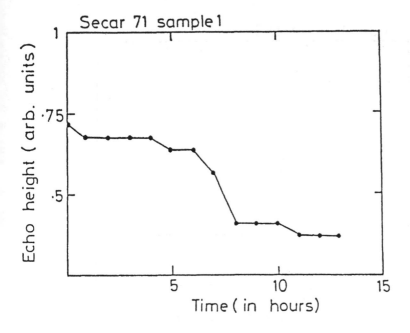

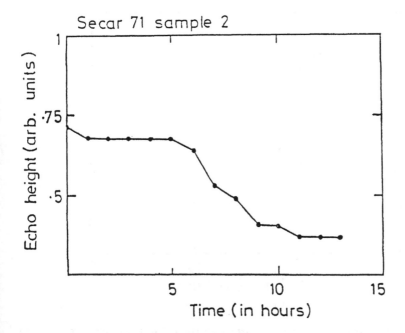

Fig. 3 Plots of spin echo height against time for hydrating Secar 71 samples 1 and 2.

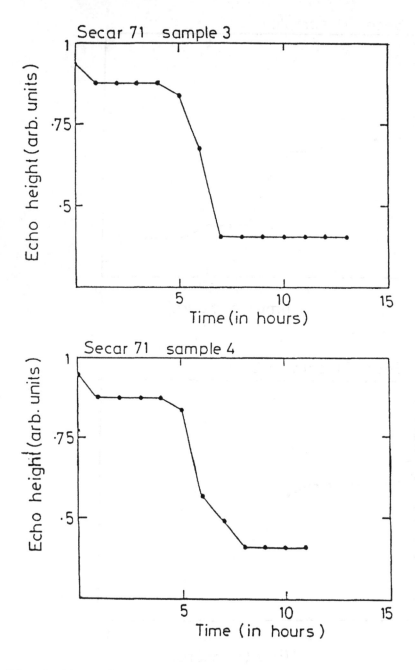

Fig. 4. Plots of spin echo height against time for hydrating Secar 71 samples 3 and 4.

The hydration behaviour of Secar 71 is known to be affected by changes in mineralogy (Fentiman). Figures 3 and 4 show the results from samples 1 and 2, and 3 and 4 of Secar 71 respectively. Samples 1 and 2 were thought to hydrate similarly but differently from 3 and 4 which were both thought to show another type of hydration behaviour. This was borne out by the hydration profiles. Samples 1 and 2 show a smaller initial echo height than samples 3 and 4 and a more gradual decrease in the step region, which also continues to a longer time, 8-9 hours, than the sharp steps of samples 3 and 4 which are complete at 7-8 hours. These results were found to be reproducible suggesting that the method is sensitive to the mineralogy of the cement.

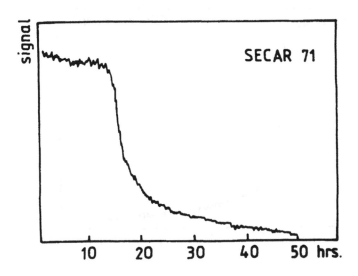

Fig.5. Plot of spin echo height against time for hydrating Secar 71 recorded on the "Magnion" spectrometer.

Figures 5 and 6 show hydration profiles recorded automatically on the "Magnion". The avalanche process is seen to occur at very similar times for these two classes of HAC, Secar 71 and Ciment Fondu, although at a later time than observed for samples of the same batch of Secar 71 recorded on the Newport N20. It is intended to further investigate this difference paying particular regard to temperature within the sample. Note that the sample sizes were different.

Figure 7 shows a profile for OPC recorded on the Magnion. The data set was sampled every six minutes over 50 hours and is therefore capable of showing more detail than the Newport N20 data which was sampled every hour at best. An increase in the signal is seen between 6 and 14 hours. This may be evidence to support the osmotic membrane hydration theory (Double (1983)) in that this

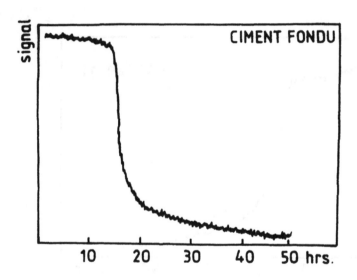

Fig. 6 A plot of spin echo height against time for a sample of hydrating Ciment Fondu recorded on the Magnion.

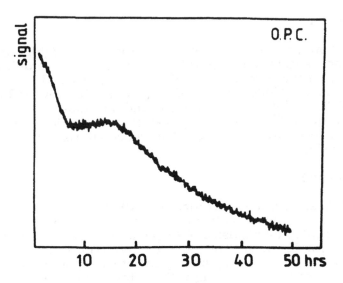

Fig. 7. A plot of spin echo height against time for a sample of hydrating OPC recorded on the "Magnion".

could be the time period during which rupture of such membranes is occurring or the water release could be consistent with the presence of ettringite (Long).

1.4. Conclusion

An NMR method for the direct observation of the hydration of cement has been described which can be performed on a relatively inexpensive spectrometer requiring little specialist knowledge for its operation or maintenance.

The results show that the method can distinguish between different classes of cements and shows a variation in hydration behaviour with mineralogy. We suggest that the method may have wide applicability in for example accelerated and retarded cements and admixtures generally. These results are of a preliminary nature, the obvious next step is to compare the hydration profiles with more accepted methods for recording setting.

^{27}Al NMR
2.1. Introduction

It has been shown, relatively recently, that the use of high magnetic fields and magic angle sample spinning can result in a useful level of line narrowing in the solid-state NMR spectra of selected quadrupolar nuclei (Muller (1981), Fyfe (1982a) Meadows (1982) and Fyfe (1982b)). ^{27}Al is such a nucleus and the suitability of cements to study by ^{27}Al NMR was proposed in an earlier paper by Lahajhar, Blinc *et al* (Lahajnar (1977)) in which an example spectrum was shown.

2.2. Experimental

Solid-state ^{27}Al spectra were recorded on a JEOL GX 400 spectrometer using a home built probe. Spectra were the result of 1000 scans recorded by digitising1024 points over a spectral width of 34 kHz using a 4µs pulse. The samples were removed, at the time intervals shown, from a mass hydrating under excess water at 333 K. Hydration was arrested by washing with acetone and drying at a filter pump.

2.3. Results

Figure 8 is a stacked plot of spectra for the samples taken. Anhydrous aluminates have chemical shifts in the region of 80 ppm, characteristic of aluminium tetrahedrally co-ordinated by oxygen, whereas the hydrated octahedrally co-ordinated aluminium phases have chemical shift in the region 0 - 10 ppm. This allows the hydration of the aluminium to be observed by the change in the ^{27}Al NMR chemical shift (Müller (1984)). From the figure it can be seen that the peak for octahedral co-ordination remains unchanged up to 15 minutes. after which time it increases and the tetrahedral peak correspondingly decreases. After one hour the relative intensities have reversed. Figure 9 shows a plot of relative fraction of octahedrally co-ordinated aluminium against time produced from the stacked plot data.

2.4. Conclusion

The solid-state ^{27}Al NMR results record the co-ordination change of the aluminium as hydration proceeds. Thus NMR provides a second technique giving direct information on another nucleus in the heterogeneous sample.

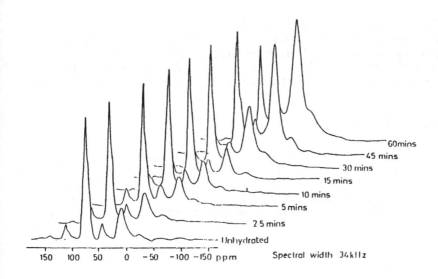

Fig. 8. Stacked plot of solid-state ^{27}Al spectra for samples of hydrating Secar 71 removed at the times shown from a mass hydrating at 333 K.

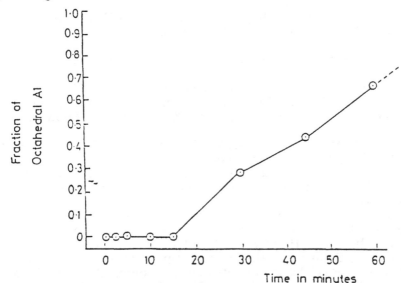

Fig. 9. Plot of fraction of octahedrally co-ordinated aluminium against time for samples of Secar 71 removed from a mass hydratings at 333 K.

This technique is more specialised than the proton measurements requiring a much more sophisticated spectrometer and suffers the additional disadvantage that samples need to be removed at intervals. Nevertheless the method appears to be capable of providing valuable information on hydration processes.

ACKNOWLEDGEMENTS

The authors wish to thank Charles Fentiman for the provision of HAC samples and useful discussions, BP for help and advice with the solid-state ^{27}Al spectra and the SERC and Dr. K. Bowden, formerly Chairman, Department of Chemistry and Biological Chemistry, University of Essex, for the award of a studentship for one of us (DJW).

3. References

Barbic, L., Kocuvan, I., Blinc, R., Lahajnar, G., Merljak, P. and Zupncic, I. (1982) The Determination of Surface Development in Cement Pastes by Nuclear Magnetic Resonance, **J. Amer. Ceram. Soc.** 65, No. 1 25-31.

Blaine, R.L. (1962),Proton Magnetic Resonance in Hydrated Portland Cements, Nat. Bureau of Standards Monograph 43 1 Paper IV-55, 501-11.

Blinc, R., Burger, M., Lahajnar, G., Rozmann, M., Rutar V., Kocuvan, I. and Urisic, J. (1978) NMR Relaxation Study of Adsorbed Water in Cement and C$_3$S Pastes, **J. Am. Chem. Soc.** 61, 1-2 35.

Double, D.D., (1983), New Developments in Understanding the Chemistry of Cement Hydration, **Phil. Trans. R. Soc. Lond.** A310 53.

Farrar and Becker, Pulse and Fourier Transform N.M.R. Chap. 1 & 2

Fentiman, C.H., personal communication.

Fyfe, C.A., Gobbi, G.C., Hartman, J.S., Lenkinski, R.E., O'Brian, J.H., Beange, F.R. and Smith, M.A.R. (1982), High Resolution Solid-State MAS Spectra of ^{29}Si, ^{27}Al, ^{11}B, and Other Nuclei in Inorganic Systems Using a Narrow-Borre 400 MHz High Resolution NMR Spectrometer, **J. Mag. Reson.** 47 168.

Fyfe, C.A., Gobbi, G.C., Hartman, J.S., Kllnowski, J. and Thomas, J.M.,(1982), Solid-State Magic-Angle Spinning Aluminium-27 Nuclear Magnetic Resonance Studies of Zeolites Using a 400 MHz High Resolution Spectrometer, **J. Phys. Chem.** 86 1247-1250.

Hahn, E.L. (1950), Spin Echoes, **Phys. Rev.** 80 580.

Lahajnar, G., Blinc R., Rutar, V., Smolej, V., Zupancic, I., Kocuvan, I. and Ursic, J. (1977), On the Use of Pulse NMR Techniques for the Study of Cement Hydration, **Cem. Concr. Res.** 7 385.

Long, G, Blue Circle, personal communication.

MacTavish, J.C., Miljkovic, L., Schreiner, L.J., Pintar, M.M., Blinc, R. and Lahajnar, G. (1985), A study of Portland Cement Hydration by Paramagnetic Iron Suppression of Proton Magnetic Resonance, **Z. Naturforsch** 40a 32-36.

MacTavish, J.C., Miljkovic, L., Pintar, M.M., Blinc, R. and Lahajnar, G. (1985), Hydration of White Cement by Spin Grouping NMR, **Cem. Conc. Res.** 15 367-377.

Meadows, M.D., Smith, K.A., Konsey, R.A., Rottgeb, T.M., Skarjvne, R.P. and Oldfield, E. (1982), High Resolution Solid-State NMR of Quadrupole Nuclei, **Proc. Natl. Acad. Sci.,** USA, 79 1351 55.

Miljkovic, L., MacTavish, J.C., Jin Jian, Pintar M.M., Blinc, R. and Lahajnar, G. (1986), NMR Study of Sluggish Hydration of Superplasticized White Cement, **Cem. Concr. Res.** 16 864-870.

Miljkovic, L., Lasic, D., MacTavish, J.C., and Pintar M.M., (1988), NMR Studies of Hydrating Cement: A spin spin relaxation study of the early Hydration Stage, **Cem. Concr. Res.** 18 951-956.

Muller, D., Gessner, W., Behrens, H.J. and Scheler, G. (1981), Determination of the Aluminium Co-ordination Number in Aluminium-Oxygen Compounds by Solid-State High Resolution[27]Al NMR, **Chem. Phys. Lett.** 79 59.

Muller, D., Rettel, A., Gessner, W., and Scheler, G. (1984),An Application of Solid-State Magic-Angle Spining ^{27}Al NMR to the Study of Cement Hydration, **J. Mag. Reson.** 57 152.

Murphy, V.R.K., Bhagat, P.K., Nickell, W.T. and Kadaba, P.K. (1980), Pulsed Nuclear Magnetic Resonance Study of Absorbed Water in Cement, **Mat. Sci. and Engineering** 43 187.

Oldfield, E. and Kirkpatrick, R.J., (1985), High Resolution Nuclear Magnetic Resonance of Inorganic Solids, **Science** 227 (4694) 1537.

Rettel, A., Gessner, W., Olieu, G. and Muller, D. (1983), On the Temperature Dependence of the Hydration of $CaO.Al_2O_3$, **Z. Anorg. Allg. Chem.** 300, 89-96.

Seligmann, P. (1960), Possible Applications of NMR in Cement and Concrete Research, **J.P.C.A. Research and Development Laboratories,** 2(3) 20-31.

Seligmann, P., ibid (1968), Nuclear Magnetic Resonance Studies of the Water in Hardened Cement Paste, (10) 53-63.

Strike, A.J. and Williamson, D.J. (1988) School of Chemical Sciences, UEA, Norwich, to be submitted.

Appendix 1

The formation of spin-echoes is best described using the concepts of the net magnetisation vector M and the rotating frame of reference (Farrar). The original spin echo method devised by Hahn (1950) utilised two 90° radio-frequency pulses of the same phase. All the experiments reported in this paper used a $90°_x$-t-$180°_y$ sequence where the 180° pulse is 90° phase shifted relative to the 90° pulse. The x and y subscripts refer to the pulses being along the x' and y' axes in the rotating frame (Farrar) respectively. Referring to Figure A1 application of a 90° pulse along the x' axis tips the magnetisation vector so that it lies along the y' axis, (Figure A1a). The observed signal is a maximum and decays due to transverse relaxation with a time constant T_2^*. T_2^* is the spin-spin or transverse relaxation time, T_2, decreased by applied magnetic field inhomogeneity. Because of the applied field inhomogeneity some of the macroscopic magnetisations m_1, which make up M arising from nuclei in different part of the field precess at rates different from that of the rotating frame. Some rotate faster and some slower than the frame, i.e. the m_l dephase in the x'-y' plane (Figure A1b). After a time τ a 180° pulse along the y' axis causes the m_l to rotate 180° about y' (Figure A1c). The result is that those m_l that rotate faster than the frame and had moved ahead are now moving towards it and those m_l that rotate slower and had lagged behind are now being "caught" by the frame, i.e. the m_l rephase, (Figure A1d). This rephasing causes a free induction signal to build to a maximum at time 2τ, dephasing again occurs and a decaying free induction signal is given back to back with the building signal, i.e. an echo. The echo height is less than that of the initial FID, due to T_2 transverse relaxation that occurs during the time 2τ and hence depends on T_2. Effects of field inhomogeneity are removed by the refocussing and the method provides a means of evaluating T_2 from a plot of echo amplitude against τ for various pulse intervals.

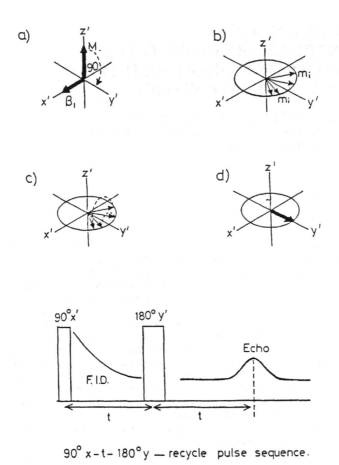

90° x - t - 180° y — recycle pulse sequence.

Fig. A1. Rotating frame diagrams depicting spin echo formation and a representation of the pulse sequence and signals produced.

8 ON THE CHANGE OF MICROSTRUCTURE DURING THE HYDRATION OF MONOCALCIUM ALUMINATE AT 20° AND 50°C

W. GESSNER, R. TRETTIN, A. RETTEL, D. MÜLLER
Central Institute of Inorganic Chemistry of the Academy of
Sciences of the GDR, Berlin, GDR

Abstract

The starting point of our investigations concerning the
hydration of High Alumina Cement (HAC) was the fact that
in contradication to the circumstances in portland cements
the relations between the parameters characterising the
microstructure and the molecular sphere, respectively, and
the macroscopic properties were object of discussion in a
few cases only. This is an unsatisfactory situation since
such investigations are essential regarding a further
development of the quality of these binding materials.
    As a contribution to a better understanding of the
development of the microstructure during the hydration of
HAC, adequate investigations on the main Clinker phase
monocalcium aluminate (CA) were carried out by measuring
the change of heat evolution, degree of hydration,
mobility of the protons of the water molecules fixed in
the hydrate phases, specific surface area, pore-size
distribution and concentrations of CaO and Al_2O_3 in the
liquid phase.

1   Experimental

The CA used for the present study was prepared by
calcining an equimolar mixture of $CaCO_3$ and α-Al_2O_3 at
1500°C. The material was then finely powdered to a
specific surface area of 0.55 m²/Kg. The purity of the
material was checked by x-ray examination. The hydration
was carried out in an atmosphere free of CO_2 with a
water/solid phase-ratio of 0.5 at 20 and 50°C. To stop
the hydration process the samples were treated with
ethanol and ether and then dried over silica gel.
    Changes of heat evolution in dependence of time were
recorded with an isothermal conductive calorimeter system
(Oliew and Wieker 1981). The degree of hydration was
determined from the ^{27}Al solid state NMR spectra using the
fact that the unhydrated CA contains AlO_4 - groups only,
the hydration products, on the other hand, AlO_4 - groups

(Müller et al 1984). The mobility of the protons of the
water molecules fixed in the hydrate phases was followed
up by measuring their resulting spin-lattice relaxation
times T_1 by means of ^1H NMR (Rettel et al 1983).
Measurement of adsorption isotherms for determining of
specific surface area and pore-size distribution were
carried out with N_2 and Kr in a device, working on the
basis of the volumetric principle (Trettin 1984 and van
Dongen et al 1971).

The pore-size distributions were calculated as reported
by Brunauer et al (1967). The concentrations of CaO and
Al_2O_3 in the liquid phase were determined
complexometrically with fluorexone as indicator and
photometrically with ferron, respectively (Schönherr et al
1987).

2   Results and discussion

2.1   The constitution of the hydration products

Before describing the course of hydration of CA at 20 and
50°C some remarks will be given on the constitution of
hydrates occurring in this process. For CA and the stable
final product of hydration C_3AH_6, constitutions with
highly
condensed $\left[Al \quad O_2^{[4]} \right] n-$ n- anions and monomeric $\left[Al \quad (OH)_6^{[6]} \right] 3-$
anions, respectively, were pointed out on the basis of
complete crystal structure analyses (Hoerkner et al 1976
and Weiss et al 1964). On the other hand, for the primary
metastable phases CAH_{10} and C_2AH_8 ambiguous data exist in
the literature.

However, the answer to the question, whether the change
of coordination number from 4 to 6 proceeds only when the
conversion of the hydrates low in lime content into C_3AH_6
takes place or in the first stages of hydration (i.e.
connected with the development of strength) is important
for discussing the reasons of the hydraulic activity of
CA.

Therefore the problem of constitution of the phases
CAH_{10} and C_2AH_8 is of particular interest. Using the
solid-state. ^{27}AL NMR spectroscopy we have shown earlier
(Geßner et al 1982) that the aluminium is sixfold
coordinated in both hydrates. Starting from this result
and using structure conceptions given in the literature
(Scheller et al 1974 and Buttler et al 1978) we have
proposed and confirmed, respectively, a constitution as
$\left[Ca_2Al^{[6]} \quad (OH)_6 \right] \left[Al^{[6]} \quad (OH)_3 \quad (H_2O)_3 \right]$ OH for C_2AH_8 and as
$Ca_3 \left[Al_4^{[6]} \quad (OH)_{24} \right]$ 18 H_2O for CAH_{10}. We consider these
results as a proof for the thesis that the hardening of
aluminates on the reaction with water is accompanied by a

change of the coordination of Al from 4 to 6. Therefore the hydration is connected with a development of strength for aluminates containing aluminium in fourfold coordination only (GeBner et al 1977).

The fact that aluminium is fourfold coordinated in CA but sixfold in the hydration products is the basis for determining the degree of hydration by means of ^{27}Al NMR spectroscopy as described above. The progress of the hydration process at a given time can quantitatively be determined from the spectra by evaluating the ratio of the intensities of the two signals typical for AlO_4 and AlO_6 groups, respectively.

2.2   Hydration at 20°C

In figure 1 the evolution of heat and the degree of hydration are plotted against the time. In the primary state of hydration a first maximum of heat evolution can be observed called the first DCA peak. Thereafter the induction period occurs without heat evolution and with an increase of degree of hydration up to lower than 1% only (Fujii et al 1986). In the following acceleration period the heat evolution arises forming the so called 2nd DCA peak. In the range of this peak the degree of hydration increases abruptly and reaches values of about 35% connected with the formation of amorphous reaction products as it has also been shown by Edmonds and Majumdar (1988). From these products after longer times CAH_{10} and C_2AH_8 crystallize, indicated - for CAH_{10} - by an abrupt decrease of the proton relaxation time T_1 (marked in Fig. 1) caused by a sudden increase of dipolar interaction of the proton of the water molecules involved in the hydration process.

After the 2nd DCA peak during the period of diffusion-controlled reactions, the degree of hydration still continues to increase but only very slowly. Also after longer reaction times C_3AH_6 and crystalline aluminium hydroxide did not occur at a hydration temperature of 20° C.

The course of specific surface area of the solid samples separated from the liquid phase in the primary range of hydration (up to 20 min) is characterised by the occurring of three maxima (Figure 2a). Afterwards only small changes of the specific surface area can be observed; it seems to be nearly constant up to the end of the induction period. A comparison of the course of specific surface area with the heat evolution shows a coincidence of the 1st DCA peak and the maxima of specific surface (Figure 2a/b).

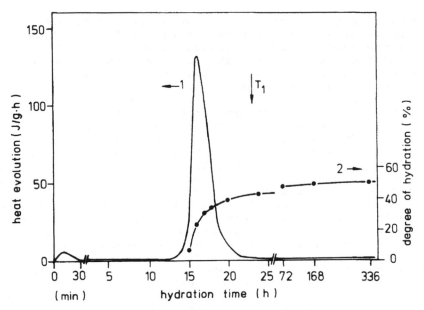

Figure 1: Course of hydration of CA at 20°C
　　　　　1)　　DCA curve
　　　　　2)　　Degree of hydration

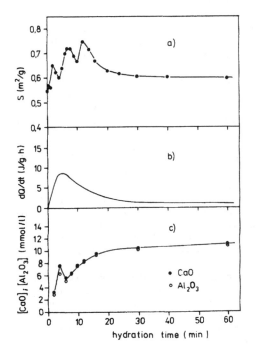

Figure 2:

Course of hydration of CA at 20°C in the primary states

a) specific surface area
b) rate of heat evolution
c) concentrations of CaO and Al_2O_3 in the liquid phase.

Obviously there is a relation between the exothermic reaction steps and the development of specific surface.

From the analysis of the liquid phase separated from the CA pastes, it follows that the concentrations of CaO and Al_2O_3 also pass through a maximum (Figure 2c) before they reach a level being nearly constant during the induction period. Only with the beginning of the acceleration period an abrupt decrease of the concentrations of CaO and Al_2O_3 occurs (not shown in Figure 2). The CaO/Al_2O_3 ratio in the liquid phase was found to be between 1.05 and 1.15 showing an incongruent solubility. The congruent solubility of CA in the primary state, discussed in literature (Rodger and Double 1984 and Barret et al 1974) could not be confirmed.

The described course of hydration in early times can be interpreted with the assumption of a first reaction layer on the unhydrated CA grain as proposed by Fujii et al (1986). Primary its occurrence may be caused by the incongruent solubility of CA leading to an enrichment of Al_2O_3 on the surface of the remaining grain. It could be supposed that the dissolving of Ca^{2+} and aluminate ions first of all takes place unequally on the CA surface, e.g. at distorted spheres of the crystal lattice, leading to a "roughing up" of the surface i.e. to an increase of the specific surface area. With this process, however, the course of hydration is explicable only within the first four minutes. The further changes of specific area observed have to be attributed to the formation and transformations of the first reaction layer. For this layer adequate changes of the chemical composition have to be supposed by reason of the course of concentrations in the liquid phase. After a hydration time of 20 min a relative stable reaction layer has been formed on the surface of the CA grains which seems to be very dense because of its small specific surface area. By this reason the reaction layer hinders the access of water and therewith the further hydration.

The finishing of the hydration period presupposes an arising of permeability of the reaction layer. A process affecting this may be the formation of nuclei within this layer. This interpretation is supported by the fact that the specific surface area is nearly constant during the last hours of the induction period. On the contrary, the formation of nuclei outside the layer i.e. via the solution as discussed elsewhere (Barret et al 1974), should be connected with an increase of the specific surface area.

The course of specific area after longer hydration times is plotted together with the known curve of heat evolution in Figure 3. It is evident that within the acceleration period the specific surface rises drastically and that the following decreases coincides with the diminishing of the rate of heat evolution. After nearly

20h the specific surface area remains constant until
between 72 and 168h a further decreasing occurs.

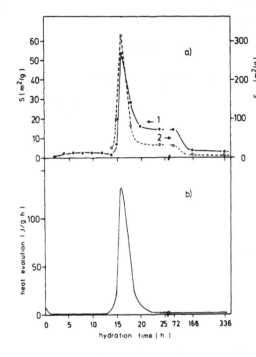

Figure 3:

Courses of hydration of
CA at 20°C at later
times.

a) specific surface
   area (curve 1);
   specific surface
   area of the
   hydration products
   (curve 2)
b) rate of heat
   evolution.

A similar course is observed for the specific surface area
of the hydration products themselves which can be
calculated using the values of degree of hydration (curve
2 in Fig. 3a). Obviously there is a relation between the
amount of specific surface area of the hydration products
and the rate of heat evolution which is correlated with
the reaction rate. That is understandable because the
hydration products cover the surface of CA and it is to be
expected that the course of hydration is influenced by the
permeability of this shell. Therefore it seems to be
useful to investigate the pore-size distribution of the
hydration products.
    The frequency of the pores of the hydration products in
the range of radii of 10 up to 80 $\overset{\circ}{R}$ increase dramatically
in the acceleration period and decreases after the second
maximum of heat evolution (Figure 4). That means that in
the acceleration period with a high reaction rate the
hydration products possess a high frequency of pores and
are more permeable.

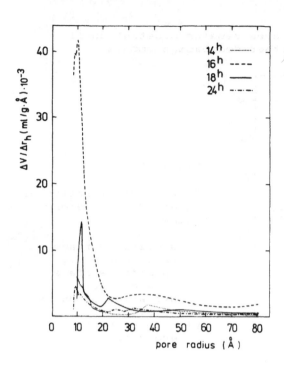

Figure 4:

Pore-size distribution of reaction products of the hydration of CA at 20°C at different times.

On the other hand, in the periods with lower reaction rate hydration products with a relatively low frequency of pores occur, which are denser and not permeable to such an extent. It is obvious to suppose the increasing densification of the hydration products as the reason for the change of the reaction to a diffusion-controlled course. A process contributing to the densification of the hydration products is the crystallization of the hydrate phases occurring primarily in an amorphous state. With this corresponds that the crystallization of CAH_{10} after about 20h - characterised also by the decrease of the proton relaxation time T_1 as described above - coincides with the first diminishing of the specific surface area. The crystallization of C_2AH_8 can be observed only after 3 days. Therewith in connection may be the second decrease of specific surface area observable at this time.

2.3 Hydration at 50°C

The macroscopic course of hydration of CA at 50°C - a temperature representative for the circumstances in the range of 30 up to 55°C (Rettel et al 1985) - is shown in Figure 5; plotted are the degree of hydration and the rate of heat evolution versus the time.

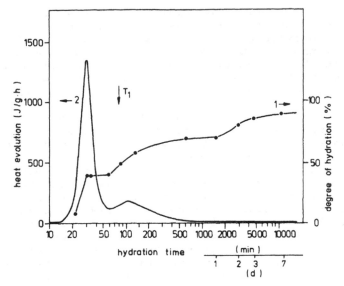

Figure 5: Course of hydration of CA at 50°C
1) degree of hydration
2) DCA curve

In comparison with the hydration at 20°C in this case the
rate of hydration increases drastically. That leads to a
shortening of the induction period and an earlier
beginning of the accelereation period. In contrast to the
circumstances at 20°C a further maximum of heat evolution
(3rd DCA peak) can be observed. In the range of the 2nd
DCA peak only C_2AH_8 occurs as crystalline hydration
product besides amorphous aluminium hydroxide. The 3rd
DCA peak is connected with the conversion of C_2AH_8 into
C_3AH_6, which proceeds very quickly under these conditions,
and with the crystallization of the amorphous aluminium
hydroxide. Obviously these effects also cause the
decrease of the proton relaxation time T_1, which takes
place at the beginning of the 3rd DCA peak (marked in
Figure 5). At the time of the 2nd DCA maximum there is a
degree of hydration of about 40%, then it remains constant
for time. With the occurrence of the 3rd DCA peak the
degree of hydration increases once more and reaches values
near 70%. Afterwards only a slight increase of the degree
of hydration is observed with a small discontinuity in the
range of 1 day hydration time.
    The described course of hydration at 50°C can be
interpreted as follows: The first increase of degree of
hydration during the acceleration period is connected with
the formation of C_2AH_8 and amorphous aluminium hydroxide.
These hydration products cover the CA grains and hinder
the access of water and therewith the further reaction.

103

Only with the conversion of C_2AH_8 into the well-crystallized C_3AH_6 and the crystallization of the aluminium hydroxide gel into gibbsite, occurring during the time of the 3rd DCA peak, the microstructure of the shell of the hydration products covering the CA grains, is so loosened that the hydration can start once more. After all this new reaction causes the 3rd maximum of heat evolution.

To get more information about the change of microstructure the development of specific surface area and the pore-size distribution was investigated also at a hydration temperature of $50°C$.

As it is shown in Figure 6, three maxima of specific surface area are observed immediately after the contact with water within the first 8 minutes of hydration. Afterwards the specific surface decreases and then changes only a little till the beginning of the acceleration period. As well this alternating course of specific surface area as the coincidence being observed also in this case with the first maximum of heat evolution and maxima of CaO and Al_2O_3 concentrations in the liquid phase could be described - as it was outlined in detail for the hydration at $20°C$ - with the suggestion of formation and conversion of a primaric reaction layer on the CA grains. The finishing of the induction period may be attributed, also at $50°C$ hydration temperature, to an arising of permeability of this layer caused by the formation of nuclei therein.

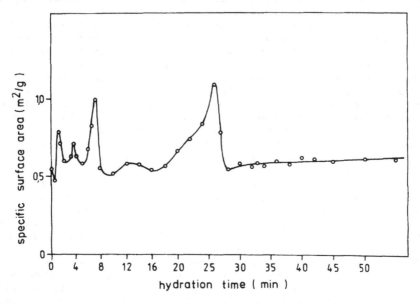

Figure 6: Course of specific surface area in the
          primary states of hydration of CA at $50°C$

With the beginning of the acceleration period the specific surface area arises reaching a maximum of 1.1 m² /g. Afterwards it decreases and remains then constant up to 60 minutes. This course is in principal adequate to those observed at 20°C. But the relative small increase of the specific surface area in the range of the acceleration period is surprising regarding the adequate value at 20°C (50m²/g) and the fact that the degree of hydration reaches about 40% at this time.

This result could be interpreted on the assumption that at 50°C the specific area after the acceleration period is determined only by the crystalline C_2AH_8 covering the amorphous aluminium hydroxide by the formation of a dense layer; in this way the surface of the hydroxide cannot be recorded. On the other hand, at 20°C hydration products are formed in this range whose exclusive amorphous character causes the large value of specific surface area observed. A drastic increase of specific surface area appears during the hydration at 50°C in the range of the third peak of heat evolution. Two maxima can be observed at 80 min (i.e. in the beginning of the 3rd DCA peak) and at 150 min (i.e. at the point of decreasing of the heat of evolution after the third maximum), respectively (Figure 7).

The maximum at 80 min has to be attributed to the loosening of the dense layer of C_2AH_8 caused by its beginning conversion into C_3AH_6. Now, the specific surface area of the amorphous aluminium hydroxide existing besides the C_2AH_8, is recorded. Afterwards its

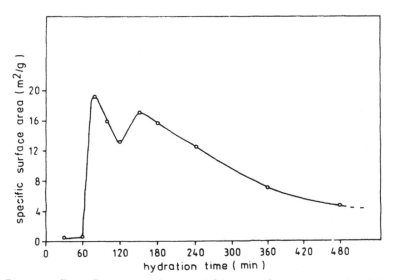

Figure 7: Course of specific surface area during the hydration of CA at 50°C in the range of the 3rd DCA peak

crystallisation into gibbsite takes place leading to a
decreasing of the specific surface area. The decrease of
the proton relaxation time T_1 mentioned above has to be
attributed to these reactions. Obviously the described
reorganisation processes of the layer of the hydration
products covering the CA grain lead to an arising
permeability which enables the increase of the degree of
hydration. By the fresh beginning hydration products are
formed are primary rich in surface and partly amorphous
again. These phases cause the maximum of specific surface
area at 150 min. With their following densification by
the crystallization processes also at last a relative low
level of specific surface area is reached (after 8h). At
this time C_3AH_6, gibbsite and only traces of C_2AH_8 occur
as hydration products.

The pore-size distributions of the hydration products
calculated at the time of the maxima and minima of the
specific surface area (80, 120, 150 min, 8h) are shown in
Figure 8.

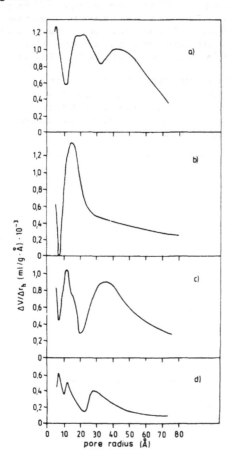

Figure 8:

Pore-size distribution
of reaction products of
the hydration of CA at
50°C at different times

a) 80 min,
b) 120 min,
c) 150 min,
d) 8h.

The frequency of pores in the range of radii between 5 and 80 $\mathring{A}$ reaches the lowest level after 8h corresponding with the densification of the reaction layer (curve d). On the other hand, after 80 min the frequency of pores shows large values (curve a). The discussed increased permeability of the hydration products may be connected with this observation. A similar large frequency of pores occurs also after 120 min (curve b) but with one maximum at 15 A only. Possibly the pores with radii > 25 $\mathring{A}$ will be covered by the crystallization of aluminium hydroxide taking place between 80 and 120 min.

The pore-size distribution in the range of the maximum of specific surface area at 150 min (curve c) which is caused by the new formation of hydration products as a result of the progress of reaction is similar to those observed after 80 min, especially with regard to the reoccurrence of the maximum at r = 40 $\mathring{A}$. Therewith the above mentioned suggestion is confirmed after which the freshly formed hydration products are similar to those occurring after 80 min with respect to their composition and degree of order. That means e.g. that the new formed aluminium hydroxide is being produced primarily in an amorphous state.

The further hydration of CA after the 3rd DCA peak is characterised by a small continuous heat evolution and by a slight increase of degree of hydration only. The observed discontinuity in the degree of hydration after 1 day (Figure 5) may be attributed to a new loosening of the dense covering layer, possibly caused by crystal growth processes of the hydration products (C_3AH_6 and gibbsite), which facilitates the access of water to the unreacted CA grain. The repeated formation of reaction products primary rich in surface during this time leads to an increasing of the specific area (Figure 9) and of the frequency of pores (not shown) once more. In the following part of hydration up to 7 days at last the densification of the reaction products takes place accompanied by a decrease of specific surface area and of frequency of pores.
Their low permeability indicated by this decrease finally leads to the diminishing of the reaction rate.

The observations described give rise to the assumption that between the 1st and the 7th day of hydration similar processes happen again as discussed for the range of the 3rd DCA peak. But obviously the effects of heat evolution per unit of time connected with these reactions are to small as they would be detectable by the differential calorimetric analysis.

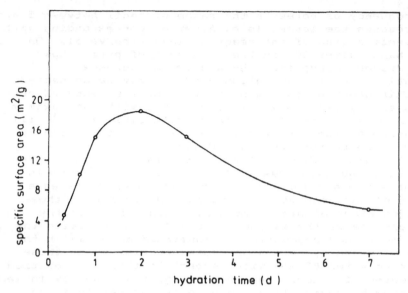

Figure 9: Course of specific surface area during
the hydration of CA at 50°C in later
times

3   Conclusion

Taking the results together one can conclude that the
course of hydration of monocalcium aluminate which is
charaterized by periods of different reaction rates would
be explicable with the suggestion of reaction layers
covering the CA grains.   The reaction rate is influenced
by the permeability of these layers which alters in
dependence of time and temperature caused by changes of
their composition and by conversion and crystallization
processes.

4   References

Barret P., Menetrier D., Bertrandie D. Cem. Concr.
    Res 4 (1974) 545

Brunauer S., Mikhail R.S., Bodar E.E., J. Colloid.
    Interf.Sci. 24 (1967) 451

Buttler F.G., Taylor. H.F.W. Il Cemento 75 (1978) 147

Edmonds R.N., Majumdar A.J. Cem. Concr. Res. 18
    (1988) 311

Fujii K., Kondo W., Ueno.H. J. Am. Ceram. Soc. <u>69</u>
   (1986) 361

GeBner W., Muller D., Behrens H.J., Scheler G., Z.
   anorg. allg. Chem. <u>486</u> (1982) 193

GeBner W., Rettel A., Winzer M., Jung. M.
   Silikattechnik <u>28</u> (1977) 232

Hoerkner W., Muller-Buschbaum H.K., J.Inorg. Nucl.
   Chem. <u>38</u> (1976) 983

Muller D., Rettel A., GeBner W, Scheler G. J.
   Magn.Reason. <u>57</u> (1984) 152

Oliew G., Wieker W. Silikattechnik <u>32</u> (1981) 152

Rettel A, GeBner W., Oliew G., Muller D.
   Z.anorg.allg.Chem. <u>500</u> (1983) 89

Rettel A., GeBner W. Müller D. Scheler G., Br. Ceram.
   Trans. J. <u>84</u> (1985) 25

Rodger S.A., Double D.D. Cem. Concr. Res. <u>14</u> (1894)
   73

Scheller Th., Kuzel H. Proc. VI. Int. Symp. Chem.
   Cem. Moscow 1974 Vol.2/1 p 217

Schönherr S., Gorz H., Bertram R. Wiss.Z. Padag.
   Hochsch. Potsdam <u>31</u> (1987) 67

Trettin R. Thesis Academy of Sciences of the GDR,
   ZIAC Berlin 1984

Weiss R., Grandjian D., Pavin J.L., Acta
   Crystallogr. <u>17</u> (1964) 1329

van Dongen R.H., Kaspersma J.H., de Boer J.H. Surf.
   Sci. <u>28</u> (1971) 237

PART THREE
ADMIXTURES

9 THE INFLUENCE OF SUPERPLASTICISING ADMIXTURES ON CIMENT FONDU MORTARS

S.M. GILL
Department of Engineering Science, Oxford University,
Oxford, UK
P.F.G. BANFILL
School of Architecture and Building Engineering, University of
Liverpool, Liverpool, UK
E. EL-JAZAIRI
Cement and Concrete Chemicals, European Technical Products
Division, Grace, Slough, UK

Abstract
Mortar cubes of Ciment Fondu containing two different
superplasticising admixtures were monitored during storage in water
at 5, 20, 30 and 40°C by measurement of the ultrasonic pulse velocity
(upv). Some mortars additionally contained an accelerating
admixture. Sufficient cubes were crushed to enable the correlation
between upv and compressive strength to be established. Hydrate
compositions were checked using derivative thermogravimetry (DTG) and
X-ray diffraction (XRD).
The presence of any of the admixtures tested reduced upv and
strength. The size of the reduction depended strongly on the
temperature of curing and storage as well as on the sample age.
Analysis of the hydrates showed that the strength loss could be
attributed to the conversion of CAH_{10} and C_2AH_8 to C_3AH_6. The same
conversion process occurred slightly later for mixes without
admixture and led to a similar loss of strength.
In conclusion, admixtures change neither the hydration processes
nor the type of hydrates formed. The influence of curing temperature
appears to be of greater significance in determining strength than
the presence of admixtures.
Keywords: Ciment Fondu, Superplasticisers, Mortar, Hydration,
Temperature, Strength, Conversion.

1 Introduction

The use of superplasticisers with Portland cement is now well
established enabling the production of extremely workable mixes or of
mixes with significantly reduced water content. Superplasticised
aluminous cement mixes with similar characteristics are an attractive
proposition. Highly workable mixes could be used where high early
strength or improved chemical resistance is required whilst reduction
of water content to less than 0.40 water/cement ratio (w/c)
significantly reduces the rate of conversion [Midgley & Midgley,
1975]. However previous studies on such systems have shown the
tendency for unacceptable set retardation [Quon & Malhotra, 1982].
The hydration of Ciment Fondu will not be described in detail in
this paper since much work has already been reported [Cottin, 1971;

George, 1983; Lea, 1970; Parker & Sharp, 1982; Robson, 1962].

The main features of Ciment Fondu hydration are:

(i) the initial hydration rate slows above 20°C until 30°C when the rate increases rapidly [George, 1983; Robson, 1962]. This results in delayed setting of pastes and concretes.

(ii) Both CAH_{10} and C_2AH_8 are metastable crystalline products which tend to convert to the stable C_3AH_6 and AH_3. The conversion reaction, which is favoured by increased temperature and w/c ratio is responsible for the loss of strength noted for some Ciment Fondu mixes [French, Montgomery & Robson, 1970; Midgley, 1967; Midgley & Midgley 1975; Neville, 1958; Neville, 1963].

This paper describes the influence of superplasticisers and an accelerator (added to overcome retardation) on Ciment Fondu mortars. Results up to 12 months are described but longer term storage is in progress and will be the subject of a later paper. Since compressive strength testing of mixes containing a large number of admixture combinations at several test temperatures and test ages would have required an unreasonably large number of specimens to be cast, a non-destructive test was preferred. The Pundit apparatus was used to monitor the ultrasonic pulse velocity (upv) of mortar cubes and the upv was correlated with compressive strength.

2 Experimental details

Physical and chemical characteristics of the Ciment Fondu used are given in the Appendix. The sand was North Nottinghamshire quartzitic, zone 2/3. The admixtures were:

Superplasticisers

S1 Sulphonated naphthalene formaldehyde condensate
S2 Development product

Accelerator (A)

Lithium salt (development product)

Twelve identical mixes containing various combinations of admixtures detailed in Table 1 were prepared at 5, 20,30 and 40°C. All mortars were 1:1.5 cement:sand at w/c 0.40. Great care was taken to ensure that materials were stored at the appropriate temperature prior to mixing. Each mortar mix (comprising 3.77kg mortar) was mixed for 5 minutes using a Kenwood planetary mixer with 3 litre mixing bowl. The admixture solution was added gradually to the dry materials during the first minute. Admixture dosages are in terms of weight % of cement. Three mixes at 40°C with extremely rapid setting times were mixed for only 3 minutes to enable cubes to be made before stiffening occurred.

To reduce the time between mixing and casting, and therefore the opportunity for temperature change the workability of these mixes was assessed only visually. An extensive investigation of the effect of admixtures on workability will be reported in another paper.

Immediately after casting the moulds and their contents were placed inside polythene bags in water at the correct temperature. A slightly different technique was used for mixes cast at 20°C which were cured in air in the laboratory, covered with polythene sheeting to prevent evaporation. Cubes were demoulded as soon as possible, typically at 6 hours after casting though considerably longer for some mixes. The first upv readings were taken 24 hours after casting with further readings taken at intervals thereafter.

Four 70.7mm cubes were cast from each mix, 2 of which were crushed for comparison with upv at randomly chosen ages as shown on Table 1. The only restriction on these timings was that the first cube was crushed at one of the initial 6 test ages and the second cube at one of the remaining test ages. This ensured that both cubes were not tested either at a very early or late age.

Table 1. Mix details and cube ages at compressive strength tests

Mix	Cube ages at compressive strength tests							
	5°C		20°C		30°C		40°C	
Control (1)	1d	12m	3d	6m	14d	3m	7d	2m
Control (2)	3d	12m	21d	2m	2d	18m	14d	3m
0.025% A (1)	1d	18m	2d	28d	7d	28d	21d	6m
0.025% A (2)	3d	6m	21d	6m	7d	28d	1d	28d
0.3% S1	3d	6m	3d	2m	1d	18m	14d	3m
0.3% S1 + 0.025% A	3d	28d	14d	18m	1d	3m	2d	12m
1.0% S1	7d	2m	2d	18m	21d	28d	21d	18m
1.0% S1 + 0.025% A	14d	12m	2d	12m	21d	3m	2d	2m
0.3% S2	7d	3m	1d	2m	14d	18m	2d	12m
0.3% S2 + 0.025% A	7d	6m	1d	3m	21d	28d	3d	12m
1.0% S2	7d	6m	14d	2m	7d	2m	21d	3m
1.0% S2 + 0.025% A	2d	18m	1d	28d	14d	12m	3d	6m

Note: d = day   m = month

The density of each cube was determined immediately prior to strength testing by weighing in both air and water. The strength was determined in general accordance with BS 915 Part 2:1972.

A sample of mortar was obtained from each cube after strength testing by drilling into the cube using a 5mm diameter masonry bit and its composition tested by derivative thermogravimetry (DTG) and X-ray diffraction (XRD).

Early results from the main programme at 20°C suggested that the presence of admixtures influenced conversion. A secondary series of mixes was therefore prepared at 20°C, using the same materials and mix proportions, to enable this behaviour to be studied in more detail. Four cubes were cast from each of the mixes listed below and upv measurements were made at intervals of 2 weeks up to an age of 4 months.

Additional mixes tested at 20°C only

Control
0.025% A
1.0% S1
1.0% S2
1.0% S1 + 0.025% A
1.0% S2 + 0.025% A

One cube from each mix was crushed at 1,2,3 and 4 months and a
sample was removed for analysis by DTG as described for the main
series.

3 Results and discussion

3.1 Workability
Both superplasticisers improved workability although S2 was more
effective than S1. The duration of improved workability is dependent
on temperature, a rapid loss of workability is observed at high
temperature.

3.2 Effect of age and hydration temperature
The differences in upv vs age relationships for mixes hydrated at
different temperatures are readily seen on Figures 1-8.
Leaving aside for the moment the effect of the admixtures it is
worthwhile to note the following general trends.
At 5°C there is a slow but continuous increase in upv up to 12
months. At test ages exceeding 28 days the highest upv measurements
are always obtained for mixes at 5°C.
At 20°C upv increases up to 1-6 months after which a steady
decrease occurs.
At 30°C there are early increases in upv with values approximately
the same as for mixes at 5 or 20°C. However after 3-7 days there is
a sharp decrease in upv and although there is some later recovery the
final upv values are much reduced compared to mixes at 5°C.
At 40°C it appears that there is a similar abrupt decrease in upv
as for 30°C but that this occurs at or before 1 day. Subsequently
there is a slight increase in upv up to 6-12 months.
The most likely explanation for the decrease in upv is the
conversion of CAH_{10} and C_2AH_8 to C_3AH_6. DTG evidence in support of
this explanation will be discussed later.

3.3 Effect of admixtures
Figures 1-8 show that the presence of either of the superplasticisers
usually causes a reduction in upv at all of the hydration
temperatures studied for all test ages. The only exceptions are 0.3%
S1 at 5°C and 0.3% S2 at 20, 30 and 40°C. The reduction in upv is
more severe with increasing dosage and is compounded by the
additional presence of the accelerator.

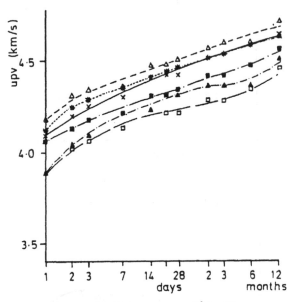

Figure 1. Upv vs age, 5°C, S1

Mixes containing only the accelerator do not show marked differences compared to the control except at 20°C where the conversion reaction appears to begin at an earlier age. The general form of the relationship between upv and age at each hydration temperature remains unaltered from the descriptions given in the previous section. The effect of both superplasticisers is very similar at all temperatures. The only notable exception occurs at 30°C when the presence of S2 extends the initial period during which upv is increasing. The subsequent minimum value of upv at 30°C is the same for all mixes, about 3.8km/s. The presence of the accelerator in addition to either superplasticiser exacerbates the drop in upv and causes it to occur earlier than for mixes containing only a superplasticiser.

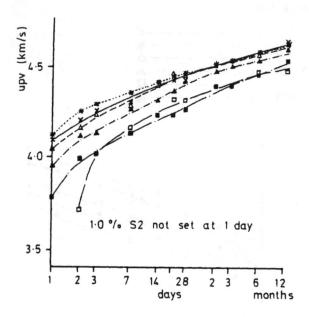

Figure 2. Upv vs age, 5°C, S2

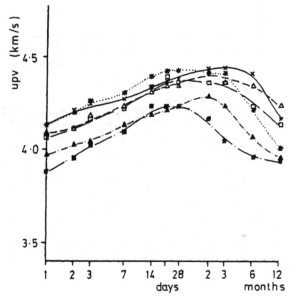

Figure 3. Upv vs age, 20°C, S1

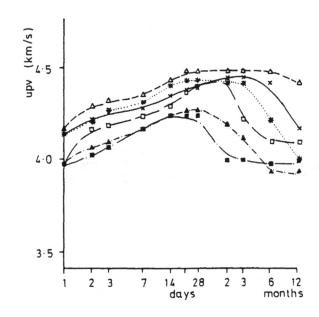

Figure 4. Upv vs age, 20°C, S2

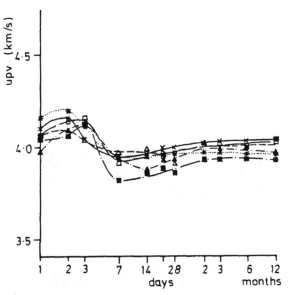

Figure 5. Upv vs age, 30°C, S1

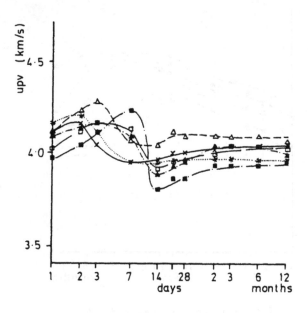

Figure 6. Upv vs age, 30°C, S2

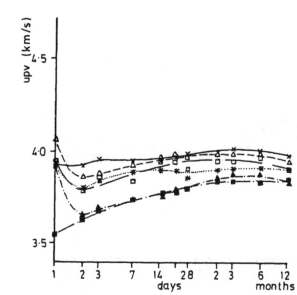

Figure 7. Upv vs age, 40°C, S1

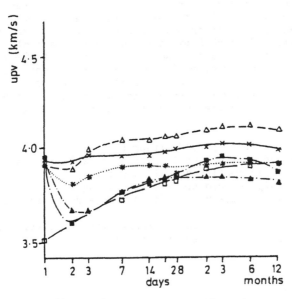

Figure 8. Upv vs age, 40°C, S2

The relationships between upv and age for the secondary series of mixes at 20°C are shown on Figure 9.

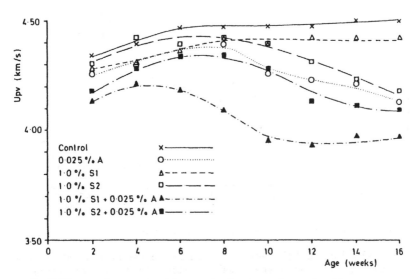

Figure 9. Effect of age on upv: secondary series of mixes

The results confirm that either plasticiser causes an early

decrease in upv. The reduction is compounded by the presence of the accelerating admixture. For the mix containing 1.0% S2 + 0.025% A the upv reduced rapidly after 6 weeks and by 16 weeks the strength of this mix had reduced to 60 N/mm^2 compared to the control mix strength. of 108 N/mm^2. Figures 3 and 4 show that the control mix at 20°C also shows a reduction in upv after 6 months and by 12 months is clearly approaching the upv of the mixes containing admixtures. Longer term testing is in progress which appears to confirm that the ultimate strength of mixes is unaffected by the presence of admixtures.

3.4 Correlation between upv and strength

It was anticipated that upv would give a reliable indication of strength and this is confirmed by Figure 10.

The relationship between upv (v) and strength (f) for mortar or concrete is generally taken to be of either a power law or exponential form.

Inclusion of all results in the regression calculation gave the following relationships between upv and strength.

$$f = 0.0025 \times v^{7.08} \tag{1}$$

$$f = 0.05 \times e^{1.70 \cdot v} \tag{2}$$

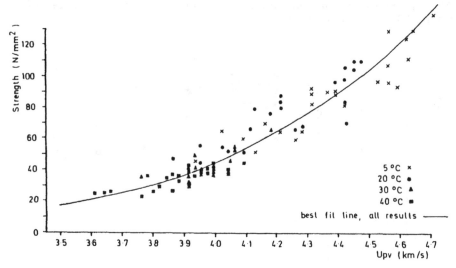

Figure 10. Correlation between upv and strength

The correlation coefficient for each of the equations above was 0.94 indicating that both are equally valid, thus only equation 1 is plotted on Figure 10. A upv of 4.5km/s therefore corresponds to a strength of about 105N/mm^2 and a upv of 3.9km/s to about 40N/mm^2.

It should be emphasised that this empirical relationship is valid only for Ciment Fondu mortars with the same mix proportions and aggregate as those used in these tests.

It might be expected that differences in hydrate composition may change the relationship between upv and strength and so the correlation at each hydration temperature was investigated. Using a power law equation it was found that the best fit curves for 20,30 and 40°C were within the 95% confidence limits calculated for 5°C. The confidence intervals for 5°C were selected since the best correlation (r = 0.92) was obtained for this temperature. There is therefore no evidence that differences in hydrate composition caused either by initial hydration temperature or by conversion have any effect on the correlation between upv and strength.

3.5 Hydrate composition
Lack of space prevents the inclusion of any DTG or XRD traces in this paper but the important features are described in this section.

In general the presence of admixtures did not change the type of hydrates produced since the major hydration products were CAH_{10}, C_2AH_8, C_3AH_6 and AH_3, as for plain mixes.

Although great care was taken to maintain the correct storage temperatures, considerable increases in cube temperature occurred as a result of the heat evolved during hydration. For example the temperature of cubes stored nominally at 20°C was found to exceed 30°C for about 3 hours during initial hydration. To check whether this is important pastes and mortars of the same composition hydrated at 5°C and 20°C were tested by DTG. The pastes were prepared at w/c 0.40 using only 10g of Ciment Fondu to ensure that there was no significant temperature rise in the sample during hydration.

The effect of the transient high temperatures in the larger mortar specimens was noticeable for both mixes. At 5°C the temperature in the mortar cube rose enough to allow formation of C_2AH_8 and AH_3 in addition to the CAH_{10} found in the paste. At 20°C the same effect was observed and was more pronouced than at 5°C.

A 70.7mm cube is comparable to the thickness of grout likely to be used in industrial applications so the effects observed during this test programme provide a reliable indication of field behaviour.

The DTG results from this programme suggest that the role of lithium in enhancing the formation of C_2AH_8 is simply to raise the hydration temperature by accelerating the reaction. This is in contrast to the suggestion that lithium ions have a specific affinity for C_2AH_8 at the expense of CAH_{10} [Rodger and Double, 1984].

For the control mix at 5°C there is little change in hydrate composition from 1 day to 1 year. The major product is CAH_{10} with small amounts of C_2AH_8 and AH_3. The very early presence of C_2AH_8 and AH_3 is indicative of the temperature rise in the cubes during hydration rather than the occurrence of conversion.

At 20°C the main initial products are CAH_{10}, C_2AH_8 and AH_3. DTG results confirmed the occurrence of conversion which proceeded most rapidly between 3-6 months. This corresponds well with the decreasing upv recorded on Figures 3 and 4.

The effect of admixtures on the earlier occurrence of conversion at 20°C has already been noted.   It was initially thought that this was again due to the increased temperature of the accelerated mixes. However 1.0% S2 causes severe set retardation but also favours initial production of C_2AH_8 and AH_3 rather than CAH_{10}.   Therefore it seems that the observed changes in hydrate production and subsequent rate of conversion are due both to temperature rises in accelerated mixes and to the presence of the superplasticisers.

At 30 and 40°C the main hydration products are amorphous gel, AH_3 and C_3AH_6.

For the control mix at 30°C C_2AH_8 is formed initially in preference to C_3AH_6 but after 14 days this has converted to C_3AH_6. The timing of the conversion reaction correlates well with the marked decrease in upv which in turn corresponds to a decrease in strength from about 65 to 40N/mm^2.

At 30 and 40°C the presence of any of the admixtures tested caused considerable set acceleration [Gill, Banfill & El-Jazairi, 1986]. This would be expected to cause higher actual hydration temperatures than for the control mix and accelerate the early conversion reaction.   Figures 5 and 6 show however that the reduction in upv (shown to correlate well with conversion) does not occur earlier for S1 mixes and indeed occurs later for S2 mixes.   It therefore appears that the presence of a superplasticiser can inhibit the conversion reaction, as was suggested by Young who proposed that organic compounds are adsorbed onto the hydrates [Young,1971].

At 40°C C_2AH_8,AH_3 and C_3AH_6 were present at 1 day.   At 7 days virtually all the C_2AH_8 had converted to C_3AH_6.   The absence of any reduction in upv after 1 day indicates that the conversion reaction must be already well advanced.   This argument is supported by the observation that the initial value of upv at 1 day is virtually the same as for mortar at 20 or 30°C after conversion.

4 Conclusions

1.   Admixtures change neither the hydration processes nor the type of hydrates formed.   They bring forward the conversion reaction to earlier ages at 20°C but strengths at later ages are no more than 10-20% lower compared to control mixes.   Larger dosages of superplasticiser cause larger strength reductions.
2.   There was good correlation (r = 0.94) between upv and strength for the mortar cubes tested.   The correlation was not significantly affected by the presence of different hydrates and the overall relationship between upv (v) and strength (f) was as follows:

$$f = 0.0025.v^{7.08}$$

3.   The occurrence of conversion, confirmed by DTG, was accompanied by a significant reduction in upv and therefore in strength.   The minimum value of upv for converted mortar was approximately 3.9km/s regardless of initial hydration temperature.   This corresponds to a strength of about 40N/mm^2.   The upv of unconverted mortars at ages exceeding 7 days was typically 4.3-4.5km/s, corresponding to a

strength of about 75-105N/mm². The strength loss associated with conversion was thus approximately 50%.
4. The heat generated during hydration can cause a considerable increase in mortar temperature sufficient to affect the type of hydrates produced. This is particularly a problem when curing specimens at about 20°C since it is in the range 20-30°C that the hydration of Ciment Fondu is most sensitive to changes in temperature. If very close control of hydration temperature is required the sample size should be minimised and an efficient water circulation system should be used to maintain the correct temperature.
5. The presence of the accelerator either alone or in combination with a superplasticiser reduces upv. Contrary to published work it was shown that the presence of lithium does not promote preferential formation of C_2AH_8 rather than CAH_{10} [Rodger & Double, 1984].

5 Acknowledgements

We gratefully acknowledge the financial support of the Science and Engineering Research Council and FEB (Great Britain) Ltd; provision of materials by Lafarge Aluminous Cement Co Ltd; and the experimental assistance of Miss P.Hooper and Mr L.Smith.

6 References

BS 915, Part 2: 1972. High alumina cement. **British Standards Institution**, London.
Cottin, B (1971) Etude au microscope electronique de pates de ciment alumineux hydrates en C_2AH_8 et CAH_{10}, **Cement and Concrete Research,**
1, 177-186.
French, P.J. Montgomery, R. and Robson, T.D. (1970) High concrete strength within the hour. **Meeting of the British Ceramic Society, Belfast**
George, C.M. (1983) in **Structure and performance of cements.** ed Barnes, P. Applied Science, London
Gill, S.M. Banfill, P.F.G. and El-Jazairi, B. (1986) The effect of superplasticisers on the hydration of aluminous cement, **8th International Congress on the Chemistry of Cement, Brazil 1986** IV, 322-327.
Lea, F.M. (1970) **The chemistry of cement and concrete.** Edward Arnold, London.
Midgley H.G. (1967) The mineralogy of set high alumina cement. **Transactions of the British Ceramic Society,** 66, 161-187.
Midgley, H.G. and Midgley, A. (1975) Conversion of high alumina cement. **Magazine of Concrete Research,** 27, 59-77.
Neville, A.M. (1958) The effect of warm storage conditions on the strength of concrete made with high alumina cement. **Proceedings of the Institution of Civil Engineers.** 10, 185-192.
Neville, A.M. (1963) A study of deterioration of structural concrete

made with high alumina cement. **Proceedings of the Institution of Civil Engineers,** 25, 287-324.

Parker, K.M. and Sharp J.H. (1982) Refractory calcium aluminate cements. **Transactions and Journal of the British Ceramic Society** 81, 35-42.

Quon, D.H.H. and Malhotra, V.M. (1982) Performance of high alumina concrete stored in water and dry heat at 25, 35 and 50°C. Journal of the American Concrete Institute, 79, 180-183.

Robson, T.D. (1962) **High alumina cements and concretes.** Contractors Record, London.

Rodger, S.A. and Double D.D. (1984) The chemistry of hydration of high alumina cement in the presence of accelerating and retarding admixtures. **Cement and Concrete Research,** 14, 73-82.

Young, J.F. (1971) The effect of organic compounds on the interconvertions of calcium aluminate hydrates. **Cement and Concrete Research,** 1, 113-122.

7 Appendix

PROPERTIES OF THE CIMENT FONDU

Setting Time (BS 915): Initial   280 minutes
                             Final   293 minutes

90 m Residue   3.8 %
Surface Area (Blaine)   276 m^2/kg

Chemical Composition (%):

SiO_2	Al_2O_3	Fe_2O_3	FeO	TiO_2	CaO	Na_2O	K_2O	SO_3	CO_2
4.14	38.67	10.43	5.63	1.98	38.5	0.06	0.04	0.19	0.42

10 THE EFFECT OF ADMIXTURES ON THE HYDRATION OF REFRACTORY CALCIUM ALUMINATE CEMENTS

J.H. SHARP, S.M. BUSHNELL-WATSON, D.R. PAYNE, P.A. WARD
School of Materials, University of Sheffield, Sheffield, UK

Abstract
The effect of four admixtures on the setting time and early hydration products of various refractory calcium aluminate cements has been investigated in the temperature range from 8 to 50°C. Although citric acid was found to be an effective retarder at all concentrations and temperatures investigated, an increase in the amount of the admixture did not necessarily cause greater retardation. Calgon, $Na_6(PO_3)_6$, was expected to act as an accelerator, but its behaviour was found to be very complex and to vary with temperature. Calcium hydroxide was found to be an effective accelerator and to favour the formation of C_2AH_8 and gibbsite. Lithium chloride was confirmed to act as a spectacular accelerator and brought about the formation of lime-rich hydrates.
Keywords: Calcium Aluminates, High Alumina Cement, Admixture, Accelerator, Retarder, Lithium Salt.

1 Introduction

Admixtures are used in the mixing and placing of refractory concretes to modify favourably the properties of the fresh or hardened concrete in such a way to make it more suitable for the job in hand (Lankard and Hackman, 1983). They are frequently added to control the working time and the flow properties of the material. Water reduction by the use of suitable admixtures is also important as this can lead to a denser material with much improved properties. Typically an admixture may modify more than one property, e.g. a specific additive may both reduce the water requirement and increase the working time for the castable. In practice placement of a refractory castable may involve the use of two or more admixtures simultaneously, leading to complex interactions.

In spite of this, there are relatively few investigations of the effect of admixtures on the hydration of calcium monoaluminate, $CaAl_2O_4$ or CA, and the commercial refractory calcium aluminate cements based on this phase.

We report here the early hydration of cements with 70-80% Al_2O_3 contents and also laboratory-synthesised samples of CA, in the presence of admixtures expected to act either as accelerators or retarders.

2 Experimental

Most of the experimental work was carried out on samples of Alcoa CA-14M, which is a typical refractory calcium aluminate cement with an Al_2O_3 content of 72% and CaO content of 27%, together with other commercial cements, described by Bushnell-Watson and Sharp (1986). X-ray diffraction showed the presence of CA and CA_2 as major phases with some alpha Al_2O_3 present and trace amounts of $C_{12}A_7$. The laboratory synthesis of the pure CA has been described by Bushnell-Watson and Sharp (1990 a). Although some batches contained trace amounts of $C_{12}A_7$, other phases could not be detected by X-ray diffraction. Calgon, calcium hydroxide and lithium chloride were laboratory grade reagents, whereas the citric acid was an Analar grade reagent. The cements were mixed with distilled water which had been boiled to remove dissolved carbon dioxide. The hydration procedure, determination of setting time parameter and the characterisation of the hydration products have been described by Bushnell-Watson and Sharp (1986). A water:cement ratio of 0.5 was used throughout this investigation as it produced a paste which could be readily poured into the setting tube containing the embedded thermocouple. Usually 30g. of cement were hydrated and the water bath was controlled to ± 0.1°C.

3 The effect of additions of citric acid

Citric acid (1%) was added to three commercial refractory calcium aluminate cements (Alcoa CA-14M, Secar 71 and Secar 80) and found to retard the setting time dramatically. Values of the setting time parameter, t_0, as defined by Bushnell-Watson and Sharp (1986) are given in Table 1 and several repeat experiments were carried out to establish that the results obtained were reproducible.
    The effect of the citric acid addition on the setting time parameter for the Secar cements was very striking. The value of t_0 at 20, 25, 30 and 35°C for Secar 71 without addition of an admixture was 3.9, 5.9, 7.3 and 3.4 hours, respectively, compared with 54.3, 67.9, 67.2 and 39.3 hours, respectively, in the presence of 1% addition. The corresponding values for Secar 80 were 2.1, 0.6, 0.3 and 0.3 hours compared with 28.1, 18.9, 7.1 and 4.7 hours. It can be seen that the addition of 1% citric acid has not changed the shape of the relationship between t_0 and temperature, discussed by Bushnell-Watson and Sharp (1986,

Table 1. Effect of citric acid on the setting time
paramenter for various cements at various temperatures

Temperature ($^\circ$C)	Setting time parameter (hours)						
	Alcoa CA-14M					Secar 71	Secar 80
	0%	0.25%	0.5%	0.75%	1.0%	1.0%	1.0%
12	2.6	60.5	60.0	40.7	52.1	48.9	-
20	6.4	62.0	65.4	59.6	55.2	54.3	28.1
25	-	-	-	-	-	67.9	18.9
28	17.1	39.2	47.3	43.8	83.2	-	-
30	-	-	-	-	-	67.2	7.1
35*	6.1	16.7	21.4	18.8	40.0	39.3	4.7
40	-	-	-	-	-	18.5	1.8
45	1.2	4.5	7.5	7.7	10.2	-	-
50	-	-	-	-	-	7.6	2.0

* 34.5°C for CA-14M

1990 a,b) in that Secar 71 shows a maximum setting time
around 28°C, whereas the setting time parameter of Secar
80 decreases continuously with increasing temperature.
   To investigate the effect of concentration of citric
acid on the setting time parameter, citric acid was added
to Alcoa CA-14M at concentrations of 0.25, 0.5, 0.75 and
1.0 weight %. Values of t_o obtained at various temperatures are shown in Table 1, from which it can be seen that
addition of citric acid in any concentration and at all
temperatures investigated caused retardation in the setting
behaviour of the cement. For example, at 20°C the setting
time parameter was 6.4 hours in the absence of citric acid,
but 55-65 hours in the presence of 0.25-1.0% citric acid.
   Surprisingly, an increase in the amount of citric acid
added did not always bring about greater retardation. The
setting time parameter was usually lower in the presence of
0.75% citric acid than in the presence of 0.5%. At 45°C
the retardation was much less dramatic than at lower
temperatures with a gradual increase in the setting time as
the concentration of citric acid was increased.
   The time taken to reach the maximum rise in temperature,
t_m, showed the same trend when plotted against the concentration of citric acid as did the setting time parameter,
t_o. In general it was found that as the setting time
increased, the maximum rise in temperature, ΔT, decreased.
   The effect of temperature on the setting time parameter,
t_o, of Alcoa CA-14M can be seen in Fig. 1. It is very
similar to data discussed by Bushnell-Watson and Sharp
(1986) for commercial cements containing CA. The setting

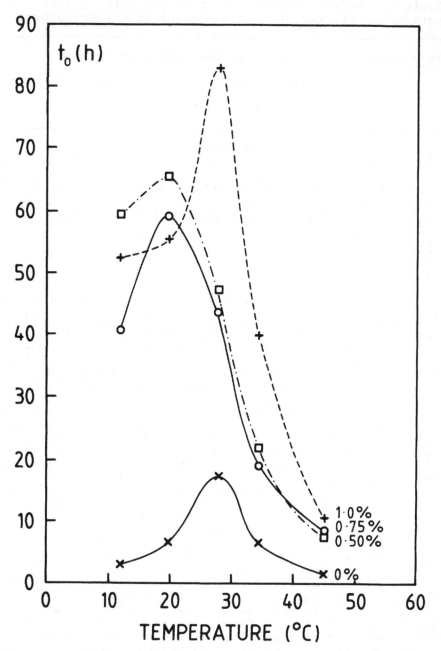

Fig.1. The effect of temperature on the setting time parameter, t_o, for various w/o additions of citric acid.

time increased with increasing temperature, reaching a maximum value at about 30°C, and subsequently decreased rapidly with increasing temperature.

The temperature dependence of the setting time parameter in the presence of additions of citric acid is also shown in Fig. 1. Addition of 1% citric acid resulted in a maximum setting time at 28°C, as in the absence of any admixture. Addition of lower concentrations of citric acid, however, resulted in maximum setting time parameters at ca. 20°C. The results obtained with 0.25% addition, which have been omitted from Fig. 1 for clarity, were closely similar to those at 0.5% addition.

3.1 Mineralogical analysis of the hydrates formed

All the samples were hydrated for t_0 + 48 hours and then quenched using acetone, and the products were examined by X-ray diffraction (XRD) and differential thermal analysis (DTA). Selected samples were also examined by scanning electron microscopy (SEM). XRD showed that the extent of hydration 48 hours after setting was similar in nearly all cases, with the height of the CA peak reduced on average to 22% of its original height and that of CA_2 reduced to about 70% confirming the greater reactivity of the CA. A substantial amount of CAH_{10} was always formed at 12°C accompanied by some gibbsite, but no C_2AH_8 was detected at this temperature. At 20°C, CAH_{10} was still the major crystalline hydrate found but some gibbsite and some C_2AH_8 were also detected. At 28°C, C_2AH_8 had become the major crystalline hydrate with some gibbsite and some CAH_{10} detected. At 34.5 and 45°C no CAH_{10} was detected, but increased amounts of C_2AH_8 and gibbsite were observed, along with trace amounts of C_3AH_6.

The results obtained at 34.5 and 45°C clearly indicate that C_2AH_8 was the major hydration product after t_0 + 48 hours. The CA remaining unreacted increased slightly as the concentration of citric acid increased, which suggests that the admixture is impeding the dissolution of calcium. It was noted that the XRD peak height of C_2AH_8 varied inversely with t_0.

These XRD results were confirmed by the DTA curves. Evidence for the presence of a "gel" phase was always observed by an endothermic peak at about 100°C. Peaks at about 150°C after hydration at 12 and 20°C were attributed to CAH_{10} and those at about 180°C after hydration at 28, 34.5 and 45°C to C_2AH_8. Gibbsite was evident in all the hydration products from its peak at about 280°C and C_3AH_6 could be seen in those formed at 34.5 and 45°C from the shoulders at about 325°C.

Whereas the DTA curves varied with hydration temperature, those obtained after hydration for t_0 + 48 hours at the same temperature but with different concentrations of citric acid were remarkably similar. This is illustrated by the five similar DTA curves shown in

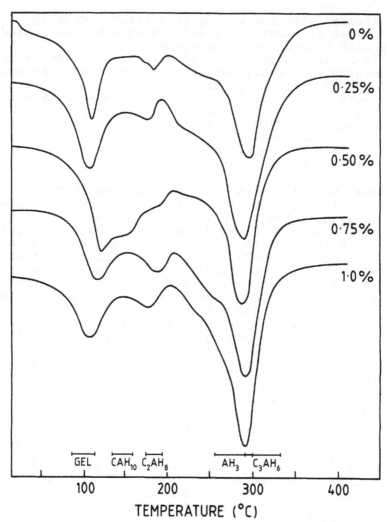

Fig.2. DTA curves of products from the hydration of Alcoa CA-14M + citric acid at 34.5°C

Fig. 2 for products hydrated at 34.5°C. Comparison of the top and bottom curves with zero and 1% admixture show closely similar traces indicating the formation of a gel phase (endotherm ca. 100°C), C_2AH_8 (180°C) and AH_3 (290°C). Only the curve obtained after hydration in the presence of 0.5 w/o citric acid is slightly different, indicating the presence of some CAH_{10} in addition to the other hydrates.

The close similarity of the XRD and DTA curves obtained after hydration for t_o + 48 hours was repeated at each individual temperature and indicates that a similar stage of hydration is always reached after this period, yet t_o

2 μm   5 μm

Fig. 3 (a) SEM of CA-14M   (b) SEM of Secar 80 + 1%
           hydrated for   citric acid hydrated
           50.6h at 12°C   for 48h at 30°C

varied from 1.2 to 65.4 hours.

SEM showed that the hydrates formed had their expected morphology. The most striking micrograph obtained was perhaps that shown in Fig. 3(a), which is of CA-14M hydrated at 12°C in the absence of any admixture. The hexagonal needles are attributed to the presence of CAH_{10}, which is not usually formed with such distinctive morphology. It is clearly distinguishable from the platey morphology of C_2AH_8 shown in Fig. 3(b), confirming that the characteristic microstructure of this phase is retained in the presence of citric acid.

4 The effect of additions of Calgon

Calgon, sodium hexametaphosphate ($Na_6P_6O_{18}$), was added to CA-14M at concentrations of 0.25, 0.5, 1.0 and 2.0 w/o and the setting time parameters were determined at the same temperatures as for citric acid additions (Table 2). The results were shown to be reproducible; thus when the 2% addition was repeated at 28°C, the values of t_o were 1.6 and 1.8h; t_m, 2.8 and 3.0h; and ΔT, 0.7 and 0.7°C.

The setting time parameter, t_o, is plotted against concentration of Calgon in Fig. 4. The complexity of the behaviour is obvious and the results are best discussed at each temperature investigated. At 12°C Calgon acted as a good retarder up to a concentration of 1%. The setting time parameter increased from 2.6h without any addition, to 11.8h with 1% Calgon, but decreased to 6.4h with 2% Calgon. At 20°C the effect of Calgon was almost neutral. A slight retardation was observed with very little change with concentration of the admixture. At 28°C, however, addition of Calgon had a marked accelerating effect, which increased with increasing concentration of the admixture. At 34.5°C, the setting time parameter was reduced from 6h

Table 2.   Effect of Calgon on the setting time parameter
           for Alcoa CA-14M at various temperatures

Temperature (°C)	Setting time parameter (hours)				
	0%	0.25%	0.5%	1.0%	2.0% Calgon
12	2.6	6.7	7.8	11.8	6.4
20	6.4	8.4	7.1	7.2	7.0
28	17.1	6.2	4.7	2.5	1.7
34.5	6.1	2.4	1.7	1.0	2.2
45	1.2	0.5	0.3	0.6	2.5

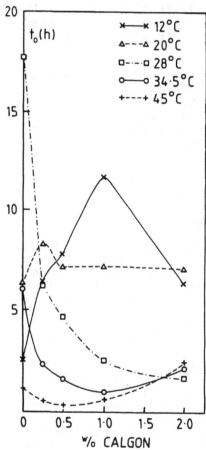

Fig.4. Effect of w/o Calgon
       on t_o of CA-14M at
       various temperatures

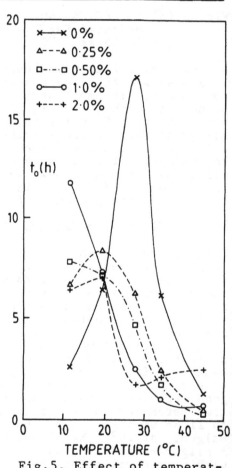

Fig.5. Effect of temperat-
       ure on t_o of CA-14M
       at various w/o add-
       itions of Calgon

without admixture to only 1.0h with 1% Calgon, but increased to 2.2h with 2% Calgon. A similar trend was observed at 45°C; t_o reduced from 1.2h without Calgon to 0.3h with 0.5% Calgon, but then increased to 2.5h with 2% Calgon.
It is evident that at 45°C an experimenter working at only one concentration would report Calgon to be an accelerator (at 0.25-1.0%), but to be a retarder at an addition of 2%. Similarly, an experimenter investigating only one temperature would report that a 1% addition of Calgon brings about acceleration at 28, 34.5 and 45°C, but slight retardation at 20°C and marked retardation at 12°C.

The variation of t_m with concentration of Calgon was of the same form as that of t_o. The highest values of ΔT were associated with the lowest values of t_o.

The effect of temperature on t_o for various additions of Calgon is shown in Fig. 5. Whereas the curve for no admixture shows maximum retardation at 28°C, additions of 0.25 and 2.0 weight % reduce this to 20°C. Additions of 0.5 and 1.0 weight % have maximum setting times below 12°C, indicating that the anomalous setting behaviour with respect to temperature has been overcome by the presence of the admixture.

Overall this remarkable variation of setting time with temperature and concentration of the admixture resulted in a surprisingly constant setting time at 20°C.

4.1 Mineralogical analysis of the hydrates formed
All the samples were hydrated for t_o + 48 hours, quenched and the products examined by XRD, which indicated that the extent of reaction 48 hours after setting was once again similar in nearly all cases. The height of the CA peak was on average reduced to 23% of its original height and that of CA_2 to 69%. These figures are remarkably similar to those obtained when citric acid was added, suggesting that the development of the hydrates in the 48 hours after setting occurs irrespective of the presence of the admixture. This observation was confirmed when the hydrates formed were examined. As in the case of citric acid additions, CAH_{10} was the major crystalline product at 12 and 20°C with some gibbsite formed at both temperatures and some C_2AH_8 at 20°C but not at 12°C. C_2AH_8 was the major hydrate observed at higher temperatures, accompan-ied by CAH_{10} at 28°C and gibbsite at all temperatures.

5   The effect of additions of calcium hydroxide

The variation of the setting time parameter, t_o, with temperature was determined for additions of 1w/o calcium hydroxide to Secar 71, Alcoa CA-14 and a laboratory-made sample of CA, which contained a trace amount of $C_{12}A_7$ but was otherwise pure. It can be seen from the results shown in Table 3 that the incorporation of calcium hydroxide

Table 3. Effect of additions of calcium hydroxide on the setting time parameter for various cements

Temperature (°C)	Secar 71 0%	Secar 71 0.5%	Secar 71 1%	Alcoa CA-14M 0%	Alcoa CA-14M 1%	Pure CA 0%	Pure CA 1%
8	1.9	1.1	-	2.6	1.5	-	-
12	2.4	-	1.0	-	-	-	-
20	3.9	2.6	2.5	4.5	2.7[+]	3.8[+]	2.7
25	5.9	3.6	5.6	9.5	3.7	10.2	6.8
28	7.8	4.5[*]	6.6	12.4[*]	4.0[*]	10.6	6.2
30	7.7	4.2	5.1	9.7	3.0	8.4	4.7
35	3.4	2.4	2.7	-	-	3.1	1.7
40	1.7	1.7	1.5	-	-	1.2	0.8
50	0.5	0.5	0.5	-	-	0.4	0.3

+ at 21°C   * at 27°C

accelerated the set in all systems. A maximum setting time was still observed at 25-28°C, but the temperature at which this maximum was observed appears to be slightly lower in the presence of added lime.

In the case of Secar 71, two different concentrations of calcium hydroxide were used. Both amounts accelerated the set at all temperatures studied, but the acceleration was greater for a 0.5% addition than for a 1% addition. This observation warrants a more thorough study of the effect of concentration of calcium hydroxide addition on the setting behaviour of refractory calcium aluminate cements.

A series of hydration products were examined by XRD and DTA with increasing time for Secar 71 hydrated at 25°C in the presence of 0.5 w/o of $Ca(OH)_2$. XRD showed that the consumption of CA was accelerated compared with the system without addition of any admixture in agreement with the reduced setting time parameter (3.6 hours). After 4.5 hours, DTA indicated the presence of a substantial amount of C_2AH_8 and some C_3AH_6, while "alumina" gel and gibbsite increased after 7 hours. Some CAH_{10} was formed, but much more C_2AH_8 was formed than in the absence of lime.

6 The effect of additions of lithium chloride

It has been known since 1937 that additions of lithium salts accelerate the set of high alumina cements (Robson, 1962), but their dramatic accelerating action was emphasised by Rodger and Double (1984). When 0.5 w/o LiCl was added to Secar 71, an exothermic effect was detected immediately the mixed paste came into contact with the thermocouple at all temperatures investigated. As a result an extrapolated onset temperature as used previously could not be determined.

Instead the setting time for Secar 71 pastes containing 0.5% LiCl was determined using a Vicat needle. The cement was mixed and poured into moulds placed in the water bath at the desired temperature. At one minute intervals a sample was removed, impressed with a 1mm x 5mm needle and a reading corresponding to the depth of penetration taken from the fixed scale. The plot of this scale reading versus time was sigmoidal. A straight line was drawn through the steepest part of the curve and its intercept with the x-axis was taken as the setting time. This method is considered to be less subjective than those based on the operator deciding whether or not a Gillmore or Vicat needle leaves an indentation mark on the surface. The setting times obtained in this manner are listed in Table 4.

Table 4. Setting times determined using a Vicat needle for Secar 71 + 0.5% LiCl

Temperature ($^{\circ}$C)	11	21	30	40
Time (minutes):	14.4	6.4, 7.0, 6.4	5.2	4.2

There was no longer a maximum in the setting time at ca. 28°C; instead the setting time decreased as the hydration temperature increased. This conclusion is in accordance with that reported by Gill et al. (1986), who showed that in ciment fondu pastes the retarded setting phenomenon is overridden by the addition of lithium chloride.

6.1 Mineralogical analysis of the hydrates formed

Hydration products of Secar 71 hydrated at 25°C in the presence of 0.5% w/o LiCl were examined by DTA and XRD after hydration for periods between 5 minutes and 1 week. These confirm that after only 5 minutes, appreciable hydration had occurred leading to rapid setting. The main hydration products were a gel phase, some AH_3 and C_2AH_8 and, most surprisingly, C_4AH_{13}. This phase is not a usual hydration product of Secar 71 or CA, although it is formed during the hydration of C_3A, especially in the presence of added lime. The C_4AH_{13} was detected in similar amounts by XRD in all the eight samples of this series, which suggests that it is formed as an initial product of hydration but not at later stages. It did not convert into C_3AH_6, although the latter was formed in increasing amounts between 1 day and 7 days hydration. No clear DTA peak which could be attributed unambiguously to C_4AH_{13} was observed, but it may overlap with the peak at ca. 190°C attributed to C_2AH_8 and $C_4A\bar{C}H_{11}$ (see Bushnell-Watson et al., 1988). A triple peak at ca. 300°C was observed in the DTA curves as previously reported (Bushnell-Watson et al., 1988).

The XRD and DTA traces corresponding to 5 minutes hydration at $25^{\circ}C$ also showed evidence for C_3AH_6, which again was unexpected. It can be concluded that lime-rich phases are favoured hydration products in the presence of LiCl, whereas little CAH_{10} was formed.

Increasing the hydration time from 5 minutes to 7 hours caused remarkably little change in either the XRD or DTA traces. This suggests that after an initial burst of hydration, which formed mainly C_4AH_{13} and AH_3, there was a dormant period of more than 7 hours. Beyond 7 hours, further hydration of the cement occurred and the amount of CA remaining decreased. The XRD and DTA peaks corresponding to AH_3 and C_3AH_6 became larger, but those due to CAH_{10}, C_2AH_8, $C_4A\bar{C}H_{11}$ and C_4AH_{13} remained remarkably constant.

A further investigation was carried out on Alcoa CA-25 cement hydrated at $25^{\circ}C$ in the presence of 0.5% LiCl for up to 7 hours. Similar observations to those just described were made; in particular the C_4AH_{13} phase was observed from the initial stages of hydration and remained more or less constant throughout. AH_3 was also formed initially, but none of the other hydrates, not even C_3AH_6, were detected in other than trace amounts. The very dramatic acceleration of the cement hydration therefore correlates with a much larger initial hydration than in the cements without addition of LiCl, accompanied by the formation of an unexpected phase, C_4AH_{13}. Once this lime-rich hydrate has formed the remaining cement undergoes a dormant stage, equivalent to that in the original cement mix. After this period further hydration occurs producing the usual calcium aluminate hydrates but perhaps still favouring the lime-rich C_3AH_6.

It has been suggested by Currell et al. (1987) that the formation of lime-rich products (such as C_2AH_8 reported by Rodger and Double (1984) and C_4AH_{13} reported above) in the presence of lithium salts is due to the increase in temperature associated with the accelerated reaction rather than to the formation of lithium aluminate as suggested by Rodger and Double. To test this hypothesis a sample of laboratory-made, pure CA was hydrated with and without addition of LiCl. The samples were maintained at $12^{\circ}C$ even during mixing by placing a sample of cement powder in the cell of a conduction calorimeter and allowing it to equilibrate at $12^{\circ}C$ over a period of several hours. Distilled water was dribbled on to the sample (by the conventional internal-mixing procedure) and the rate of heat evolution was monitored. The maximum temperature rise was observed after only 6 minutes in the presence of LiCl, indicating a setting time of 4-5 minutes, and the maximum increase in the temperature of the sample was only $0.7^{\circ}C$. After 24 hours the sample was examined by XRD and DTA and the results obtained are shown in Fig. 6, where they are compared with a similar sample hydrated under identical

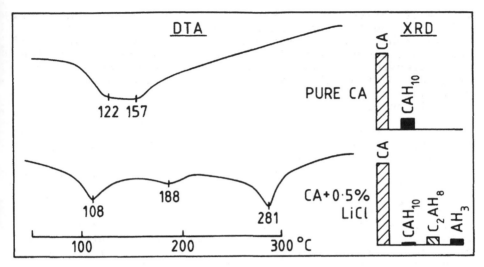

Fig. 6.   DTA curves and XRD peak heights of the reaction
          products of pure CA hydrated with and without
          LiCl after 24 h. at 12°C.

conditions but without addition of LiCl. The amount of CA
consumed was similar in both systems, but the hydration
products were different. Without any addition of LiCl, the
only products were crystalline CAH_{10} and an amorphous "gel"
phase, whereas in the presence of LiCl, much less CAH_{10} was
formed, but C_2AH_8 and AH_3 were also observed in significant
quantities. It must be concluded that lime-rich products
are formed because the Ca:Al ratio in solution is much
higher in the presence of LiCl, leading to formation of
C_2AH_8 at 12°C and C_4AH_{13} at 25°C. Neither of these are
normal hydration products (except perhaps in trace amounts)
under the conditions of these experiments in the absence of
lithium salts.

7 Discussion

The action of small additions of the admixtures studied can
bring about enormous changes in the setting behaviour of
calcium aluminate cements. Thus addition of citric acid
can result in setting time paramenters of 50 - 80 hours for
cement that would set after 2 - 17 hours in its absence,
whereas addition of LiCl can lead to a flash set for the
same cements.
      The effect of concentration of an admixture is, however,
often much more complex. It is shown in section 4 that at
12°C Calgon acted as a good retarder at concentrations up
to 1%, but was less retarding at 2%. Similarly in section
5 it is shown that a 0.5% addition of calcium hydroxide is

a more effective accelerator than a 1% addition. When this effect is coupled with the effectof temperature, very complex behaviour is observed. Thus at 20°C, the effect of Calgon was almost neutral, but at 28°C and above it became an accelerator. At 45°C, however, a 2% addition, while bringing about acceleration, was much less effective than a 0.5% addition.

It is surprising that in spite of the frequent use of admixtures to control the setting characteristics (e.g. Lankard and Hackman 1983) of refractory cements, there are few papers about their effects, with notable exceptions such as those by Rodger and Double (1984) and Currell et al. (1987), yet Robson (1962) was able to indicate that the accelerating or retarding properties can vary significantly with changes in concentration. This phenomenon is worthy of further study.

The presence of an admixture affects the appearance of plots of setting time parameter versus temperature. Some are merely displaced, e.g. when 1% citric acid is added to Secar 71 or Secar 80, but in others the anomalous setting behaviour with a maximum setting time parameter ca. 28°C is shifted to a lower temperature or even over-ridden, e.g. by the presence of Calgon or LiCl. This supports the suggestion made by Bushnell-Watson and Sharp (1986, 1990a) that the observed setting behaviour of Secar 80 is affected by the presence of an admixture or admixtures.

The progress of hydration beyond setting is remarkably consistent in the presence of both accelerators and retarders and seems to depend principally on temperature.

We have not made a thorough study of the mechanism of acceleration and retardation but some observations may be made. The amount and composition of the amorphous phase formed is probably significant and may account for the retarding effect of citric acid, as suggested by Rodger and Double (1984), by blocking access to the surface of anhydrous grains. The ratio of calcium to aluminate ions in solution is of critical importance. Admixtures that increase this ratio act as accelerators, whereas those that decrease it act as retarders. Thus the presence of calcium hydroxide increases the amount of lime in solution and has an accelerating effect, while the spectacular acceleration brought about by addition of a lithium salt may be due to the removal of aluminate ions as lithium aluminate (Rodger and Double, 1984). Support for this mechanism is provided by our observation of the formation of the lime-rich phases, C_2AH_8 and C_4AH_{13}, in the presence of LiCl at 12°C and 25°C, respectively.

One way to reduce the concentration of calcium ions in solution is to chelate them using retarders such as sugars and citric acid. The complex behaviour of Calgon, sodium hexametaphosphate, is probably related to its ability to act as a water-softener. It reacts with calcium ions in solution to form a complex salt:

$$2Ca^{2+} + (Na_4P_6O_{18})^{2-} = 4Na^+ + (CaP_6O_{18})^{2-}$$

Such activity will not only modify the ratio of calcium to aluminate ions in solution, but will also change the rate of dissolution of CA.

8 Acknowledgments

SMBW and DRP were supported by SERC studentships, the former under the CASE award scheme in collaboration with ICI Agricultural Division.

9 References

Bushnell-Watson, S.M. and Sharp, J.H. (1986) The effect of temperature upon the setting behaviour of refractory calcium aluminate cements. **Cem. Concr. Res.**, 16, 875-884.

Bushnell-Watson, S.M. and Sharp, J.H. (1990 a) Further studies of the effect of temperature upon the setting behaviour of refractory calcium aluminate cements. **Cem. Concr. Res.**, in the press.

Bushnell-Watson, S.M. and Sharp, J.H. (1990 b) On the cause of the anomalous setting behaviour with respect to temperature of calcium aluminate cements. **Cem. Concr. Res.**, in the press.

Bushnell-Watson, S.M., Winbow, H.D. and Sharp, J.H. (1988) Some applications of thermal analysis to cement hydrates. **Anal. Proc.**, 25, 8-10.

Currell, B.R., Grzeskowiak, R., Midgley, H.G. and Parsonage, J.R. (1987) The acceleration and retardation of set high alumina cement by additives. **Cem. Concr. Res.**, 17, 420-432.

Gill, S.M., Banfill, P.F.G. and El-Jazairi, B. (1986) The effect of superplasticers on the hydration of aluminous cement. **Proc. 8th Intl. Congress Chem. Cem.**, Rio de Janeiro, vol. IV, 322-327.

Lankard, D.R. and Hackman, L.E. (1983) Use of admixtures in refractory concretes. **Bull. Amer. Ceram. Soc.**, 62, 1019-1025.

Robson, T.D. (1962) **High alumina cements and concretes.** Contractors Record, London

Rodger, S.A. and Double, D.D. (1984) The chemistry of hydration of high alumina cement in the presence of accelerating and retarding admixtures. **Cem. Concr. Res.**, 14, 73-82.

11 PROPERTIES OF FRESH MORTARS MADE WITH HIGH ALUMINA CEMENT AND ADMIXTURES FOR THE MARINE ENVIRONMENT

N.C. BAKER, P.F.G. BANFILL
School of Architecture and Building Engineering, Liverpool University, UK

Abstract

HAC mortars, made at 5°, 20° and 40°C, were mixed using seawater, de-ionised water and reconstituted seawater. The admixtures used were: an accelerator, a superplasticiser, anti-washout, air-entrainment, water-proofer and an EVA polymer latex dispersion. The mortars were tested for initial and final setting times, workability and anti-washout characteristics. Setting times at 5° and 40°C were faster than at 20°C, with seawaters having a retarding action. The EVA mixes were particularly slow-setting. There was not much variation in the workability, except for the anti-washout mixes which were very stiff and sticky. There was a wide range of anti-washout characteristics, with slightly better results at lower temperatures.
Keywords: Mortars, Workability, Setting Time, Admixtures.

1 Introduction

HAC is known to have superior qualities in resisting attack by seawater and many other hostile chemical environments. HAC is widely recommended as being more durable than OPC in seawater (Gjorv, 1971, George, 1983). Conversion does continue but is usually very slow (up to 15% in 34 years) except in the tidal zone or in warm waters. HAC made with w/c < 0.4 has been found to be particularly resistant.

A research programme was set up to investigate the durability of HAC mortars, mixed with sea and fresh waters and with various common types of admixtures. Some authors have suggested that HAC mixed with seawater was more likely to lead to problems, due to formation of chloroaluminates, (Neville, 1975) and some that it might even be better (Gjorv, 1971). Halse and Pratt (1986), for instance, found that although there was early retardation with seawater and HAC, the later microstructure was very similar to that with fresh waters. Some of this work is relatively elderly and so, by using modern analytical methods and with contemporary knowledge of the conversion process, it is hoped to investigate this aspect of HAC use. Also, little previous research has looked at the use of admixtures, whose use is becoming increasingly common, particularly for repair work.

This paper presents early results of this programme and concentrates on the characteristics of the fresh mortars.

2 Experimental programme

2.1 Materials

HAC (Ciment Fondu) from a single batch supplied by Lafarge Aluminous Cement Co Ltd was used throughout. Graded sands were mixed in proportions set by AFNOR (see Table 1) so as to be comparable with other mortars made by Lafarge. Three different mixing waters were used: de-ionised water (di), Irish Sea water, settled but not filtered (sw), and reconstituted seawater (corrosion test mixture - BDH Chemicals (rsw)), see Table 2. Each mix was made with one of the above waters alone and also containing additional sea salts in an amount intended to simulate the use of unwashed sea-won aggregates (di + salt, sw + salt and rsw + salt). The amount of corrosion test mixture needed to achieve this was 6.6g total solids per kg of sand in the mortar. In every case the total water/cement ratio was 0.4 and the sand/cement ratio was 1.5. The admixtures used were chosen to represent a range of types such as might be used in marine work, whether for bulk work or for repairs. The admixtures and their dosage rates were:

    lithium citrate accelerator at 0.025% by weight of cement
    superplasticiser (FEB SP3) at 0.6% by weight of cement
    anti-washout (Conplast UW) at 1% by weight of cement
    air-entrainer (Cormix AE1) at 45ml/50kg of cement
    water-proofer (Palace "Intrapruf") at 1:30 in the mixing
    water
    EVA dispersion polymer (Vinamul 3281) at 5% solids by
    weight of cement.

Table 1. Sand particle size distribution in mortar (AFNOR)

Grade	A	B	C	D	E
Size fraction mm	1.2-2.4	0.6-1.2	0.3-0.6	0.15-0.3	0.09-0.15
Proportion by weight	7	10	4	4	2

In each case the proportions used were the maximum recommended by the manufacturer. Including the nil admixture this programme resulted in a factorial design of 126 mixes (7. admixtures x 3 waters x 3 temperatures x 2 aggregate salts content).

Table 2. Composition of sea waters (ppm by weight of each species)

Species	Cl^-	Na^+	$SO_4^=$	Mg^{2+}	Ca^{2+}	K^+	$CO_3^=$	Br^-
Irish Sea Water	19000	10500	2650	1350	400	380	160	65
Reconstituted	16200	9960	1830	570	440	-	-	-

2.2 Mixing method

The materials, except for admixtures, for each mix were brought to the temperature (5, 20, 40°C) at which they would eventually be cured. Each batch of about 5kg was dry mixed, in a Kenwood Chef domestic mixer, at 120 rev/min for 1 minute, then 1 minute of hand mixing with the mixing water plus admixture, followed by 1 minute at 250 rev/min.

2.3 Test methods

Workability was indicated by the consistence (Dropping Ball) and the Flow Table methods, in accordance with B.S.4551 (B.S.I., 1980). Times of Initial and Final Set were established using a thermocouple test. This semi-isothermal method consists of a copper-constantan thermocouple wire inserted into a 400g sample of mortar in an insulated container as soon as possible after mixing. A chart-recorder records the changing temperature with time with respect to 0°C, using an electric zero-reference. Initial set is taken to be the intersection of 2 "best" straight lines: through the early flat part of the curve and through the rising slope of temperature due to the main peak of heat production, see Fig. 1 (Bushnell-Watson and Sharp, 1986). We take final set as being at the point of maximum heat production. Anti-washout characteristics, for those mortars containing anti-washout admixture, were found by using a version of a method outlined by Davies (1987). An expanded metal mesh basket containing about 1kg of mortar is repeatedly dropped vertically through a 1.3 metre high column of water (see Fig. 2). The basket of mortar is weighed following each drop and the test criterion is taken as the % weight loss per drop. Davies used a longer column and 5 drops in his apparatus are equivalent to 8 in our apparatus.

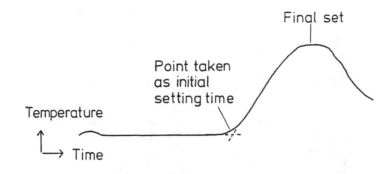

Fig. 1. Use of temperature profile for setting time

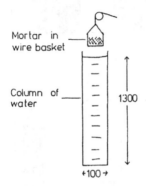

Fig. 2. Washout test apparatus (schematic)

3 Results

A large mass of data has been accumulated and space permits
only selected significant features to be presented here.
Analysis of Variance was carried out to assess the
significance of the factors in the factorial design and only
those factors found to be significant are described.

3.1 Workability
Fig. 3 shows the effect of admixtures on the flow of mixes
made with deionised water at each temperature. Similar
trends were noted for the other waters and Fig. 4 shows the
variation in the flow of mixes made with accelerator and
superplasticiser at each temperature. The workability of
mixes containing the other admixtures was unaffected by the
composition of the mixing water. The consistency results
showed a larger experimental error but generally confirmed
these trends.

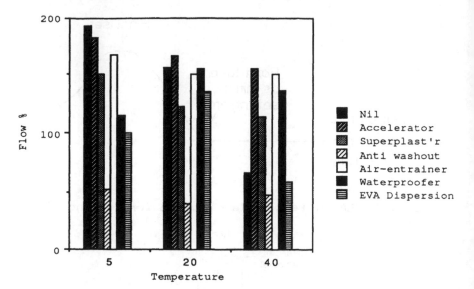

Fig. 3. Effect of temperature on the flow of mixes made with deionised water.

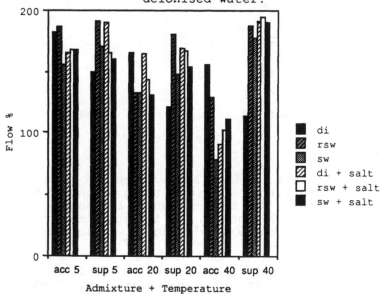

Fig. 4. Effect of temperature on the flow of mixes containing accelerator and superplasticiser with different mixing waters.

3.2 Initial setting time

Figs. 5-11 show the effect of temperature and mixing water on the initial setting time for each admixture.

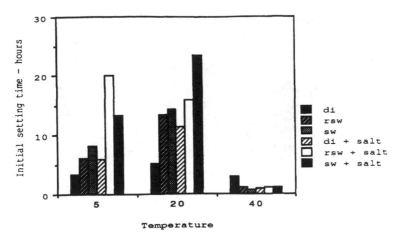

Fig. 5. Influence of temperature on setting time of mixes containing no admixture.

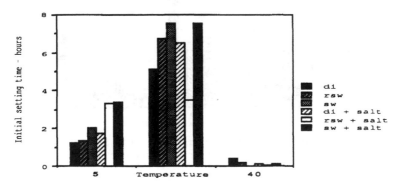

Fig. 6. Influence of temperature on setting time of mixes containing accelerator.

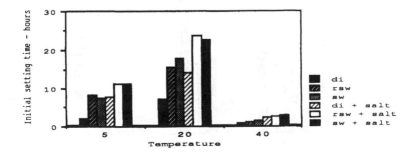

Fig. 7. Influence of temperature on setting time of mixes containing superplasticiser.

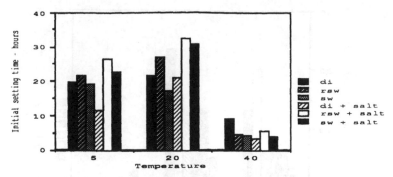

Fig. 8. Influence of temperature on setting time of mixes
containing anti-washout admixture.

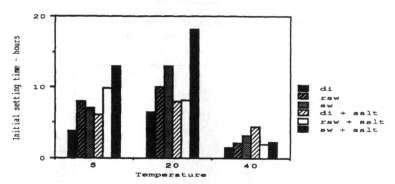

Fig. 9. Influence of temperature on setting time of mixes
containing air-entrainer.

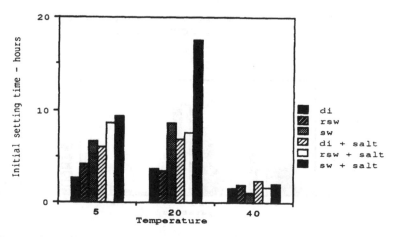

Fig. 10. Influence of temperature on setting time of mixes
containing water-proofer.

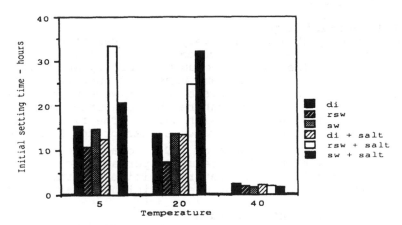

Fig. 11. Influence of temperature on setting time of mixes
containing EVA dispersion.

3.3 Anti-washout capability

Fig. 12 shows the effect of the antiwashout admixture on the
weight loss at 20°C. The weight loss after 8 drops at 5°C
was 27-70% and at 40°C was 45-70% but within this range the
order of influence of the mixing water also varied.

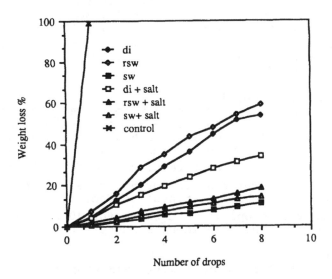

Fig. 12. Rate of weight loss with successive drops at 20°C.

4 Discussion

4.1 Workability
Workability showed relatively little variation, except for the anti-washout mixes, which were very stiff and sticky to work. EVA mixes had a medium workability and were easily worked but were at the same time self-supporting until trowelled. Superplasticised mortars were more workable with added salt at all temperatures but the other mixes were generally less workable with salts. The flow% varied little with the different temperatures, except for EVA, which was noticeably more workable at 5°C. Consistency did not always correlate with the flow results, perhaps because the dropping ball method is more operator sensitive for the range of mortars used and susceptible to finishing variability.

4.2 Initial setting time
The general pattern of setting times was that mixes at both 5° and 40°C were significantly faster than at 20°C, with seawaters being slower than de-ionised. This effect of temperature is the same as that already well known for HAC. Final set was generally about 2-3 hours after initial set. Variations from this pattern occurred when using anti-washout admixture, when the salt water mixes were faster, especially at 40°C. EVA mixes were very slow setting, especially at 5° and 20° C. With water-proofer, 5°C mixes were only very slightly faster than 20°C, but at 40°C initial set was much faster. Setting times with water-proofer were faster than other admixtures for all temperatures, with the contrast particularly noticeable at 5°C. It is possible that this is due to the extra nucleation sites provided by the silicate particles in the admixture.

4.3 Anti-washout capability
The rates of washout varied over a wide range. HAC mortar without the anti-washout admixture is completely washed away during the first drop. The anti-washout admixture was more effective at the lower temperatures and when salts were present in the mix. The best results were for 20°C mixes with salts. The poorer results for 5°C mixes may be because the cold coagulates the "gelatinous" effect of the admixture before it can work properly, and those for 40°C mixes due to the warmth melting the gelatinous structure too much.

4.4 General
The influence of admixtures on long term properties will be the subject of a future paper, but it is evident that the choice of an admixture to improve durability may have such a profound effect on the fresh properties that the long term performance may be impaired either through poor compaction or

through damage before setting.

5 Conclusions

Admixtures may have a great effect on the fresh properties of HAC mortars. These effects are further modified by the use of seawater in the mix and may also be affected by the presence of salts in the aggregate.

Users of HAC in the marine environment would be well advised to carry out trials on any admixtures which are being considered using conditions which will be found in real life.

6 References

B.S.I. (1980) **B.S. 4551: Methods mortars, screeds and plasters**, British Standards Institution.

Bushnell-Watson, S.M. and Sharp, J.H. (1986) The effect of temperature on the setting behaviour of refractory calcium aluminate cements. **Cem. Concr. Res.**, 16, 875-884.

Davies, B. (1987) Laboratory methods of testing concrete containing polymers for placement underwater, in **The production, performance and potential of polymers in concrete** (ed. B. Staines), Brighton Polytechnic, 375-378.

George, C.M. (1983) Industrial aluminous cements, in **Structure and Performance of Cements** (ed. P. Barnes), Applied Science Publishers.

Gjorv, O.E. (1971) Long term durability of concrete in seawater. **J. Amer. Concr. Inst.**, 68, 60-67.

Halse, Y and Pratt, P.L. (1986), The development of microstructure of high alumina cement, in **8th International Conference on Chemistry of Cement**, Communications theme 3, vol IV, 317-321.

Neville, A.M. and Wainwright P.J., (1975) **High Alumina Cement Concrete**, Construction Press, London.

7 Acknowledgements

We are grateful to the Science and Engineering Research Council and Lafarge Aluminous Cement Co. for financial support under the CASE award scheme.

PART FOUR
CALORIMETRY

12 ROLE OF FOREIGN CATIONS IN SOLUTION IN THE HYDRATION KINETICS OF HIGH ALUMINA CEMENT

M. MURAT, El. H. SADOK
Institut National des Sciences Appliquées de Lyon,
Villeurbanne, France

Abstract
The theoretical modelling, previously proposed to describe the homogeneous nucleation step during the hydration reaction of calcium sulfates (orthorhombic anhydrite and β-calcium sulfate hemihydrate) in the presence of foreign cations in solution, does not hold for hydration of high-alumina cement in the presence of foreign cations with z/r parameter (z electrical charge and r ionic radius) higher than 2.5 Å^{-1}. The noticeable increase of nucleation rate observed (investigation by isothermal calorimetry) is explained by the hypothesis of an heterogeneous nucleation process : potential nuclei responsible for the phenomenon are generally precipitated metallic hydroxide grains, the hydroxide precipitation being due to the increase of pH of the intersticial liquid phase that results from the dissolution of anhydrous calcium aluminate. For hydration in the presence of lithium or nickel ions, potential nuclei are solid particles resulting from rapid precipitation of an hydrated Li- or Ni-aluminate.
Keywords: High-alumina cement, hydration, kinetics, foreign cations, heterogeneous nucleation.

1 Introduction

It is well known that foreign ions in solution modify the hydration kinetics of mineral binders, but the nature of the exact step involved in the 'dissolution-nucleation-growth' process associated with non congruent dissolution of anhydrous phases was never clearly identified
   To clarify the situation, a first approach was attempted in the case of hydration of a mineral binder leading to the precipitation of one hydrate only. The binder chosen was orthorhombic anhydrite, the hydration of which leads to precipitation of calcium sulfate dihydrate ($CaSO_4 . 2H_2O$) and needs the addition to water of inorganic salts which act as 'chemical activators' to accelerate the reaction. Foreign metallic cations used as 'chemical activators' were introduced in the liquid water phase in condition of anion (sulfate) isoconcentration. Two important observations were made by Murat and El Hajjouji (1987): firstly, both dissolution rate of the anhydrous phase and growth rate of the reaction product are practically not affected by the presence of foreign cations in solution, and secondly, the nucleation rate 'Vn' of dihydrate diminishes very noticeably with the increase of a z/r

parameter characterizing the foreign cation (z and r, electric charge and ionic radius, respectively).

The theoretical modelling proposed to explain experimental results (Murat and El Hajjouji, 1987; Murat et al., 1987) was based on a modification of the expression of nucleation free enthalpy (hypothesis of an homogeneous nucleation): in the 'dilute solutions approximation' it leads to a single equation taking into account the z/r parameter characterizing the foreign cation:

$$Ln(Vn) = A + B(z/r) \tag{1}$$

where Vn is the nucleation rate assimilated to ($1/to$) where the time 'to' corresponds to the induction (or latent) period determined from experimental determination of $\alpha = f(t)$ curves (α, conversion rate).

Expression (1) was experimentally verified and extended to hydration of calcium sulfate hemihydrate plaster (Murat et Sadok, 1988; Sadok, 1989).

It is possible that the proposed model needs certain modifications when applied to solid-liquid systems implying an increase of pH of the liquid intersticial phase during hydration, the precipitation of more than one hydrate, the impossibilty to introduce the foreign cation with a common soluble anion (e.g. silicate or aluminate in the case of OPC or HAC, respectively), or the intervention of a 'non-homogeneous' nucleation process, all conditions which were not encountered in the case of our preceding investigations on hydration of orthorhombic anhydrite.

These considerations have led us to attempt the extension of the nucleation model to hydration reaction of pure monocalcium aluminate and high alumina cement (Murat and Sadok, 1988; Sadok, 1989)). Effectively, with the actual data we were unable to decide exactly what is the concerned step in the 'dissolution - nucleation -growth' process involved in the hydration reaction. Parker (1952) and Robson (1967) mention a decrease of setting time with basic additives increasing the pH, when acids promote some delays in the hydration process. Rodger & Double (1984) have pointed out the particular role of lithium cation which promotes an heterogeneous nucleation process by very rapid precipitation of a lithium aluminate hydrate whose grains play the role of potential nuclei for growth of hexagonal calcium aluminate hydrate. More recently, Currell et al. (1987) have schown that setting time obeys the following increasing order according to the nature of ions introduced in hydration water :

$Li(+) < Na(+) < $ Pure Water $< K(+) < Ca(2+) < Mg(2+) < Sr(2+) < NH_4(+)$
and $\quad OH(-) < $ Pure Water $< Cl(-) < NO_3(-) < Br(-) < $ Acetate

These authors have proposed an interpretation of their results partially based on the 'ligand theory' of Misono et al. (1967) and a schematization of the change in hydrate formation by modification of aluminum hydration sphere.

Results described and discussed in the present paper will concern only the role of foreign cations introduced as chlorides in hydration water. Effectively, when foreign cations are introduced as sulfates, one can help by cation concentration, either an acceleration or a retardation of hydration reaction (Robson, 1967; Banfill, 1986), and, as shown by Sadok (1989), two successive hydration peaks may appear on calorimetric curves, the first one corresponding to precipitation of ettringite. In such conditions, kinetical interpretation becomes difficult when based on the determination of the time corresponding to the end of the latent period.

2 Experimental

Experimentation have concerned hydration of 3 samples:

Two pure CA samples synthetized by high-temperature solid-solid reaction: the first one (called 'Turin sample' in the text) was provided by Bachiorrini (1985), the second one, called 'CA Lafarge', by Sorrentino from Lafarge Coppée Recherche (Viviers sur Rhône, France)
A high alumina Secar 71 cement (called 'Secar 71' in the text), also provided by Lafarge Coppée Recherche: its mineralogical composition, determined by XRD, is approximatelly: CA 68.6 % ; CA_2: 27.1 % ; α-Al_2O : 3.2 %; β-Al_2O : 1.1 %

Hydration kinetics was investigated by isothermal calorimetry (use of the calorimeter developped by Karmazsin and Murat (1978)) to obtain kinetical data (Fig.1). XRD analysis was used to identify the precipitating hydrates.

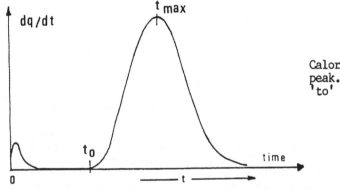

Fig.1.

Calorimetric hydration peak. Détermination of 'to' and 'tmax'.

The rapid hydration of CA and Secar 71 samples allows an accurate determination of both the latent period 'to' and the time 'tmax' of the exothermal hydration peak (Fig.1). Nucleation rate 'Vn' and growth rate 'Vg' were approximated by the following expressions:

$$Vn \simeq 1/to \qquad (2)$$
$$Vg \simeq 1/[tmax - to] \qquad (3)$$

XRD investigations have concerned samples (1 g) hydrated in small glass tubes (like for calorimeter experiments), after stopping the reaction using a process derived from that described by Bachiorrini and Guilhot (1982): three successive washing with a 0.5/0.5 Ethanol/Ether mix, and drying of the solid for 2 hours in air before grinding. Calcium aluminate hydrates were identified by the measure of intensity of their two more characteristic lines (d in Å) : 7.16 and 14.20 for CAH_{10} ; 10,70 and 2.86 for C_2AH_8; 2.30 and 2.04 for C_3AH_6.

The liquid/solid ratio (l/s) used for both hydration experiments was unity (0.5 cm^3 of solution for 0,5 g of solid for hydration in the calorimeter; 1.0 cm^3 of solution for 1.0 g of solid for hydration in glass tubes).

Hydration reaction were carried out either with pure water (PW) or with different foreign cations (Monovalent : Cs, NH_4, K, Na, and Li; Bivalent : Ba, Sr, Hg, Ca, Cd, Mn, Zn, Cu, Co, Ni, Mg; Trivalent : La, Fe, Al; Tetravalent : Zr) introduced in the hydration water as chlorides in conditions of anionic isoconcentration (0.025 Mole/l). This concentration was chosen for a comparison with results of Rodger & Double (1984) who have used a liquid/solid ratio of 0.5 with 0.050 M chloride solutions. So, the 'Chloride/Aluminate (or Secar 71)' ratio is identical in both experiments.

Compressive strength was measured on hydrated Secar 71. Samples were gauged at w/c = 0.4, then poured into plexiglass moulds to obtain minicylinders of diameter 2.0 cm and height 4.0 cm, and cured in saturated water vapour up to 7 or 28 days. The concentration of foreign cation in the solution was fixed at 0.0625 M in chloride ion for comparison with calorimetric and XRD analysis.

3 Results and discussion

Hydration calorimetric curves are reproducible and give rise to one peak only. Generally nucleation rate (Vn) and growth rate (Vg) are increased by introduction of foreign cations in the hydration water and their values may differ according to the origin of the investigated samples, but whatever it is, growth rate does not greatly vary versus the z/r parameter characterizing the foreign cation, contrary to nucleation rate which decreases up to z/r $\simeq$ 2.5 A^{-1}, then suddenly increases for 2.5 $Å^{-1}$ < z/r < 3.0 $Å^{-1}$ and slowly diminishes again for z/r > 3.0 $Å^{-1}$ (Fig.2).

The Ln(Vn) = f(z/r) diagram, proposed to explain results obtained for hydration of orthorhombic anhydrite, presents the same discontinuity for 2.5 A^{-1} < z/r < 3.0 A^{-1} (Fig.3). Consequently, the hydration process of monocalcium aluminate or high alumina cement is quite different from that of calcium sulfates, at least for sample with z/r parameter higher than 2.5 $Å^{-1}$.

Lithium cation gives rise to the more rapid hydration, as observed by Rodger & Double (1984): no latent period is measurable on the calorimetric curve. So we must consider that nucleation rate is infinite (arrow on Fig.2 and Fig.3) as a consequence of the heterogeneous nucleation process consecutive to an instantaneous precipitation of hydrated lithium aluminate.

Hydrates formed during hardening after one day are essentially CAH_{10} and C_2AH_8. However, the variation, versus z/r, of the intensity of the characteristic lines of these two hexagonal hydrates shows that except for hydration with lithium cation (which leads to formation of C_2AH_8 only + a small quantity of lithium aluminate hydrate $LiH(AlO_2)_2.5H_2O$, observed by Rodger & Double), CAH_{10} is the predominant hydrate, but its

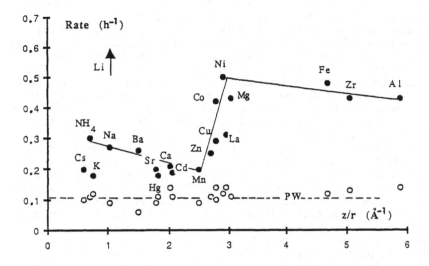

Fig.2. Variation, versus z/r, of nucleation rate (solid points) and growth rate (open points) for hydration of Secar 71. Horizontal line gives nucleation rate in pure water.

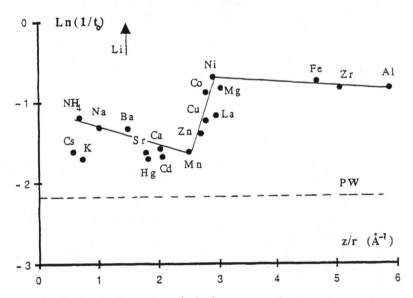

Fig.3. Variation of Ln(1/to) versus z/r for Secar 71. Horizontal line : value for hydration in pure water.

content decreases with cation at high z/r parameter.

On the other hand, there is no direct correlation between CAH_{10} content and the z/r parameter characterizing the foreign cation. That confirms the limited intervention of foreign cations in the growth process during formation of calcium aluminate hydrates.

Compressive strength (Rc), measured at 7 and 28 days, does not present any correlation with the z/r parameter characterizing foreign cations. Independently of the decrease of strength in relation to hydration in pure medium, this behaviour is quite different from that of anhydrite binders (El Hajjouji and Murat, 1987). In the later case, compressive strength decreases when z/r increases which agrees with the hypothesis of an homogeneous nucleation largely influenced by the presence of foreign cations in solution. The higher the nucleation rate (hydration with cation at low z/r), the higher the strength because at constant Solid-Liquid weight, the higher the number of nuclei formed, and consequently the lower the crystal size after hardening, then the lower the porosity of the hardened material. In the present case, the absence of correlation between Rc and z/r let us to believe that the nucleation process is quite different from that occuring during hydration of calcium sulfate binders, and is certainly heterogeneous in type.

Effectively, and not taking into account the case of very rapid hydration in presence the of lithium ion, the fact that nucleation rate can be multiplied by a factor of up to ten with certain foreign cations (e.g. nickel or zirconium) let us suppose that the increase of supersaturation caused by the addition of cations to water in the conditions of low concentration adopted for the experimentation, is not concerned. So, in the hypothesis of an heterogeneous nucleation, we had to determine what was the nature of potential nuclei.

It is well known that pH does not stay constant during hydration of HAC (Lea, 1970; George, 1980). The pH of the intersticial phase effectively increases up to about 12 as it was pointed out by measurement during dissolution of CA Lafarge and Secar 71 samples (Sadok, 1989). In these conditions the following reaction can be written :

$$M^{n+} + n \ (OH)^- \rightarrow M(OH)_n \qquad (4)$$

and the grains of the precipitated hydroxide may play the role of potential nuclei.

Two types considerations have confirmed the possible precipitation of solid metallic hydroxides for cations with z/r parameter $\geqslant 2.5 \ \text{Å}^{-1}$:

a. The nucleation rate (Vn = 1/to), drawn versus the values of solubility product (Ks) of metallic hydroxides, given by Charlot (1974), decrease up to -logKs = 12 corresponding to bivalent manganese ion ($z/r = 2.5 \ \text{Å}^{-1}$), then increase up to -logKs = 40 (corresponding to trivalent Fe ion ($z/r = 4.68 \ \text{Å}^{-1}$) (Fig.4). Nickel and cobalt ions, on one hand, and mercury ion on the other, stay largely out of the drawn average curve.

b. To explain the role of foreign ions in the hydration kinetics of monocalcium aluminate, Currell et al. (1987) have refered to the paper of Misono et al. (1967), concerning the theory of 'ligands': after these authors, the pK of the reaction of metallic complex formation :

$$M^{n+} + mL \rightarrow MLm^{n+} \qquad (5)$$

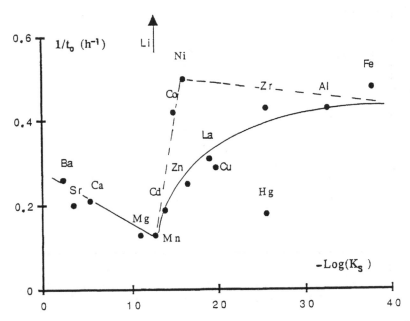

Fig.4. Variation of nucleation rate (hydration of Secar 71) versus solubility product of hydroxide of foreign cations.

can be given by the expression :

$$pK = - logK = X + Y + \gamma \qquad (6)$$

where X and Y refer to the electronegativity and the acidity of the metallic ion, respectively. Currell et al.(1987) have proposed that their experimental results presented a certain trend with the Y parameter of eq.(6). The drawing, versus Y, of nucleation rate ($Vn \simeq 1/to$) calculated from calorimetric results does not point out any correlation (Sadok 1989) (Fig.5). On the contrary, the representation, versus X (Fig.6), leads to approximately the same curve than the $Vn = f(z/r)$ one (Fig.2). Now, following Misono et al.(1967), the parameter X is directly related to the solubility product (Ks) of metallic hydroxides

$$- logKs = A.X \qquad (7)$$

where A is a constant.

Consequently, the increase of nucleation rate is related to the decrease of Ks (or the increase of -log.Ks). This observation confirms the direct correlation between Vn and Ks, that is to say, the increase of hydrated calcium aluminate nucleation rate with the trend of the foreign metallic ion to precipitate as hydroxide in the intersticial phase during the hydration of the binder.

Another hypothesis must be advanced to explain the particular role of cations such as nickel and cobalt, which lead to a nucleation rate lying above the the Vn = -logKs curve. Pourbaix's works (1963) which

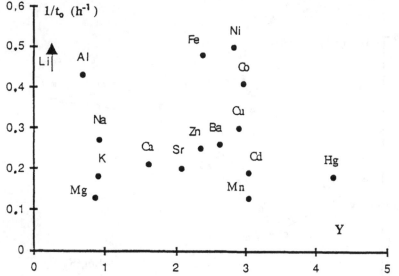

Fig.5. - Variation of the nucleation rate (hydration of Secar 71) versus the parameter Y of Misono et al. (1967)

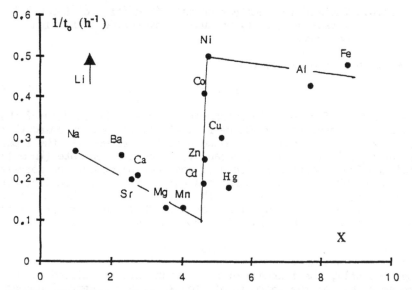

Fig.6. Variation of the nucleation rate (hydration of Secar 71) versus the parameter X of Misono et al. (1967)

concern the precipitation of metallic hydroxides, indicates that when pH increases, many hydroxides can pass again through the solution to form soluble complex metallic anions (for example aluminate anion AlO_2^- in the case of aluminum hydroxide). That may occur with Al-, Zn-, Cu-,

Hg-, Ni-, Zr- and Cr-hydroxides, all metallic cations investigated as modifiers in the present research. But with experimental conditions previously defined (use of metallic chlorides in solution at isoconcentration 0.025 M in Cl⁻) the possibility of formation of a soluble complex metallic anion concerns the trivalent aluminum cation only (Sadok, 1989).

Consequently, the hypothesis of rapid precipitation of an hydrated metallic aluminate other than calcium aluminate (such as occurs for hydration with lithium ion) must be extended to the case of some cations such as nickel and cobalt which give rise to a particularly high nucleation rate.

To verify experimentally these two hypothesis, metallic hydroxides of cation with $z/r \geqslant 2.5$ Å⁻¹ have been prepared separately by precipitation, using a saturated calcium hydroxide solution poured on metallic chloride solution at chloride isoconcentration 0.025 M. Gels obtained were washed four times with demineralized water to eliminate residual calcium and hydroxyl ions. Then, one cubic centimeter of the fresh gel suspension was introduced upon one gram of CA (or Secar 71) sample in the cell of the calorimeter, and the hydration curve was recorded. Results of Table 1 clearly show that, with respect to hydration with pure water, freshly prepared gels diminish the induction period and now promote a noticeable increase of nucleation rate (although not entirely exact in the case of an heterogeneous process, nucleation rate was approximated by $1/t_o$).

These experiments verify the hypothesis of the precipitation of metallic hydroxide during hydration in the presence of foreign cations introduced as chlorides, the grains of precipitated hydroxide playing the role of potential nuclei. But nickel ion, introduced as freshly prepared hydroxide gel, leads to the higher nucleation rate, such as when introduced as chloride solution, although its solubility product is higher than for foreign cations giving lower nucleation rate. Keeping in mind the particular role played by lithium ion (precipitation of an hydrated lithium aluminate, whatever the nature of lithium compound used, and particularly lithium hydroxide) and referring to some papers related to the formation of hexagonal hydrated nickel aluminate (Longuet, 1948; Longuet-Escard, 1951; Gallezot, 1969), the XRD analysis of hydration product stopped at 4 h, then dried, have given results listed on Table 2. (X-ray lines of non-hydrated calcium aluminate are not listed on the Table). Comparison of these results with characteristic X-ray lines of both two common hydrated nickel aluminates and nickel hydroxide (Table 3) points out firstly the absence of nickel hydroxide in the hydration product and secondly the presence of two aluminates: hexagonal C_2AH_8 alone as calcium aluminate hydrate (line at 2.87 Å), and hydrated nickel aluminate $Al_2O_3.2NiO.nH_2O$ (lines at 3.9; 2.5 and 1.51 Å), but the more intense line at d = 7.55 Å is not present on the XRD diagram and a complementary line appears at about 29 Å.

These results suggest that as soon as the aqueous suspension of nickel hydroxide gel is introduced on solid monocalcium aluminate (or Secar 71 cement), the dissolution of which is very rapid, the formation of a hydrated nickel aluminate occurs as the result of interaction between nickel hydroxide gel and dissolved aluminate anions. This hypothesis is certainly better when we refer to the work of Longuet-

Table 1. Calorimetric data obtained for hydration of Secar 71 with pure water or a washed suspension of metallic hydroxides.

Cation	to (h)	1/to (h^{-1})	tm (h)
Mn(2+)	3.5	0.29	16.0
Zn(2+)	5.6	0.18	11.9
Co(2+)	5.5	0.18	13.1
Cu(2+)	5.1	0.20	13.7
Ni(2+)	0.5	2.00	2.1
La(3+)	2.5	0.40	13.6
Fe(3+)	4.4	0.23	14.9
Al(3+)	4.0	0.25	17.2
Zr(4+)	4.0	0.25	23.5
Pure Water	8.9	0.11	16.0

Table 2. X-ray lines (other than lines of anydrous phases) observed after hydration (4 h) of Secar 71 with a whased suspension of nickel hydroxide

d ($\overset{\circ}{A}$)	I (%)	Identified phase
20.2069	20.81	hydrated Ni-aluminate ?
12.1416	8.80	C_2AH_8 ?
3.9027	7.02	hydrated Ni-aluminate*
3.5958	15.60	
3.1911	12.57	
2.8721	19.16	C_2AH_8
2.5114	30.05	hydrated Ni-aluminate*
2.3805	10.47	
2.1895	13.40	
1.9119	12.46	
1.5312	17.49	
1.5168	12.04	hydrated Ni-aluminate*
1.4567	13.51	
1.3714	10.47	
.....	.....	

* $2NiO.Al_2O_3.nH_2O$

Table 3. Characteristic X-ray lines of hydrated Ni-aluminates and nickel hydroxide.

$2NiO.Al_2O_3.nH_2O$		$5NiO.5Al_2O_3.18H_2O$		$Ni(OH)_2$	
d (A)	I (%)	d (A)	I (%)	d(A)	I (%)
7.8	100	7.55	100	4.61	100
3.9	< 100	3.76	40	2.71	45
2.5	< 100	2.55	30	2.33	100
1.51	< 100				
Longuet-Escard		ASTM file 22-452		ASTM file 14-117	

Escart (1951) who have prepared the nickel aluminate hydrate by mixing a solution of nickel chloride and the easily hydrolisable solution of aluminum chloride.

The precipitation of hydrated nickel aluminate promotes an heterogeneous nucleation with growth of C_2AH_8. In the hypothesis of a rapid precipitation of hydrated nickel aluminate, the explanation of the formation of C_2AH_8 alone (and not CAH_{10}) as calcium aluminate can be easily understood (Sadok, 1989) by reference to the Al_2O_3-CaO-H_2O diagram, as Rodger & Double did (1984) to explain the role of lithium ion in the hydration reaction. On the other hand, the structure of the short-time precipitating hydrated nickel aluminate would be only sligthly different from that of the hexagonal $Al_2O_3.2NiO.nH_2O$ hydrate (absence of line at $d = 7.55$ Å, but presence of a line at $d = 29$ Å, which corresponds to a parameter 'c' multiplied by about 4). This point could be investigated more precisely.

4 Conclusion

The present investigation has shown that during hydration of both pure monocalcium aluminate and high-alumina cement, foreign cations with z/r parameter higher than 2.5 Å^{-1} essentially modify the nature of the nucleation process which becomes 'heterogeneous', the potential nuclei formed being grains of metallic hydroxide (or hydrated aluminate for Li and Ni ions). The phenomenon leads to an increase of the nucleation rate. This result lead us to believe that the hydration process of both CA and HAC is quite different from that of C_3S and portland cement for which the precipitation of metallic hydroxide does not promote any acceleration of the hydration kinetics, due to formation of a protective layer on the anhydrous phases, as pointed out by many authors and reported by Skalny (1980). Only Zamorani et al. (1989) have recently observed an increase of hydration rate in the case of OPC by a solution containing nickel ion. So, it should be interesting to investigate again and more precisely the role of foreign cations in the nucleation step associated with the hydration of tricalcium silicate.

5 References

Bachiorrini, A. (1985) Physico-Chemical Interactions between Monocalcium Aluminate and Different Carbonates during Hydration Reaction. **Doctorat Thesis**, Univ. de Lyon, France. (in Fr.)

Bachiorrini, A. and Guilhot, B. (1982) First Echeances of Monocalcium Aluminate Hydration: Influence of the Stopping Procedure. **Cem. Concr. Res.**, 12, 559-567. (in Fr.)

Banfill, P.F.G. (1986) The Effect of Sulphate on the Hydration of High Alumina Cement. **Cem. Concr. Res.**, 16, 602-604.

Charlot, G. (1974) **Quantitative Analytical Chemistry.** Masson, Paris, 6th Edition, (in Fr.), p. 37

Curell, B.R. Grzeslowiak, R. Midgley, H.G. and Parsonage, J.R. (1987) The Acceleration and Retardation of Set High Alumina Cement by Additives. **Cem. Concr. Res.**, 17 [3] 420-432.

El Hajjouji, A and Murat, M. (1987) Strength Development and Hydrate Formation Rate. Investigation on Anhydrite Binders. **Cem. Concr. Res.**, 17 [5] 814-820.

Gallezot, M. (1969) The Hydrated Nickel Aluminate. **Comptes-Rendus Acad. Sci.**, Paris, 268, Ser.B, 329-331. (in Fr.)

George, C.M. (1980) Aluminous Cements. A synthesis of Recent Publications, Proceed. **7th. Intern. Congr. Chemistry of Cement**, Septima, Paris, France ,Vol. I, pp. V-1/3 - V-1/26. (in Fr.)

Karmazsin, E. and Murat, M. (1978) Study of a 'Solid + Liquid → Solid' Reaction (Calcium Sulfate Hemihydrate Hydration) by Simultaneous Isothermal Calorimetry and Electrical Resistivity Measurement. **Cem. Concr. Res.**,8 [5] 553-558.

Lea, F.M. (1970) **The Chemistry of Cement and Concrete.** Edward Arnold Ltd., London, 3rd Edition, p. 512.

Longuet, J. (1948) On an Hydrated Nickel Aluminate. **Comptes-Rendus Acad. Sci.**, Paris, 226, 579-580. (in Fr.)

Longuet-Escard, J. (1951) Study of an Hydrated Nickel Aluminate. **Mém. Serv. Chim. de l'Etat**, 36 [2] 187-193. (in Fr.)

Misono, M. Ochai, E. Saito, Y. and Yoneda, Y. (1967) A New Dual Parameter Scale for the Strength of Lewis Acids and Bases with Evaluation of Their Softness. **J. Inorg. Nucl. Chem.**, 29, 2685-2691.

Murat, M. and El Hajjouji, A. (1987) Role of Electric Potential Created by a Foreign Cation in Solution in the Nucleation Process of a Solid Phase from a Solution. **J. Chim. Phys.**, 84 [4] 209-212. (in Fr.)

Murat, M. El Hajjouji, A. and Comel, C. (1987) Investigation on Some Factors Affecting the Reactivity of Synthetic Orthorhombic Anhydrite. I. Role of Foreign Cations in Solution. **Cem. Concr. Res.**, 17 [4] 633-639.

Murat, M. and Sadok, El H. (1988) Role of Foreign Cations in the Hydration Kinetics of β-Calcium Sulfate Hemihydrate. Poster at the **A.C.S. 90th Annual Meeting**, Cincinnati; Annual Meeting Abstract, The ACS Inc., p. 90.

Parker, T.W. (1952) The Constitution of Aluminous Cement. Proceed. **3rd. Intern. Symp. Chemistry of Cement**, Cement and Concrete Association, London, 5, p. 512.

Pourbaix, M. (1963) **Atlas of Electrochemical Equilibrium at 25°C.** Masson, Paris (in Fr.)

Robson, T.D. (1967) **High Alumina Cements and Concretes.** John Wiley & Sons Inc.,New York.

Rodger, S.A. and Double, D.D. (1984) The Chemistry of Hydration of High Alumina Cement in Presence of Accelerating and Retarding Admixtures. **Cem. Concr. Res.**, 14 [1] 73-92.

Sadok, El H. (1989) Compared Roles of Foreign Cations in Solution in the Nucleation Step Associated to Calcium Sulfates and Monocalcium Aluminate Hydration," **Doctorat Thesis**, Institut National des Sciences Appliquées, Lyon, France. (in Fr.)

Skalny, J. (1980) Hydration Mecanism of Portland Cement. Proceed. **7th. Intern. Congr. Chemistry of Cement**, Septima, Paris, Vol.1, p. II-1/3 - II-1/47. (in Fr.)

Zamorani, Z. Ibrar Sheikh and Serrini, G. (1989) A Study of the Influence of Nickel Chloride on the Physical Characteristics and Leachability of Portland Cement," **Cem. Concr. Res.**, 19 [2] 259-266.

13 CALORIMETRIC STUDIES ON HIGH ALUMINA CEMENT IN THE PRESENCE OF CHLORIDE, SULPHATE AND SEAWATER SOLUTIONS

D.L. GRIFFITHS, A.N.F. AL-QASER
Department of Chemistry, University of Salford, Salford, UK
R.J. MANGABHAI
TH Technology Ltd R & D Centre, Rickmansworth, UK

Abstract
The hydration of high alumina cement (HAC) at various temperatures and in the presence of chloride, sulphate and sea water solutions at water/cement ratio (w/c) of 0.40 has been studied using an isothermal conduction calorimeter.

The maximum rate of heat evolution of the second peak (Q_{max}) and the time taken to reach the maximum (t_{max}) have been used to compare the hydration process in the various solutions studied. For all solutions the position of the second peak at (t_{max}) at 30°C is greater than at either 15°C or 40°C. This is in agreement with previously reported work.

Of all the chloride solutions studied the most interesting result is that $CaCl_2$, a well known accelerator for OPC, retards the hydration of HAC. Among the sulphate solutions, magnesium produces a greater accelerating effect than either sodium or potassium. The sea water solutions had a greater influence on Q_{max} than any of the individual solutions at all temperatures.

Keywords: Acceleration, Calorimetry, Chloride, High alumina cement, Retardation, Sea water, Sulphate, Temperature.

1 Introduction

The hydration reaction of High Alumina Cement (HAC) has been extensively reported in the literature (Robson (1962), Lea (1976), Neville (1975) and George (1983) and will not be discussed in detail here.

Parker (1952) reported the effect of various additives on the setting time of HAC and suggested that the major influence was the pH of the solution, the higher the pH the faster the setting. Robson (1962) discussed the effects of additives and concluded that dilute solutions of sodium, potassium and calcium hydroxide and organic bases such as triethanolamine acted as accelerators but magnesium and barium hydroxide could cause retardation possibly due to the buffering action.

Rodger and Double (1984) studied the effect of accelerating and retarding admixtures (lithium salts, citric acid and other metal chlorides). They showed that lithium salts accelerated the hydration by precipitation of a lithium aluminate hydrate which acts as a heterogeneous nucleation substrate. It was suggested that retardation by citric acid is due to the precipitation of protective gel coatings around the cement grains which impede hydrolysis or inhibit growth of the hydration products.

Currell et al. (1987) studied the acceleration and retardation of HAC by additives. They showed that both cations and anions have a profound effect on the chemical reactions which cause the hardening of HAC. The effect on the time of setting was summarised as follows:

for cations $Li + << Na^+ < none < K^+ < Ca^{2+} < Mg^{2+} < Sr^{2+} < NH_4^+$

for anions $OH^- << none < Cl^- < NO_3^- < Br^- < acetate$

Banfill (1986) studied the hydration with sulphate and superplasticizers and showed that sulphate retards the hydration but only at concentrations above 0.50 %. The superplasticizers themselves have a greater effect on the hydration than the free sulphate they contain.

This study investigates the hydration reaction of HAC in the presence of chlorides, sulphates and sea waters at 15°C, 30°C and 40°C using an isothermal conduction calorimeter.

2 Materials, apparatus and experimental

2.1 Materials
Cement : High alumina cement (HAC) Fondu complying to BS 915 was obtained from Lafarge Aluminous Cement Co Ltd (1985) and used in the investigation. The chemical composition and surface area are given in Table 1.

Table 1. Chemical composition and ssa of HAC.

% Oxide composition
CaO :38.53, SiO_2 :4.14, Al_2O_3 :38.67, Fe_2O_3 : 10.43, FeO :5.63,
SO_3 :0.186, MgO :0.02, Na_2O :0.06, K_2O :0.036, TiO_2 :1.98, CO_2 :0.42

Specific surface area : 276 m^2/kg

Chlorides and sulphates : 0.05 M solution of $BaCl_2$, $CaCl_2$, KCl, $MgCl_2$, NaCl, K_2SO_4, $MgSO_4$ and Na_2SO_4 were prepared using laboratory grade reagents. The concentration of 0.05M was chosen to allow for comparison with similar work reported by Rodger and Double (1984) and Currell et al. (1987).

Sea waters: Synthetic samples of ASTM (1975) and North Sea (Moxon, 1981) sea water were prepared in the laboratory using laboratory grade reagents. The composition of sea water samples is given in Table 2, as the concentration of the individual salts present expressed both in terms of g/l and molar solution.

Table 2. Composition of sea waters.

| Salt | ASTM | | North sea | |
	g/l	M	g/l	M
NaCl	24.53	0.422	25.00	0.431
$MgCl_2.6H_2O$	5.20	0.0255	3.00	0.014
$MgSO_4.7H_2O$	4.09	0.028	2.00	0.008
$CaSO_4.2H_2O$			1.50	0.008
$CaCl_2$	1.16	0.005		
KCl	0.695	0.009		
$NaHCO_3$	0.201	0.0024		

2.2 Apparatus
The conduction calorimeter used was of the type developed by Forrester (1970) and modified by Wexham Developments Ltd (1979). Details of the operation of this unit are given in instruction manual (1979).

2.3 Experimental
Mixing was carried out external to the calorimeter using the following technique. Ten grammes of HAC were placed in a polythene bag and four mls of the appropriate solution added. Excess air was squeezed out of the bag, sealed using a heat sealer and the excess plastic cut off. The bag was kneaded for two minutes by hand and placed in the calorimeter. The hydration reaction was studied at 15°C, 30°C and 40°C for 24 or 48 hours depending on the rate of heat evolution.

3 Results and discussion

3.1 Heat evolution
The hydrating cement pastes show a characteristic two peak profile with an intermediate dormant/induction period (Figure 1).

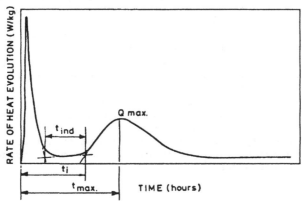

Figure 1. Typical heat evolution curve.

Explanations for the form of the curves have been discussed by Edmonds and Majumdar (1988). Under normal conditions (20°C), the sharp highly exothermic first peak derives from the heat of wetting of the cement powder and the early hydration of the clinker phases. The second lower peak which reaches a maximum after some 8 hours derives mainly from the hydration of the aluminates.

The heat evolution pattern obtained from the conduction calorimeter for de-ionised water (Control) at three temperatures are shown in Figure 2. The mixing was carried out externally and therefore the heat of wetting is not observed.

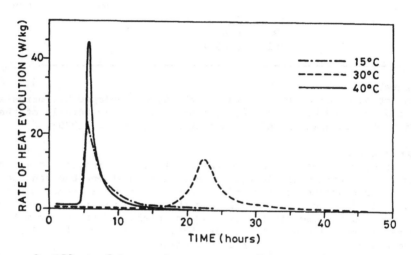

Figure 2. Effect of temperature on rate of heat evolution of HAC (control).

The initial setting time (t_i) was calculated as described by Bushnell-Watson and Sharp (1986) as the intersection of 2 "best" lines: through the early flat part of the curve and through the rising slope of temperature due to the main peak of heat production (see Figure 1). The induction period (t_{ind}) was also calculated as shown in Figure 1. At 15°C the t_i and t_{ind} values are 4.5 and 3.5 hrs and reaches maximum rate (Q_{max}) at 5.53 hours after mixing, whilst at 30°C and 40°C t_i is (19 and 4.5), t_{ind} (18.5 and 3.5) and t_{max} is (22.3 and 5.6) respectively. The results of the t_i are in agreement with those of Bushnell-Watson and Sharp (1986).

The heat evolution profile results are similar to those of Edmonds and Majumdar (1988, 1989), who studied the hydration of individual aluminates and Secar 71 at various temperatures.

Figure 3-5 shows a typical rate of heat evolution curves obtained when the hydration was carried out in the presence of chlorides, sulphates and sea waters respectively at 30°C.

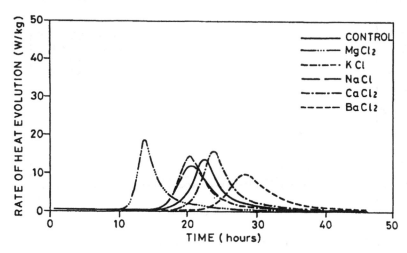

Figure 3. Effect of chlorides on rate of heat evolution of HAC at 30°C.

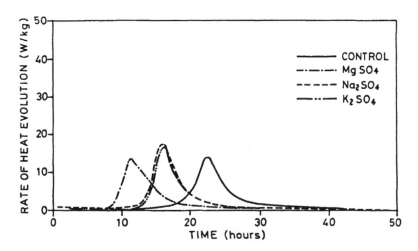

Figure 4. Effect of sulphates on rate of heat evolution of HAC at 30°C.

The t_i and t_{ina} values for the additives were calculated for the additives but are not reported here since t_{max} and Q_{max} values have been used to assess the acceleration/retardation process of HAC.

The heat evolution data obtained for all the solutions studied is summarised in Table 3, which shows for each solution at each of the temperatures the value of t_{max} and Q_{max}.

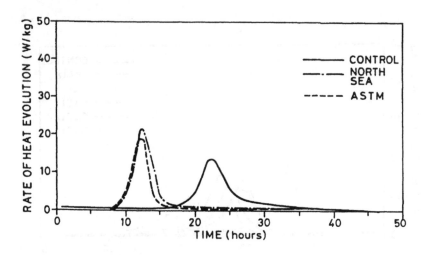

Figure 5. Effect of sea waters on the rate of heat
evolution of HAC at 30°C

Table 3. t_{max} and Q_{max} values for various solutions
at 15°C, 30°C and 40°C.

	15°		30°		40°C	
	t_{max} (hrs)	Q_{max} (W/kg)	t_{max} (hrs)	Q_{max} (W/kg)	t_{max} (hrs)	Q_{max} (W/kg)
control	5.53	23.92	22.27	13.76	5.62	44.50
$BaCl_2$	5.31	22.02	28.04	9.72	7.42	32.03
$CaCl_2$	5.58	26.20	23.31	15.73	5.81	47.17
KCl	4.17	29.15	20.01	14.51	5.17	36.99
$MgCl_2$	3.91	29.09	13.52	19.37	3.47	80.09
NaCl	4.32	28.97	20.29	11.95	5.22	40.05
K_2SO_4	6.06	19.85	16.02	16.79	3.91	51.99
$MgSO_4$	4.87	19.77	10.81	14.32	2.93	57.49
Na_2SO_4	5.34	20.21	15.94	17.42	3.68	45.11
North sea	9.20	15.75	12.11	21.58	2.95	65.53
ASTM sea	18.44	10.70	11.81	19.09	2.77	212.62

Using t_{max} to assess the acceleration/retardation process of HAC by
chlorides shows the following order at various temperatures:-

$MgCl_2$>KCl>NaCl>$BaCl_2$>control>$CaCl_2$ at 15°C

$MgCl_2$>KCl>NaCl>control>$CaCl_2$>$BaCl_2$ at 30°C

$MgCl_2$>KCl>NaCl>control>$CaCl_2$>$BaCl_2$ at 40°C

$MgCl_2$ accelerates the hydration process more at 30°and 40°C than at 15°C. At all three temperatures there is only a slight difference between the acceleration caused by NaCl and KCl. $CaCl_2$ behaves as a slight retarder at all three temperatures. $BaCl_2$ acts as a slight accelerator at 15°C but at 30 and 40°C acts as a stronger retarder.
In the presence of sulphate the following order is shown :-

$MgSO_4$>Na_2SO_4>control>K_2SO_4 at 15°C

$MgSO_4$>Na_2SO4>K_2SO_4>control   at 30°C

$MgSO_4$>Na_2SO_4>K_2SO_4>control   at 40°C

$MgSO_4$ acts as an accelerator at all three temperatures, being much more effective at 30°C and 40°C than at 15°C. Na_2SO_4 always acts as an accelerator but at 15°C the effect is slight. K_2SO_4 acts as an accelerator at 30 and 40°C but shows retardation effect at 15°C. Both the ASTM and North Sea sea waters show similar effects. At 15°C the hydration process is retarded while at 30 and 40°C the process is accelerated. At 15°C the retardation is much more pronounced than that shown by any of the individual salt solutions. The ASTM sea water retards more than North Sea sea water. At 30 and 40°C the hydration process is accelerated, both sea waters showing similar behaviour.
The behaviour of the sea waters follows more closely the effect shown by the sulphate solutions than the chlorides. All the sulphates accelerate the hydration at 30 and 40°C whilst at 15°C the K_2SO_4 acts as a retarder and $MgSO_4$ and Na_2SO_4 only slightly accelerate the process.
Wilding et. al. (1984) used a plot of Q_{max} vs $1/t_{max}$ to assess the acceleration/retardation of Portland cement by various admixtures. In this study the above plot was normalised (ie. Q_{max} solution/ Q_{max} water vs t_{max} water/ t_{max} solution) to allow a direct comparison at the various temperatures. Figure 6 shows the plot for chloride and sulphate solution at the three temperatures studied.
The value of a plot such as Figure 6 is that it shows both the accelerating/retarding properties of a salt and its effect on the maximum rate of heat evolution.
$MgCl_2$ at 40°C both accelerates and increases the maximum rate of heat evolution. $BaCl_2$ at 30 and 40°C retards the reaction and reduces the maximum rate of heat evolution. Among the chlorides at 30 and 40°C the monovalent NaCl and KCl act as accelerators whilst the divalent $CaCl_2$ and $BaCl_2$ both act as retarders. The position of $MgCl_2$ is anomalous and represents the ease with which it can replace calcium ions in the cement hydrate (as in diffusion of ions). At 15°C it is difficult to identify clear cut trends. At 30 and 40°C the divalent $MgSO_4$ is positioned further to the right than the monovalent Na_2SO_4 and K_2SO_4, which are closely grouped.

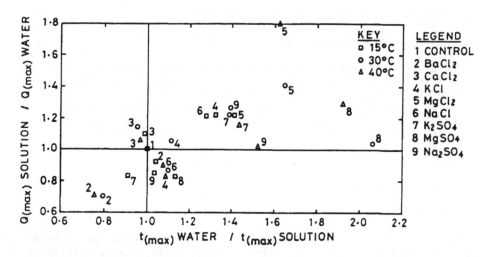

Figure 6. Q_{max} solution/Q_{max} water vs t_{max} water/t_{max} solution.

3.2 Total heat of hydration

Figure 7 shows the total heat of hydration (THH) against time for HAC at various temperatures and is obtained from the rate of heat evolution curve Figure 2. These results show that at 15°C, the THH rises after 5 hours and reaches a maximum of 250 kJ/kg within 10 hours, whilst at 30°C there is a slow gradual increase upto 20 hours followed by a more rapid rise to reach a maximum of >300 kJ/kg after 30 hours and at 40°C rises after 5 hours and reaches a maximum of 350 kJ/kg in 15 hours.

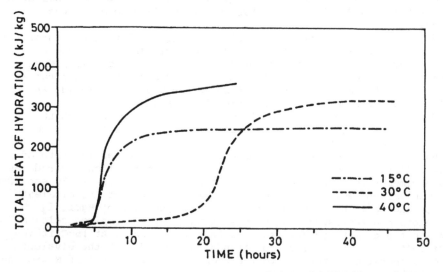

Figure 7. Effect of temperature on total heat of hydration of HAC (control)

The total heat of hydration of the completed process at 15°C for control sample was 251 kJ/kg after 48 hours whilst at 30 and 40°C were 320 kJ/kg (after 48 hrs) and 360 kJ/kg (after 24 hrs) respectively. The range of total heat of hydration for the additives for the completed process are given in Table 4. The average of all the additives indicates that the total heat of hydration is not significantly altered.

Table 4. Total heat of hydration for the completed process.

Solution	15°C	30°C	40°C
	48 hrs (kJ/kg)	48 hrs (kJ/kg)	24 hrs (kJ/kg)
Control	251	320	360
Chlorides	298-316	304-343	337-376
Sulphates	254-318	318-336	336-354
Sea water	236-266	294-338	332-347

3.3 Differential Thermal Analysis

Exploratory DTA analysis were carried out using Stanton Redcroft DTA 673-4 (with heating rate 10°C/min and Nitrogen as flowing gas) on various samples with control samples at 15°C, 30°C and 40°C after 24 hours hydration and shown in Figure 8. These results show the presence of gel and small amount of AH_3 at 15°C whilst at 30 and 40°C shows gel, C_2AH_8 and AH_3. Figures 9 and 10 shows the DTA curves obtained in the presence of various chlorides solution at 15 and 30°C after 24 hours of hydration. The presence of the additives does not significantly alter the pattern of DTA traces. These results shows the presence of gel at 15°C and very little AH_3, whilst at 30°C shows the presence of gel, C_2AH_8 and AH_3. In both cases there is no evidence of the presence of CAH_{10} or C_3AH_6. SEM and XRD analysis were not carried out on these samples and should be investigated.

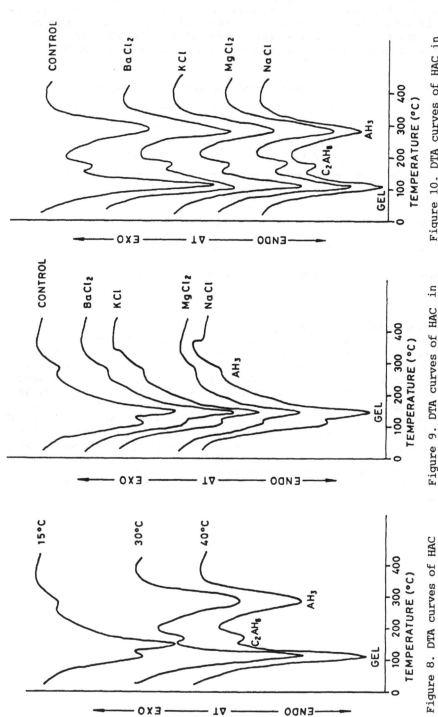

Figure 8. DTA curves of HAC at various temperatures of hydration after 24 hours (control).

Figure 9. DTA curves of HAC in the presence of chloride solution at 15°C after 24 hours hydration

Figure 10. DTA curves of HAC in the presence of chloride solution at 30°C after 24 hours hydration.

4 Conclusions

The hydration of HAC using pure water has been shown to be slower at 30 · C than at either 15 or 40 C. The presence of a salt does not change this pattern of behaviour, but will either accelerate or retard the process. A relationship between the salt used and its effectiveness in affecting the speed of the reaction and the rate at which heat is evolved has been demonstrated.

The total heat of hydration of the completed process was not significantly altered by the presence of additives.

5 References

ASTM standard (1975). D1141-75, Standard specification for substitute ocean water. ASTM book, Part 13, 949-951.

Banfill, P.F.G. (1986) The effect of sulphate on the hydration of high alumina cement. **Cem. Concr. Res.** 16, 602-604.

Bushnell-Watson, S.M. and Sharp, J.H. (1986). The effect of temperature upon the setting time of refractory calcium aluminate cements. **Cem. Concr. Res.** 16, 875-884.

BS 915, Part 2 (1972) High alumina cement, British Standard Institution London.

Currell, B.R., Grzeskowlak, R., Midgley, H.G. and Parsonage, J. R. (1987). The acceleration and retardation of set high alumina cement by additives. **Cem. Con. Res.** 17, 420-432.

Edmonds, R. N. and Majumdar, A. J. (1988). The hydration of monocalcium aluminate at different temperatures. **Cem. Concr. Res.** 18, 311-320.

Edmonds, R. N. and Majumdar, A. J. (1988). The hydration of 12 CaO.7Al$_2$O$_3$ at different temperatures. **Cem. Concr. Res.** 18, 473-478.

Edmonds, R. N. and Majumdar, A. J. (1989). The hydration of Secar 71 aluminous cement at different temperatures. **Cem. Concr. Res.** 19, 289-294.

Forrester, J. A. (1970). A conduction calorimeter for the study of cement hydration. **Cement Technology.** 1, 95-99.

George, C.M., (1983) in **Structure and performance of cements.** Ed. Barnes, P. Applied Science, London.

Lea. F. M. (1976). **The chemistry of cement and concretes.** Edward Arnold, London.

Lafarge Aluminous Cement Co. Ltd., (1985). England.

Moxon, S. (1981). Wimpey Laboratories Ltd., Hayes, Middlesex, England. Private communication.

Neville, A.M. (1975). **High Alumina Cement and Concretes.** Construction Press. Lancaster.

Parker, T.W. (1952). The constitution of aluminous cement. **Proc. Third. Int. Symp. Chemistry of Cement,** Cement and Concrete Association, London (1954), 5, 485-529.

Robson, T. D. (1967). **High Alumina Cements and Concretes.** John Wiley & Sons. Inc. New York.

Rodger, S.A. and Double, D.D. (1984). The chemistry of hydration of high alumina cement in the presence of accelerating and retarding admixtures. **Cem. Concr. Res.** 14, 73-92.

Wexham Development Ltd. (1979). Instruction manual for conduction
    calorimeter. Wexham Springs, Slough, SL3 6PL.
Wilding, C.R., Walter, A., and Double, D.D. (1984). A classification of
    inorganic and organic admixtures by conduction calorimetry.
    Cem. Concr. Res. 14, 185-194.

PART FIVE
DURABILITY

14 MANUFACTURE AND PERFORMANCE OF ALUMINOUS CEMENT: A NEW PERSPECTIVE

C.M. GEORGE
Lafarge Calcium Aluminates, Chesapeake, Virginia, USA

Abstract
Information is assembled from a variety of sources, many
not previously published, with the object of providing a
coherent technical portrait of this increasingly widely
used cement. Particular emphasis is given to the question
of reliability. To this end the method of manufacture and
the results achieved are examined in some detail.
Confidence in a material must be based not only on test
data but on an understanding of the properties displayed.
A simple model is used to discuss the criticality of
water/cement ratio to long-term performance of concrete,
and explain the recommended conditions for application.
The inevitability of conversion is stressed. Problems
arising from failure to respect these considerations are
shown to be predictable.
Keywords: Aluminous Cement, Manufacture, Quality Control,
Predictability, Critical Water/Cement Ratio, Hydration,
Porosity, Compressive Strength, Durability, Conversion,
Prestressed Concrete.

1   Introduction

When the first successful patent for manufacture of
aluminous cement appeared, the rotary kiln process for
Portland cement was already established. Calcium
aluminates - the essential ingredient of aluminous cements
- were however, by no means scientific newcomers, as Lea,
(1970) records:

1846   Vicat recommends sulfate resistance of alumina rich
        compositions.
1853   French society for the Encouragement of National
        Industry stimulates research on sulphate attack of
        concrete.
1865   Fremy reports good hydraulicity of calcium aluminate
        melts.
1869   Michaelis confirms Fremy's results.
1888   British patent for limestone - bauxite cement.

1890	Increasing severe problems of
1898	sulphate attack of Portland cement
1900	concrete, particularly of railway
1902	construction in the South of France. Bied, amongst others, investigated.
1908	Bied-Lafarge patent for the manufacture of aluminous cement.

The rotary kiln process did not lend itself to the manufacture of the aluminous cement composition targeted by Bied, and after some early difficulties a continuous process evolved in which the raw materials were completely fused. The product was called "Ciment Fondu" - french: melted cement. Today this process has become a large scale, fine-tuned technology.

The first practical use of this cement (1916), was for the construction of gun emplacements, The criterion of choice was the rapid strength development of aluminous cement rather than its sulphate resistance. General commercialization began in 1918, since which time applications for aluminous cement have been in continuous development. The single largest current use is in refractory and heat resistance applications, yet over 60% of shipments find uses in other areas. Amongst these, rapid hardening remains a prime feature with corrosion resisitance and suitability for cold weather concreting as complementary advantages. Often a given application will utilize more than one of these special properties. Examples of this now traditional use of aluminous cement mortar and concrete can be found all over the world. As with Portland cement, the use of chemical additives, particularly to enhance workability has become increasingly widespread. The virtually instant setting characteristics of mixtures of aluminous and Portland cements have also long been known. However, the development of complex formulations in which calcium aluminate is combined with calcium silicate and sulphates as well as a wide variety of chemical additives is a sophisticated industry of comparatively recent origin, now consuming large amounts of aluminous cement. Most of this is to be found in the form of premixed grouts, screeds and renderings.

2   Manufacture

Fully integrated plants for the manufacture of this cement, by far the most widely used internationally, are located in the north and in the south of France. Today these two plants produce more cement than three plants provided in the mid 1980's. This has been possible in part due to improved quality/production control.

The technology resembles that of the Blast Furnace for the production of iron with the differences that the furnace operates in a predominantly oxidising condition and that the molten product decants continuously from the furnace. Thus the (reverberatory) furnace - see figure 1 - used for fusing the raw materials produces a slag with no metallic phase as in the Blast Furnace. However, the principle of combustion gases traversing a permeable burden of lump, natural raw materials remains common to both. The temperature of fusion (about 1450°C) is somewhat higher than in the Blast Furnace. It is virtually the same as that used to manufacture Portland cement but substantially lower than steel making temperatures.

Figure 2 shows the entire process schematically and summarizes the production and quality control procedures.

A   Raw materials reception. Lump material ranging in size from a few mm to several hundred mm is difficult to sample representatively. Theoretically a minimum sample size of over 50 tons is needed. This is quite impractical. A compromise is made by taking combinations of 50 kg sized samples. The samples are analysed chemically, providing precise but not necessarily accurate information, about each stock pile of raw material. Individual deliveries of raw materials are kept in segregated stock piles which are not released for use until the chemical analysis has been established.

B   After a preliminary homogenisation, the raw materials are transferred to silos from which they are individually weighed to a common conveyor. The proportioning is based on the raw materials analysis as a first level of control.

C   The conveyor transports the raw materials to the top of the furnace stack. This, and all other materials movement in the process are monitored by closed circuit T.V. from the central control panel, from which all electrical and mechanical functions are operated.

D   Melting at the bottom of the stack causes the burden to continuously descend against the counter current of combustion gases. It is for this reason that the raw materials are in lump form, to maintain permeability in the stack. In the furnace, combustion conditions are monitored and melting temperature controlled. Liquid product in the horizontal section decants continuously into iron conveyor pans being replenished by further raw

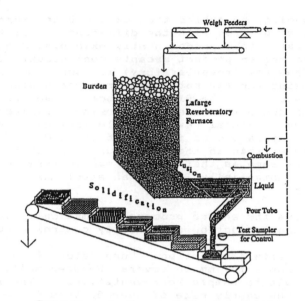

Figure 1. Reverberatory Furnace for Aluminous Cement Production.

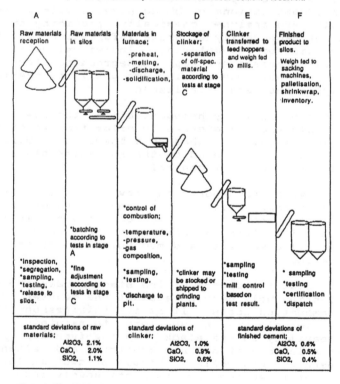

A	B	C	D	E	F
Raw materials reception	Raw materials in silos	Materials in furnace; -preheat, -melting, -discharge, -solidification,	Stockage of clinker; -separation of off-spec. material according to tests at stage C	Clinker transferred to feed hoppers and weigh fed to mills.	Finished product to silos. Weigh fed to sacking machines, palletisation, shrinkwrap, inventory.
		*control of combustion; -temperature, -pressure, -gas composition,			
*inspection, *segregation, *sampling, *testing, *release to silos.	*batching according to tests in stage A *fine adjustment according to tests in stage C	*sampling, *testing, *discharge to pit.	*clinker may be stocked or shipped to grinding plants.	*sampling *testing *mill control based on test result.	* sampling *testing *certification *dispatch
standard deviations of raw materials; Al2O3, 2.1% CaO, 2.0% SiO2, 1.1%		standard deviations of clinker; Al2O3, 1.0% CaO, 0.9% SiO2, 0.6%		standard deviations of finished cement; Al2O3, 0.6% CaO, 0.5% SiO2, 0.4%	

Figure 2. The Melting Process for the Manufacture of (Fondu) Aluminous Cement.

materials fusion.   It is thus a truly continuous
process operating 24 hours per day in long
campaigns.   Samples of liquid product are taken
regularly at intervals of a few hours and analysed
for main chemical constituents.   Good representivity
is achieved at this stage and the results are used
to fine tune the weigh feeder settings for raw
materials proportionings.

It is here that an intrinsic advantage of the
process is seen.   As a result of complete melting of
the feedstock, homogenization automatically occurs,
a feature not observed in rotary kiln sintering.
The standard deviations for the main constituents
before and afte melting in figure 2 clearly show
this effect.

E   The material solidifies on the hot pan conveyor and
discharges at the end of the conveyor.   From there
it is stock piled by overhead cranes.   During start
up procedures, when some off-specification clinker
may be generated, this stage in the process allows
rejects to be separated for retreatment.   Once the
target composition has been attained the control
procedures described above rarely fail to maintain
that composition within specified limits.   The
equipment is designed to enable cooling rate to be
held constant.   The result is a cement of
predictable performance.

An additional feature of the process deserves
comment.   The newly solidified "clinker" is glass
like in aspect although crystalline, appearing as
irregular lumps, and 50mm by down, with smooth shiny
surfaces and a very dense, impervious structure.
After a few days the surfaces become dull and later
take on a grey-white appearance as weathering
proceeds.   From this stage very little further
change occurs.   The material having been
superficially hydrated (the loss on ignition is
typically below 0.1%), is virtually immune from
further attack.   As a result this "clinker", is
unground form, may be stored outdoors for months or
even years, without significant deterioration.   The
practical advantages, including quality, are many.

F   In the integrated manufacturing plants the clinker
is ground in ball mills to cement fineness on the
same site.   Other plants carry out only these
grinding operations being supplied with clinker by
ship from the integrated plants.   In both cases, the
rehandling of the clinker from stock to feed silos
and the mixing that takes place during the grinding
and stocking operations, provides a small but useful
further homogenisation of composition.   This is also

185

apparent from the data in figure 2.

Most features of the grinding of aluminous cement clinker are similar to Portland cement operations. The principal control parameter for cement quality is the surface area of the finished product exiting the mill, measured on frequent, regular samples by the air-permeability technique. A typical american SPC control chart for this stage in the process is shown, in figure 3.

The level of control achieved in this special aluminous cement manufacturing process can be judged by an examination of the consistency of the finished product.

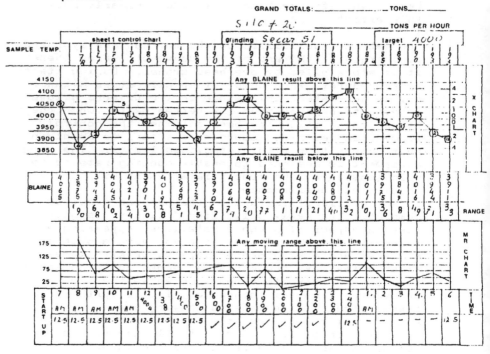

Figure 3. SPC Control Chart for Surface Area of Grinding.

3 Finished Product Characterisation

Each manufacturing plant has a control laboratory with trained staff and equipment to measure the chemistry, mineralogy, physical and mechanical properties of the finished product. These laboratories also monitor the reliability of on-line process sample taking and testing, and provide the chemical information needed to adjust materials flow. A control laboratory monitors the plant laboratories.

The aluminous cement discussed in this paper is supplied, where the relevant codes apply, according to British Standard Specification 915 and the requirements of the official French VP classification. Testing by the control laboratories ensures compliance. The French Classification, being somewhat more comprehensive than the British one, will be discussed.

While factory monitoring of cement quality is carried out on a daily basis, two additional monthly checks are carried out on a random sample taken by a representative of the official government laboratory. This sample is subdivided, part being tested by the official laboratory (usually referred to as VP=Ville de Paris) and part tested in the manufacturer's central laboratory. A comparison of some annual results is shown in table 1.

While there is apparently a slight measurement bias between laboratories the results are statistically indistinguishable.

The 6 hour and 24 hour compressive strengths of standard mortars (total w/c 0.4 @ 20°C) have minimum values specified officially in France and guaranteed by the manufacturer. Using SPC procedure we may thus evaluate the process capability and the process performance relative to these specifications. The same comparison is also possible for setting time. Based on typical annual data in each case (12 monthly random samples) the results of these calculations are shown in table 2.

The capability indices and performance indices (based on short-term variation and long-term variation respectively) are calculated from the relationship:

$$\text{Index} = \frac{\text{specification limit - average}}{3 \times \text{standard deviation}}$$

Thus a performance index of 1.0 or greater indicates a minimum of 99.7% conformance of the measured property to the specified tolerance. A capability index that exceeds the performance index implies scope for further improvement in quality control. Based on these criteria the process is operating well and close to optimum.

Other properties such as chemical composition, mineralogy, standard consistency, etc., are not the subject of standard specifications. However, if measured against the manufacturer's own internal standards these generate indices well above unity.

All the test data described in this section has been for measurements carried out at normal laboratory temperatures, 20 +/- 2°C. For the aluminous cement which is the subject of this paper some evidence of predictability at higher temperatures is important because of the well known phenomenon of conversion.

Routine tests at 38°C, (100°F), again on random monthly

Table 1. Comparison of Monthly Testing of Random Cement Samples.

factory laboratory			A		B	
			VP	M	VP	M
compressive strength MPa (mortar prisms)	6 hrs	av.	48.6	43.4	48.9	44.5
		s.d.	1.8	3.0	1.6	2.9
	24 hrs	av.	63.8	60.8	65.5	63.7
		s.d.	3.0	3.7	3.8	4.0

Table 2. Capability Indices (CI), and Performance Indices (PI), for Guaranteed Specified Cement Properties.

factory property		specification	A		B	
			CI	PI	CI	PI
compressive strength, MPa	6 hrs	>30	1.7	1.5	1.8	1.7
	24 hrs	>50	1.2	1.0	1.1	1.1
initial set, minutes		>90	1.5	1.5	1.6	1.1

samples have been carried out for many years on concretes gauged at total water/cement ratios of 0.35, 0.4 and 0.5. In each case, two curing conditions are used: 24 hours at 20°C and 95% RH, and 5 days under water at 38°C. This latter, at first sight apparently arbitrary set of conditions results from a great deal of past study and has been shown to provide a relatively rapid and reliable estimate of the minimum long-term strength which these concretes develop.

In table 3, the compressive strength achieved after 5 days curing under water at 38°C for a total water/cement ratio of 0.4 is taken as the reference value. The strengths achieved for the other test conditions are then expressed as a ratio over the reference strength. This is a convenient way of displaying the coherency or reproducibility of the data.

Table 3. Strengths obtained under the conditions shown as a ratio on the reference strength.

curing	5 days under water at 38°C			24 hours at 20°C and 95% RH		
w/c	0.35	0.4	0.5	0.35	0.4	0.5
average	1.26	1.0	0.66	1.69	1.56	1.39
standard deviation	0.07	/	0.03	0.10	0.06	0.10
coefficient of variation,%	5.5	/	4.5	5.9	3.8	7.2

The average reference strength for the period 1980-1989 was 40.6 MPa with an s.d. of 3.5.

The last 25 monthly results for the reference condition are plotted in control chart form in figure 4.

When reporting long-term test data on this cement, Teychenne (1975), expressed the long-term, 38°C strengths as a percentage of the strength at 24 hours standard curing, as a function of water/cement ratio. In figure 5, his 1965 data is compared with the results shown in table 3.

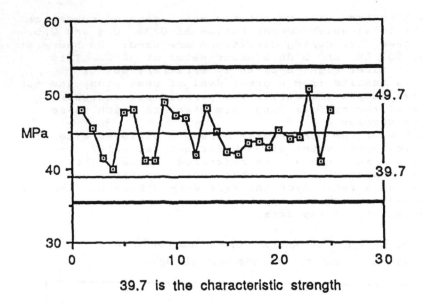

39.7 is the characteristic strength

Figure 4. Minimum strengths at 38°C and 0.4 w/c on random monthly samples.

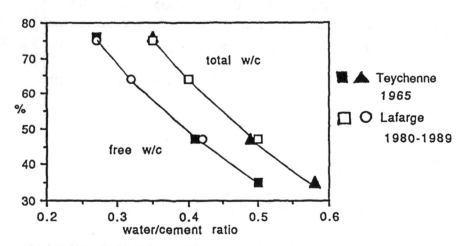

Figure 5. Ratios of 38°C strength to 20°C strength as a function of w/c ratio.

4   The Criticality of Water/Cement Ratio

4.1   Hydration and Porosity - A Simple Model

The reaction between cement and water to produce crystalline hydrates is a chemical one and therefore takes place in fixed proportions. The ratio of water/cement required for complete reaction is therefore a constant. This leads to the notion of the theoretical w/c ratio (also referred to as the critical w/c ratio).

In fact, from a chemical standpoint it is the stochiometric water/cement ratio.

The value of this ratio depends on the specific reaction involved. In the case of the aluminous cement under discussion, two extreme values exist depending on the reaction temperature. At low temperature, (typically below about 20°C) a low density hydrate with a high water content is formed. At high temperature, (typically above 50°C) a denser hydrate system containing less water is formed. When the cement is initially hydrated at low temperature and subsequently exposed to higher temperatures, the low density system is converted to the high density system. The higher the temperature, the faster this change, known as conversion, occurs. This effect is not reversible. Thus, on a long-term view, only the high temperature system can be considered stable.

It will be evident from the foregoing that the stochiometric w/c ratio for the low temperature hydrate system will be greater than that for the high temperature system. These ratios are usually expressed by weight. It is however, instructive to express them as volume ratios. Several relevant w/c ratios are recorded in table 4.

Expressed on a volume basis the ratio, for complete reaction of aluminous cement to its long-term stable condition, is close to unity.

In other words the porosity of the fresh paste is about 50%, (52%). For Portland cement the corresponding initial porosity is little different, at about 46%.

As the reaction proceeds, hydrates are produced and cement and water consumed. At 50°C the principal reaction of the aluminous cement can be written:

$$CA + 4H = 1/3 \ C_3 AH_6 + 2/3 \ AH_3$$

The density of the hydrate system is 2.45 whereas that of the fresh paste is 2.1. On the other hand, the density of the cement is 3.3. As a consequence, the hydrates fill some of the space originally occupied by water but never all of it: porosity is reduced but not eliminated. This very important result is illustrated in figure 6 along the lines of the pattern used by Cottin and Reif (1970).

At 20°C the essential reaction is: $CA + 10H = CAH10$ and the hydrate density is 1.7 while the initial density of the paste (for complete reaction at 20°C), is 1.6.

Thus once again porosity is reduced but not eliminated. Again, see figure 6. Similar considerations apply to Portland cements.
To summarize:

initial volume of   volume of   initial volume
cement + water   >   hydrated cement   >   of cement

   This description of hydration is the basis of our simple model, which can be given quantitative form if known values for chemical compositions and densities are used. These are well documented. See Lea (1970) for example, reference (4).

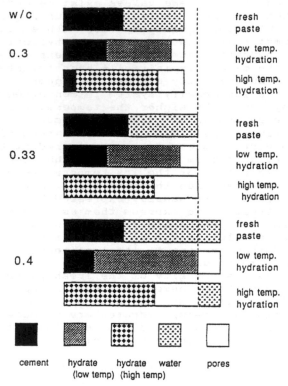

Figure 6. Hydration, water/cement ratio and porosity.

   For good predictive capability some refinements to the model are needed. These include:

      •allowance for the fact that the industrial aluminous cement does not consist only of calcium aluminate, (CA) so that the water/cement ratio (by weight) needed to totally react with the cement is

192

0.33 (at 50°C) compared to the theoretical values for pure CA shown in table 4.

• allowance (for low temperature hydration) for the fact that not all the water added to the cement reacts, even at water/cement ratios well below the stochiometric value, Cottin and Reif 1970.

• allowance (when predicting the properties of mortars and concretes) for the water associated with aggregates and for air entrapped during mixing.

Table 4.   Stochiometric Water/Cement Ratio for Complete Hydration.

Cement	Ratio By Weight		Ratio By Volume	
Portland	0.28		0.85	
Aluminous 50°C		0.33*		1.08
Pure CA 50°C	0.46		1.37	
Aluminous 20°C		0.83		2.74
Pure CA 20°C	1.14		3.40	

*Determined experimentally by prolonged hydration of the cement at 70°C, removal of excess free water by treatment with acetone and ether followed by measurement of loss on ignition, (*).

4.2   Predictive Capabilities of the Model

Neat pastes prepared at less than the stochiometric water content will contain unhydrated cement after all the water has reacted. This residual cement does not, of course, contribute porosity.

Pastes prepared above the stochiometric water content will contain unused water after all the cement has reacted. In this case porosity associated with hydration will be further increased by the excess water. The penalty of exceeding the stochiometric w/c is thus increasingly severe. As a consequence the curve of porosity versus water/cement ratio displays a discontinuity as it passes through this critical water/cement. The model described in the previous section accurately predicts the porosity versus w/c curve including its discontinuous nature, (see figure 7), for high temperature hydration.

Calculation from the model porosities for the low density hydrate system are hampered by the need to allow for incomplete reaction at low water/cement ratios. For Portland cements the problem is more complicated.

Overall, the model describes well the dependence of porosity on water/cement ratio. There is also a very precise relationship between porosity and compressive strength for both aluminous and Portland cement. For

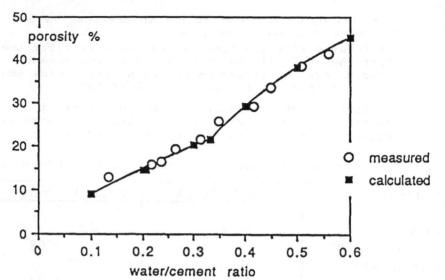

Figure 7. Measured and Calculated Porosities of Aluminous Cement Paste.

aluminous cement the relationships,

low temperature hydration (10°) Ln C= 5.37 - 0.085P
high temperature hydration (70°C) Ln C= 6.25 - 0.105P
(C expressed in MPA and P in volume %).

were established, (Cottin and Reif 1970) for a w/c range
of 0.1 to 0.6. Coefficients of correlation are 0.997 and
0.976 respectively.
    Inspection of these two equations reveals that for a
given level of porosity, the stable long-term hydrate
system is always stronger than the low density hydrate
system.
    It is also worth noting that while the coefficients in
these equations show a greater sensitivity of strength to
porosity (and hence w/c), for the high density system, the
corresponding coefficient for Portland cement closely
approximates that for the aluminous cement high density
system -see also reference (5).
    The intercorrelations between water/cement ratio,
porosity and strength have been reported previously,
(George 1983).
    Because the model reliably predicts porosity and since
strength is strongly correlated with porosity it follows
that the model can be used to predict strength, over a
range of water/cement ratios that encompass those
generally encountered in service.
    Some data on concrete is presented in table 6. Since
the type of aggregate used and the mix designs for these

Table 5. Specific volumes at theoretical water/cement ratio.

	fresh paste	fully hydrated	cement
Aluminous	0.477	0.370	0.304
Portland	0.475	0.398	0.328

Table 6. Cube crushing strengths (MPa), and Porosities (P%), of Concretes.

total water/ cement ratio	curing temperature, °C (48 hours)							
	10		20		70		90	
	MPa	P%	MPa	P%	MPa	P%	MPa	P%
0.33	91.0	9.7	90.0	8.8	46.5	12.0	55.5	12.5
0.4	81.5	9.6	77.0	9.8	34.5	11.4	34.5	14.0
0.5	70.0	11.6	72.0	11.7	22.0	13.5	19.5	15.3
0.67	60.5	13.4	55.0	13.8	9.0	18.1	7.0	18.0
1.0	23.0	17.9	23.0	19.0	2.0	22.6	2.0	22.9

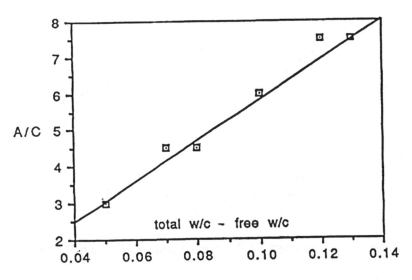

Figure 8. Total minus free w/c as a function of aggregate/cement ratio.

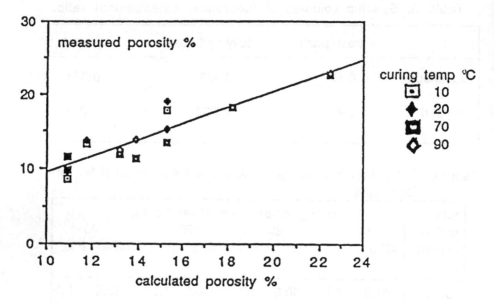

Figure 9.   Comparison of measured and calculated porosities of concrete.

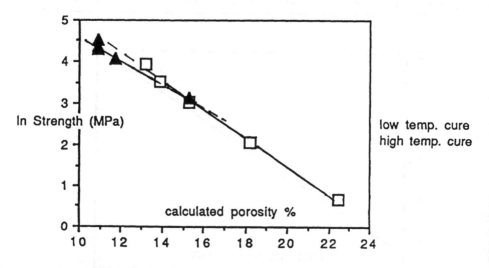

Figure 10. Compressive Strength of Concrete versus Calculated Porosity,%.

concretes are recorded the model can be used to calculate
porosity. Absorption of some of the gauging water by
aggregate and sand reduces the amount available to react
with the cement, hence the concept of free water/cement
ratio which is lower than the total water/cement ratio-
only in a neat cement paste are the two ratios equal. It
is the free water/cement ratio that is used when
translating the model to concrete. The difference between
free and total water/cement ratio can be substantial as
figure 8 shows.

The calculated porosities obtained are compared with
the measured values in figure 9. The agreement is very
reasonable. Next, it is interesting to see if the same
exponential type of relationship exists between concrete
strength and porosity as was found for neat paste.
Certainly if a semi log plot is made of the experimental
data in table 6, a fairly clear straight line
relationship, allowing for scatter in the data is
obtained. When the measured strength results are plotted
against the porosities calculated from the model the
correlation is excellent: figure 10.

The difference in the curves for the low density and
high density systems is still detectable although less
pronounced than in neat cement pastes. Note that for the
lowest porosities the high temperature hydrate system
provides the greater strengths. These low porosities in
the high density stable hydrate system are obtained only
at free water/cement ratios at or below the stochiometric
value, (0.33) for full hydration.

The model provides an obvious justification of a well
known prescription for the reliable long term performance
of aluminous cement concrete, namely that total
water/cement ratios should not exceed 0.4. For typical
concrete mixes a total w/c of 0.4 will provide a free w/c
virtually equal to the stochiometric value, thus avoiding
the inevitable excess porosity for water additions above'
this limit and ensuring good long-term performance.

Strength is not the only significant property of
concrete. Durability, or resistance to a variety of
destructive forces is highly important. This aluminous
cement, as related, was conceived to provide enhanced
resistance to sulphate attack and has a proven record for
this purpose. It also shows superior resistance to a
number of industrial products and effluents, particularly
mild acids. But despite its intrinsic suitability a key
requirement is still using w/c's at or below the critical
0.33. The same remarks apply to other potential forms of
attack, such as freezing and thawing and alkaline
hydrolysis.

Maintaining this critical water/cement ratio in
concrete requires using sufficient cement to achieve the
desired placeability. For most practical purposes this
demands a cement content in excess of 400 kg/m^3, (George

1980). Water/cement ratio can be driven lower without
increasing cement content by the use of workability aids.
They should not be used as a device for reducing cement
content. A weakness of the model is that it looks only at
total porosity and not permeability which would be a
better predictor of durability. However a comparison of
some long-term data with estimates of porosity is not
without interest.

Long-term tests in sea water were reported by George in
1980, (8). Seven series of semi-immersed specimens at
total water/cement ratios from 0.285 to 1.0 were examined
for signs of visual attack up to 7 years (4 series), and
up to 20 years, (3 series). Three specimens were totally
destroyed after 15-20 years at a total water/cement ratio
of 1.0, while others remained intact. These were the only
specimens cured and stored at 20°C. For specimens cured
at 50°C or 70°C some attack occurring at 0.6, and in one
case at 0.5 w/c ratio. At 0.4 and below all specimens
remained undamaged. Porosities estimated from the model
for the high temperature cured series, ranged from 7.4% to
14%. The lowest porosity for which corrosion was seen was
11.7% and the highest porosity exhibiting no attack was
12.4%. Similar tests in MgSO$_4$ solution were carried out
on concretes from the programme referred to in table 6.
Attack occurred at porosities above 12% and within one
month at 19 to 22%. The same concrete series was used to
evaluate freeze-thaw resistance. Estimates were based on
either the percentage of strength retained after 25 freeze-
thaw cycles or the number of cycles needed to
disintegrate the test specimens. In figure 11 both types

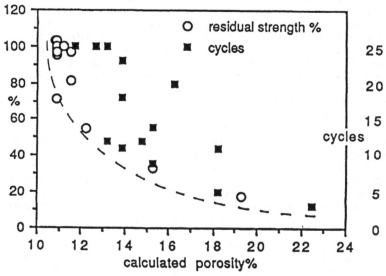

Figure 11. Porosity and Freeze-thaw Resistance of Concrete.

of data are displayed against the porosities of the
specimens. Once again performance clearly improves as
porosity decreases, with the limit for complete immunity
occurring at about 11%.

Thus in each of these studies, the cut-off below which
there is no apparent attack occurs at an estimated
porosity of 11 to 12%. This corresponds to a free
water/cement ratio equal to the stochiometric 0.33 or
less.

Alkaline hydrolysis is a form of attack specific to
aluminous cements. Some structures built with the cement
in the very early years, (when official guidance
recommended not less than about 0.6 w/c) deteriorated
rapidly through this form of attack. By 1935, the
phenomenon had been recognised, the mechanism elucidated
and published, (Rengade et al. 1936) and recommendations
changed to a maximum limit for water/cement ratio. Today
incidents of alkaline hydrolysis are rare. Such attack is
encountered in light weight refractory concretes when the
normal drying out period is inadvertently delayed.

For alkaline hydorlysis to occur, water, soluble
alkalis and carbon dioxide are needed. These can be
brought together with the cement only in a concrete of
high porosity. Attempts to provoke this form of attack in
the laboratory have been made on numerous occasions, using
cement rich mortars with a variety of sands. In these
tests porosities ranged from an estimated 11.5% to 26.3%.
Attack took place (after one to two months) in 6 out of 48
test conditions. The minimum porosity at which alkaline
hydrolysis was observed was 20.9% and the highest porosity
where attack was absent was 22.7%. No deterioration
occurred below a total water/cement ratio of 0.7. The
mortar containing the sand with the highest soluble alkali
content was not attacked. That with the sand containing
the least alkalis was among those affected. This anomaly
may be explained by the fact that more cement was used,
for the same water/cement ratio, in the mortar which
suffered attack, thus increasing its overall porosity.

Deterioration of concretes due to alkalis can also
happen with Portland cement concrete. This form of attack
known as alkali-aggregate reaction was first recognised in
structures in 1940. The problem is not totally resolved.
As in aluminous cement concrete, maintaining low
porosities, hence low water/cement ratios, may be
beneficial, however, if portlandite is a factor, high
cement contents could be detrimental. The aluminous
cement discussed in this paper does not liberate calcium
hydroxide when hydrated.

5   Conversion and Long-Term Performance

In the mid 1970's, this cement became the focus of

199

considerable controversy, triggered by failures in structures containing aluminous cement concrete. These occurred at schools in Camden (1973) and Stepney, London and at the University of Leicester (1974). At the centre of attention was the phenomenon of conversion.

H. G. Midgley has shown (10) that the conversion of the low density hydrate system to the high density, stable form occurs within minutes at temperatures approaching 100°C and takes many years at temperatures approaching 0°C. Concretes cured and stored in outdoor conditions typical of Western Europe will reach their minimum long term strengths in about 5 years, (11). It is difficult to be precise about the time-scale because it depends on the detailed thermal history of the concrete. For example the long-term minimum strength is reached within a week by curing continuously at 38°C, (100°F). Three months is needed if exposure to 38°C is preceded by 24 hours curing at 20°C.

Again, Teychenne, 1975 showed that concretes cured in water at 18°C for nearly 9 years and still well above the long-term minimum strength, when placed in water at 38°C, reached a stable value in 3 months at strengths close to, though slight greater than those achieved by initially curing for 3 months at the same temperature.

Recently, the final results of this comprehensive test programme were published, (Collins and Gutt 1988). Two storage temperatures were used, 18-20°C and 38°C. Total water/cement ratios spanned the range 0.3 to 0.73. Estimates of degree of conversion (by DTA), as well as cube crushing strengths were made at ages from 24 hrs to 20 years. The results clearly demonstrate that concretes reach their minimum long-term strengths before conversion is complete at 38°C and after conversion is complete at 18-20°C. Although the minimum long-term strength at the lower storage temperature was only reached after 20 years and at the highest water/cement ratio, this strength was virtually the same as that achieved in 3 months storage at 38°C. The authors conclude that at lower water/cement ratios a minimum strength will eventually be reached at or slightly above the corresponding value for the higher curing temperature.

Cottin (1982) has shown that intermittent curing at 30°C brings about the same result on a cumulative basis as continuous curing at the same temperature for an equivalent time.

There is thus good evidence that conversion is inevitable, even at "ordinary temperatures", but that the term "converted strength" is ambiguous. The phrase- **long-term minimum strength**- used in this paper is preferred. Moreover the level of long-term strength will depend, not on when it is achieved but on the prevailing temperature, (and the water/cement ratio of course). In passing is it worth summarising some other factors that have caused

confusion in the discussion of conversion.   By introducing
temperature variations during the initial curing of
aluminous cement concrete, it was shown that large, and
apparently anomolous changes in subsequent strength
development patterns could result, (Teychenne 1975).
These effects were later shown to be predictable according
to whether the temperature changes were introduced before,
during or after the period of initial setting of the
concrete, (Cottin 1982).

   Lack of completeness in reporting conditions can also
make results difficult to interpret, for example, failing
to distinguish between free and total water/cement ratios
and average and characteristic strengths, (14).   The
difference between dry and wet curing is also often
ignored.   Collins and Gutt, (1988), show that dry curing
gives higher strengths than wet curing and data published
by Quon and Malhotra, (1978) show that the effect is more
pronounced at higher w/c.   Chassevent and Domine, (1950),
have shown that this effect also occurs with portland
cement and is probably purely mechanical in origin, a
conclusion supported by the reversibility of the phenomena
as demonstrated by Rabot and Stiglitz, (1957) - Figure 12.

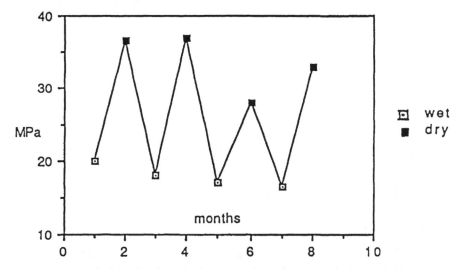

Cured in monthly cycles alternately wet then dry.

Figure 12. Strength differences as a result of storing wet or dry.

   Attempts to avoid conversion have been only partially
successful, by delaying it.   When eventually the change
takes place the minimum strength may be somewhat greater
than otherwise but at the sacrifice of a significant
degree of the very early strength the cement normally
achieves.   Furthermore the long-term stability of some of

the complex hydrate systems proposed for overcoming
conversion remains to be demonstrated.

In figure 13 the long-term minimum strengths achieved
over a range of curing temperatures, (French et al 1971),
are plotted on the same (logarithmic) time-scale as the
short-term strengths obtained when curing specimens for
product quality assurance, (the long-term data was
generated at total water/cement ratio of 0.38, whereas the
routine testing is conducted at 0.4 total w/c. This
relatively small difference does not invalidate the
comparison). The graph illustrates the strikingly early
strength potential of this type of aluminous cement
concrete - over a very wide range of exposure temperatures
the long term stable strength is achieved within 6 hours
of standard curing. This phenomenon will be well known to
users of aluminous cement, but it is helpful to see it
demonstrated in a systematic and comprehensive manner.
Furthermore it is evident that the often quoted statement
that aluminous cement concretes achieve in 24 hours what
Portland cement concrete only obtains in 28 days is, while
true, a somewhat inadequate oversimplification.

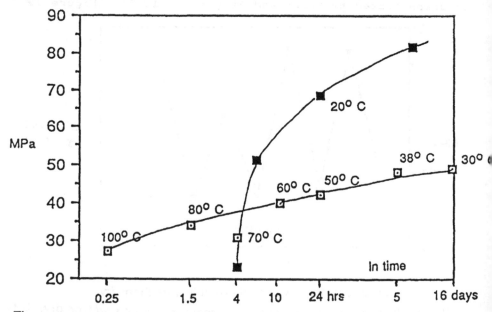

Figure 13. Long-term minimum strengths for the curing temperatures
shown versus short-term strengths for 20°C curing.

Figure 13 has other implications. Since the minimum strength potential is exceeded, (temporarily) at 20°C within a few hours, concretes designed for long-term applications can be tested and put into service very quickly, if design strengths are chosen to reflect the long-term properties. Indeed, this approach is the only way to take full advantage of the time savings offered by this cement.

Once the principle of the inevitability of conversion has been understood and accepted, the design of aluminous cement concretes based on the metastable low density hydrate system will become of historical interest only and phrases such as 'the harmful effects of conversion', will lose their meaning.

This was not the situation in 1973. At that time many structures existed in the United Kingdom built with prestressed aluminous cement concrete. Aluminous cement had been used because of its rapid hardening and the concretes were designed on the assumption (incorporated in the Code of Practice 116) that the very high 24 hour strengths achieved by normal curing would be permanent. After most intensive investigations, official guidelines for the appraisal of such buildings recommended assuming a minimum compressive strength for the concrete of marginally above 20 MPa. The minimum strength was less than half the control limit during manufacture of the concrete. Most of the structures were 5 to 10 years old. Numerous DTA tests showed that conversion had occurred. Further investigation revealed, unexpectedly, that temperatures in buildings in the UK could readily reach 30°C, and as high as 35°C in summer. Conversion therefore could be induced, not at a comfortable 20 to 25°C where its impact on strength would be small, but at temperatures in the mid 30's, where the effect is far from negligible.

Ignorance of the service temperature to be encountered however, would not have been disastrous had the provisions of the Code of Practice been fully respected. In a report on the failure at Stepney, (Bate 1974) it is stated that 'free water/cement ratios may have exceeded the maximum of 0.4 aimed at in production'. Inspection of CP116 reveals that the limit for prestressed concrete was 0.4 total. With aggregate/cement ratios as high as 9:1, a glance at figure 8 will show how important the distinction is: 0.4 free w/c implies more than 0.5 total w/c, and even that may have been exceeded. In fact, laboratory measurements of the total w/c of concrete samples taken from the prestressed beams at Stepney showed an average of 0.54 with minimum of 0.45 and a maximum of 0.64, (George 1975).

Minimum compressive strength for a total water/cement ratio of 0.54 is 20 MPa.

Part of the procedures for determining the water/cement ratio in a hardened aluminous cement concrete involves a determination of porosity. Results on a number of

concrete samples taken from structures in the UK,
including Stepney, are portrayed in figure 14. Few
results are below the limit of 0.4 total w/c and most are
above 0.4 free w/c. The correspondingly high measured
porosities come as no surprise based on the model
described earlier in this paper.

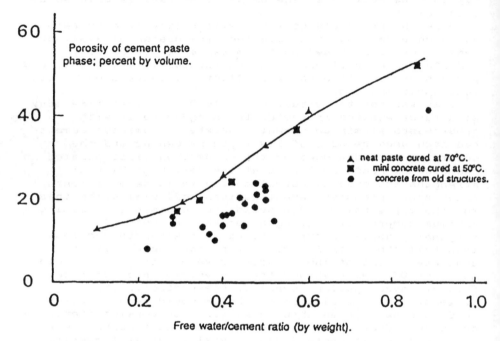

Figure 14. Porosity of Aluminous cement as a function of
water/cement ratio.

In retrospect, there is no mystery about the incidence
of low strength aluminous cement concrete: if service
temperatures and water contents had been diagnosed the
results could have been predicted. How could this happen?
Consider the following:

• quality control in the manufacture of prestressed
concrete was based on batch tests of cubes cured at
20°C, for 24 hours. It has already been noted that,
compared to long-term minimum strengths, the
metastable 20°C strengths are relatively insensitive
to water/cement ratio. In fact the lower limit
for these strengths could be achieved at
water/cement ratios greater than recommended in
the Code.

204

● Official recommendations placed an upper limit
on the cement content of concrete to prevent
initial overheating, and the consequent 'harmful
affects of conversion', because it was assumed that
it would not be significant in service.

● British Codes of Practice are not mandatory.

   It is noteworthy that in 1973 the British Standards
Institution had begun considering changing the Code of
Practice to require design of aluminous cement concrete
only on the basis of its minimum long-term strengths.
This is precisely the approach adopted in the 1979
official French circular, (21) for the structural use of
this type of concrete that includes a minimum cement
content of 400 kg/m^3, virtually identical to the maximum
recommended in CP116. Water/cement ratios must not exceed
0.4 total. However, before CP116 was changed the decision
was taken (in 1975), to discontinue its use, as not
satisfying the official UK Building Regulations. As a
result it is no longer used in Britain in public works,
even for non-structural applications such as sulphate
resistance, where its performance is unrivalled.
   When problems occur with Portland cement concrete,
thorough investigations are made to discover and eliminate
the cause rather than to incriminate the cement.

6   Summary and Conclusions

The cement described in this paper is the longest existing
and most largely used aluminous cement world-wide, with
applications in refractories, building and construction.
Its method of manufacture by melting, which has not
changed in principle since it began 80 years ago, is today
a large scale continuous process. There are intrinsic
advantages to this process from the point of view of
homogenisation which, combined with systematic and
rigorous control, provide a highly consistent and reliable
product.
   A simple physico-chemical model of hydration allows the
effect of both water/cement ratio and the conversion
phenomenon on the porosity of cement paste to be
quantitatively predicted. This in turn can be related to
the strength and durability of mortars and concretes thus
providing a justification of the 0.4 total water/cement
ratio limit normally recommended for practical
applications of this cement.
   The predictive capabilities offered by an understanding
of cement behaviour, coupled with the demonstrable
consistency of the product provide a sound basis for
confidence in its use as well as a full explanation of
earlier problems encountered with prestressed concrete.

The approach to aluminous cement concrete that seeks to avoid conversion is unrealistic. Describing the effects of conversion as "harmful" is to credit the unconverted state with a potential for permanence it does not possess. Conversion is the normal and predictable route by which the material reaches its stable long-term characteristics. The unique advantage of this cement is that the long-term performance potential can be attained (and exceeded temporarily) within a few hours of placing the concrete under standard curing conditions.

Current restrictions on the use of the cement in some countries need to be re-examined in the light of these considerations.

7   References

(1)   Lea F.M., The Chemistry of Cement and Concrete, 3rd Edition, pp. 490-491. 1970.

(2)   Teychenne D.C., Long-Term research into the characteristics of high alumina cement concrete. Mag. Conc. Res. Vol 27, No 91: June 1975.

(3)   Cottin B. and Reif P., Parameter Physiques regissant les properties mechaniques de pates pures de liants alumineux, Rev. Mat. Constr., 661, 291 (1970).

(4)   See Reference 1, p 270

(5)   See Reference 2, p 88, figure 5.

(6)   George C.M., Industrial Aluminous Cement, (in Structure and Performance of Cement, edited by P. Barnes), 1983.

(7)   George C.M., Aluminous Cement, sub-theme V-1, Vol. 1, 7th International Conference on The Chemistry of Cements, Paris, 1980.

(8)   Long-term and accelerated tests of the resistance of cements to sea water, with special reference to aluminous cements. Amer. Concr. Inst. SP-65. pp 327-349, 1980.

(9)   Rengade E., L'Hospitalier P. and Durand de Fontmagne P. Rev. Mat. Constr. Trav. Publ., **318**, 52; **319**, 78,, 1936.

(10)   Midgley A. and Midgley H.G., The Conversion of High Alumina Cement Concrete, Mag. Concr. Res., 27 (91), 59 (1975).

(11)   See Reference 7. figure 2.

(12)   Collins R. J. and Gutt W., Research on Long-Term
        Properties of High Alumina Cement Concrete, Mag.
        Concr.Res., Vol. 40, No.145, P. 204, December,
        1988.

(13)   Cottin B, et al, Reactivity of Industrial Aluminous
        Cements: An Analysis of the Effect of Curing
        Conditions on Strength Development.
        International Seminar on Calcium Aluminates,
        Turin, 1982.

(14)   See Reference 7, figure 6.

(15)   Quon D. H. H. and Malhotra M. V., Performance of High
        Alumina Cement Concrete Stored in Water and Dry
        Heat at 25, 35 and 50°C, CANMET Report 78-15,
        June 1978.

(16)   Chassevant and Domine, Sur les variations de
        resistance de liants hydrates. Rev. Mat. Constr.
        No. special p. 44 October 1950.

(17)   Rabot R. and Stiglitz P., Central Laboratory notes
        O/S 6990, April 1954 and February 1957.
        Unpublished.

(18)   French P. J., Montgomery R. G. J. and Robson T. D., High
        Concrete Strength Within the Hour, Concrete, p. 3
        August 1971.

(19)   Bate   S. C. C., Report on the Failure of Roof Beams
        at Sir John Cass's Foundation and Red Coat Church
        of England Secondary School, Stepney. Building
        Research Establishment Current paper, CP 58/74.

(20)   George C. M., The Structural Use of High Alumina
        Cement Concrete. Published by Lafarge Special
        Cements, Grays, Essex, UK. 1975.

(21)   Circular No 79-34, 27 March 1979. Ministere de
        l'Environment, France.

15 LONG-TERM PERFORMANCE OF HIGH ALUMINA CEMENT IN SULPHATE-BEARING ENVIRONMENTS

N.J. CRAMMOND
Building Research Establishment, Watford, UK

Abstract
In 1970, a wide variety of concretes in the form of 150mm diameter precast cylinders and insitu piles were buried in the sulphate-bearing clay soil at Northwick Park, London. 100mm cubes made from the same batches of concrete mixes were immersed in controlled sulphate solutions to provide an accelerated laboratory test.

The programme assessed the long-term sulphate resistance of concretes made using a variety of different cement types including high alumina cement (HAC). The buried precast cylinders and insitu piles made from HAC concretes have so far shown good sulphate resistance after 15 years in the sulphate-bearing soils of Northwick Park (0.26% SO_3 in the ground water). However, 100mm concrete cubes made from the same mixes have not performed as well in the accelerated laboratory test.

The sulphate resistance of HAC concrete can be improved by incorporating a water-reducing admixture.
Keywords: High alumina cement, concrete, sulphate attack, conversion, X-ray diffraction, ettringite, gypsum.

1 Introduction

It was the belief in the good chemical resistance of High Alumina Cement in sulphate-bearing soils and groundwaters which first encouraged the use of High Alumina Cement Concrete (HACC) in this country at the beginning of the century. This type of concrete has the added advantage of a high early strength and it was this quality which led to its extensive use in the precast beam industry during the nineteen fifties. Its success in reducing demoulding times ensured its continued use in the nineteen sixties and early seventies.

However a few roof collapses in 1973 and 1974 (Bate, 1984) led the engineering community to question the use of precast HACC components. Some of the failures and near failures were caused by poor detailing and practice but others were a direct result of the excessive loss in strength in association with chemical attack or inferior durability connected with the misuse of HACC.

The main problem connected with the use of HAC is that once it hydrates, it undergoes the conversion reaction, which can be expressed in its simplest form as:

$$CAH_{10} \longrightarrow C_3AH_6 + 18H \qquad\qquad (1)$$

Conversion is accompanied by a decrease in solid volume and an increase in water content, resulting in a more porous cement paste matrix. However, a high degree of conversion does not automatically lead to poor durability which is governed by the rate of conversion and water/cement ratio (W/C) as follows:

1. If a concrete converts quickly, large C_3AH_6 crystals form leading to greater permeability and lower concrete strength. Elevated temperature is the main factor which produces a fast rate of conversion but other factors such as high humidity, high W/C ratio and the presence of free alkalies also accelerate the conversion rate.
2. CAH_{10} needs more water of crystallisation to form compared with C_3AH_6. Low total W/C ratios ensure that not all the CA hydrates initially to form CAH_{10} but continues to hydrate after the conversion reaction starts to release extra water into the system. Therefore the pore space produced during conversion is reduced as it becomes filled with newly-formed CAH_{10}.

Slow rates of conversion and low W/C ratios guarantee that the increase in permeability is kept at a minimum and that a higher quality converted concrete ensues.

Although HACC has good sulphate resisting properties, this fact is attributed to the low permeability of good quality unconverted concrete and to the inert behaviour of the mineral CAH_{10}. Sulphate attack will occur more readily in poor quality converted concrete because:

a. the concrete has increased permeability to sulphate solutions
b. C_3AH_6 readily reacts with sulphates to produce ettringite ($3CaO.Al_2O_3.3CaSO_4.31H_2O$)
c. the additional free water formed can mobilise the sulphate anions

In 1970, a long-term performance test was initiated in which a series of different concrete mixes were buried in the sulphate-rich soil at Northwick Park, London (0.26% SO_3 in the ground water). The 5 year results have been reported (Harrison et al, 1981) and the general conclusion at that stage was that very little sulphate had penetrated the buried concrete or the laboratory-stored HACC specimens described below. A BRE Report reviewing the 15 year exposure results is being prepared for publication early in 1990 (Harrison). The detailed results on HACC are discussed separately in this paper. The paper contains two main sections; the first reports on the 15 year results from the buried concretes on site and the second reports on both the 10 year and 15 year findings from the accelerated laboratory tests carried out on identical concrete mixes stored in sulphate solutions at BRE.

Other long-term sulphate resistance tests have been reviewed by Bate (1984) and Lea (1970), who have concluded that good quality HACC, not subjected to elevated temperatures, performs very well in sulphate-rich environments.

In 1979, Lafarge Fondu International issued guidelines, which still apply today, on the use of HACC in structural units (1979). Two of their main recommendations for its use are:

1. The total W/C ratio should be less than or equal to 0.4 and a minimum cement content of 400 Kg/m^3 should be used.
2. Provision should be made to keep cast concrete cool and prevent surface drying.

Provided these are compiled with and that the aggregates used and the chemical composition of the HAC are satisfactory, Lafarge claim that HACC can produce excellent resistance to many corrosive environments including groundwaters containing sulphates.

The Northwick Park HAC concretes were made with cement contents of 335 and 375 kg/m^3 and total water/cement ratios ranging from 0.47 to 0.60; mixes which do not comply with the current manufacturer's recommendations.

2 Description of the Northwick Park specimens

The chemical analysis of the HAC used and details concerning the Thames Valley aggregate have been recorded in the 5 year and 15 year Northwick Park Reports (Harrison, 1981, 1990). These reports also include a complete account of the 150mm diameter precast cylinders and insitu piles buried at Northwick Park, the HACC basement wall panels (not investigated in the current programme of work) and the 100mm precast cubes used in the accelerated laboratory tests. Table 1 shows the HACC mix designs relevant to the results discussed in this paper.

It should be noted that although all the piles were poured with no compaction, the surface of one side of the pile was generally found to be well formed against the clay face whereas its opposite side was poorly formed. This appeared to be the result of air trapped at the surface due to the method of casting and the lack of any vibration during casting. Samples from each side of the pile were analysed separately.

For the accelerated tests, 16 cubes of each mix were examined at 5 years, 10 years and 15 years. Four cubes were kept stored in water, four in solution I containing 1.5% SO$_3$ as Na$_2$SO$_4$, four in solution E containing 1.5% SO$_3$ as MgSO$_4$ and four in solution C containing 0.35% SO$_3$ as MgSO$_4$.

3 Methods of testing
Compressive strength testing was carried out on five precast cylinders and three sections of each insitu pile after 15 years exposure to sulphate-bearing groundwater and four cubes from each of the four solutions were tested after 10 and 15 years of immersion. At the same time, one cylinder, one pile section and one cube from each solution were analysed using a variety of chemical and mineralogical techniques (Harrison, 1990). The total sulphate contents were determined using the LECO method of sulphur analysis and the amounts of ettringite and

Table 1 Mix details of the HACC Northwick Park Specimens

Mix No	Cement content (Kg/m3)	Total W/C	Free W/C	Workability	Precast Cylinders	Insitu piles	Precast Cubes
23					X		X
	335	0.53	0.46	Medium			
23P					X		
24						X	X
	335+A	0.60	0.53	Extra high			
24PS						X	
25						X	X
	335	0.60	0.53	Extra high			
25PS						X	
26	375	0.47	0.40	Medium			X
27	375	0.54	0.48	Extra high		X	X

P = Partly compacted
PS = Poorly formed surface
A = Admixture

gypsum ($CaSO_4.2H_2O$) present were determined using quantitative X ray diffraction analysis (Crammond, 1985). The degrees of conversion (DC) were calculated in accordance with procedures followed previously (BRE Information Paper, 1981) using both X ray diffraction (XRD) and differential scanning calorimetry (DSC). Additional mineralogical information was obtained using XRD and optical microscopy.

4 Results of the tests on the HACC specimens buried at Northwick Park for 15 years

4.1 Physical Condition
The HAC concretes from all the different specimens (including the basement walls) have performed satisfactorily after 15 years exposure to sulphate-rich groundwaters.

4.2 Sulphate ingress and degree of conversion

The amount of sulphate penetration into the buried concretes was established by removing selected samples and determining their sulphate content. The samples were removed in 2mm-thick consecutive layers until a depth was reached where negligible sulphate ingress was detected. An additional centre sample, 75mm from the surface was also analysed. Figure 1 shows the amount of sulphate ingress with depth for one precast, partly compacted cylinder, the well formed surface sections of three piles and two poorly formed surface sections of the lower cement content MIX24 and MIX25 piles. The degrees of conversion have been calculated for each sub-sample and are shown in Figure 1 in the form of low, medium or high conversion values (BRE Information Paper, 1981). Negligible amounts of sulphate had penetrated the fully

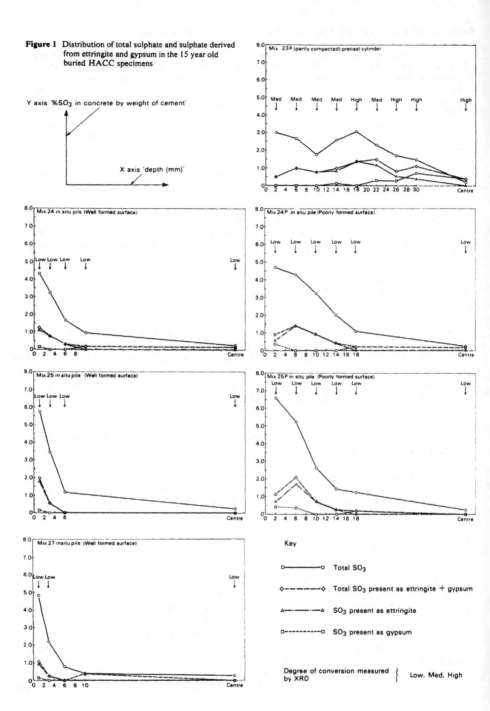

Figure 1 Distribution of total sulphate and sulphate derived from ettringite and gypsum in the 15 year old buried HACC specimens

212

compacted MIX23 cylinder, so the results have not been represented graphically. The DC values at the surface and centre of the MIX23 and MIX23P cylinders are very similar.

There is a significant difference between the degrees of conversion found in the precast cylinders and those found in the insitu piles. A low DC was found throughout each of the piles, irrespective of mix type, whereas the two precast cylinders were highly converted (as determined by XRD; medium conversion detected by DSC) at their centres with a less converted outer skin.

The main sulphate mineral present was found to be ettringite, although the discrepancy between total sulphate and sulphate derived from ettringite and a gypsum shown in Figure 1 indicates that substantial amounts of the absorbed sulphate is present within amorphous phases which are undetectable using the XRD technique. Apart from the fact that sulphates penetrate farther into partly compacted concrete or concrete with a poorly formed surface, the overall sulphate content is not significantly greater in the highly converted cylinders compared with the slightly converted piles.

4.3 Strength development

The rate of conversion has not affected the extent of sulphate penetration but has influenced the compressive strengths of the two specimen types. At 5 years, the precast cylinders had lost about 20% of their 3 month strength and no further reduction in strength was found after 15 years. The 15 year pile strengths range between 60 and 70 N/mm^2, and are about twice the values obtained for the highly converted precast cylinders.

5 Results of the tests on the cubes stored in sulphate solutions and water for 10 and 15 years

5.1 Physical Condition

Each cube comprises a mid-brown interior typical of converted concrete surrounded by a darker brown and much harder surface skin up to 10mm thick. The layer of outer skin varies in thickness and can be as narrow as a couple of millimetres especially on the 'struck' face. This outer skin is less porous because of its low conversion and is also carbonated so it forms a very hard protective coating.

After 10 years five out of the eight MIX25I cubes shown in Figure 2, two MIX25E cubes, two MIX25C cubes and one MIX27I cube had started to expand. These cubes were cast using a high W/C ratio and their surfaces appeared to be substantially intact apart from cracking which had been caused by explosive expansion from within the cube. After 15 years the number of exploded cubes has greatly increased and their distribution in relation to solution type and mix number is shown in Figure 3. The extra high workability mixes MIX25 and MIX27 with W/C ratios greater than 0.54 have shown the worst performance especially in strong sodium sulphate solution. After 15 years, only three of the four MIX25I cubes remain (Figure 4) and they have completely disintegrated possessing zero compressive strength.

Out of the four identically stored MIX25E cubes, one cube has

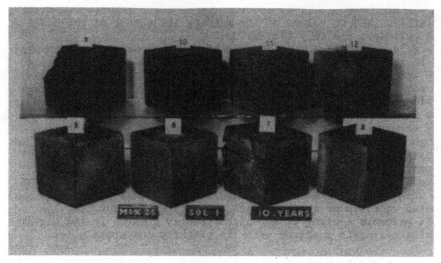

Figure 2 MIX25 HACC cubes immersed in 1.5% Na_2SO_4 solution
for ten years

remained totally intact whereas the other three have completely
deteriorated. This effect is shown in Figure 5 and is probably due to
the variable thickness of the harder more impervious outer skin; a
route was found for sulphate penetration through to the more
vulnerable centre in three of the cubes but not the fourth. Another
example showing this variability is shown in Figure 3 in the case of
the high cement content (375 kg/m^3) and low water cement ratio (0.47)
MIX26 cubes immersed in weak magnesium sulphate solution, in which
sulphate had penetrated into the cente of one cube but not the other
three. Many other workers have also found that using cubes to monitor
sulphate attack gives variable results because their corners and edges
are more suseptible to unavoidable differences in casting and testing
procedures.

The MIX25 cubes were the most seriously deteriorated but all the
MIX24 cubes, an identical mix apart from the incorporation of a
water-reducing admixture, have remained unattacked, although no
reduction in the water/cement ratio was actually made (Harrison et al,
1981).

The surfaces of the cubes from 4 of the 5 mixes stored in weak
solution C have a very eroded appearance as shown in Figure 6 with
aggregate pieces clearly exposed. Figure 3 shows that this effect
could loosely be connected with a total W/C ratio in excess of 0.5.

5.2 Degree of conversion
After 5 years, the cubes made from all the mixes with the exception of
MIX24 had highly converted centres and low to medium converted surface
skins. The only difference found in the 10 and 15 year results was
that the centres had become fully converted.

The rate of conversion for the MIX24 cubes containing an admixture
was much slower than the other mixes as their centres had only reached

214

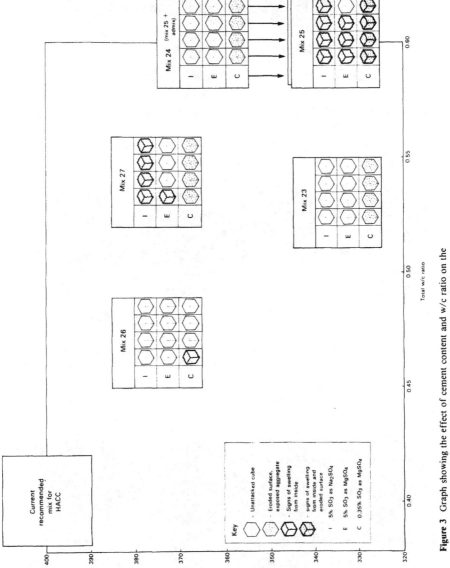

Figure 3 Graph showing the effect of cement content and w/c ratio on the sulphate resistance of the 15 year old laboratory-stored cubes

Figure 4 MIX25 HACC cubes immersed in 1.5% Na_2SO_4 solution for fifteen years

Figure 5 MIX25 HACC cubes immersed in 1.5% $MgSO_4$ for fifteen years. One out of the four cubes remains completely unattacked

Figure 6 MIX23 HACC cubes immersed in 0.35% $MgSO_4$ solution for fifteen years. Surface erosion can clearly be seen

low to medium conversion after 5 years, high conversion after 10 years and full conversion after 15 years. The slower rate of conversion has clearly affected the overall performance of this mix as shown in Figure 3.

The presence of sodium ions should enhance conversion but in fact the cube surfaces in contact with water were found to be more converted than those in contact with solution I.

5.3 Sulphate ingress

The main compositional difference between the deteriorated and intact cubes is the presence of excess sulphate at the centres of the former. The sulphate can be present as crystalline ettringite, crystalline gypsum or sulphate-bearing amorphous material. It is impossible to ascertain which of these phases is responsible for the initial cube expansions as the reactions at the centres of all the exploded cubes tested will have reached different stages in their expansion development. Optical microscopy studies have shown that two distinct types of ettringite grow within the pores at the centre of the disrupted cubes. The first type of ettringite comprises a straw-yellow colloidal variety and the second comprises acicular (needle-shaped) crystals which are clear coloured under plane polarised light. Both types are shown in Figure 7. Crack formation probably preceded the secondary deposition of well-formed ettringite crystals as many open cracks and spherical voids are filled with this material. It is possible that the precipitation of the colloidal ettringite was responsible for the expansive disruption seen in many of the cubes.

An equal amount of ettringite and gypsum was detected at the centre of the 10 year MIX25E cube but only gypsum was detected after 15 years. Gypsum was found to a much lesser extent in other $MgSO_4$-stored cubes but was not detected in any of the Na_2SO_4-stored cubes. The reason for this is that ettringite becomes unstable and decomposes in strong $MgSO_4$ solution resulting in the formation of gypsum as a sulphate-bearing reaction product.

Ettringite was found precipitated in the surface voids of half of the 5 year old cubes and in practically all the 10 year old and 15 year old cubes. After 15 years, large amounts of total sulphate, in the region of 10% by weight of cement, were found in the surface of many intact cubes.

5.4 Strength development

After 5 years, the water-stored cube strengths for the 4 mixes cast without an admixture had dropped to a value of about half the 28-day strength. At 5 years, all the cubes made with MIX24 were found to be less converted than the other mixes and showed a corresponding high strength of 50 N/mm^2 as opposed to the 20 N/mm^2 found for the other mixes.

The strengths at 10 and 15 years have remained about the same as the 5 year strengths apart from two exceptions:

1. All the disrupted cubes had very low strengths and some of those showing severe disintegration had zero strength.

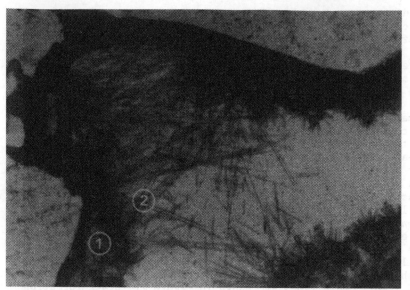

Figure 7 Photomicrograph showing the two types of ettringite found
          lining cracks and voids
(1) represents the deposition of primary amorphous ettringite
(2) represents the depostion of secondary acicular ettringite

2. Without exception, all the 15 year old intact cubes stored in
   solution C (0.35% SO_3 as $MgSO_4$) had lower strength than those
   stored in the two stronger solutions and in water. The MIX26C
   cubes with non-eroded surfaces were also weaker than the other
   MIX26 cubes. Further research is needed in order to explain this
   phenomenon.

5.5 Carbonation

The outer 1 to 2mm of both the struck and cast (opposite to struck)
faces of each 5 year, 10 year and 15 year cube are carbonated
irrespective of solution type. There is an appreciable increase in
the amount of calcite detected in the surface of the Na_2SO_4-stored
cubes, probably because of the presence of alkalis.

The centres of the completely disintegrated cubes are also
carbonated and small spherical accumulations composed of aragonite
were detected within the many cavities and cracks in the disrupted 15
year old MIX25E cube. These 'blobs' are off-white in colour, are
translucent and comprise radiating needle-shaped aragonite crystals
<5mm in diameter. Why aragonite should form and why it should form
with such an unusual habit is not known.

6 Discussion of the results

There is a clear distinction between the high DC values found in the
centres of the precast cylinders and precast cubes compared with the

low values found in the cast insitu piles. All the moulded samples also possess a lower converted outer skin which can vary in thickness from a couple of millimetres to over ten. A possible explanation is given in the following paragraphs.

The early strength of HAC is accompanied by a large heat of hydration and when dealing with site concrete any shuttering or moulding should be removed as soon as possible and if necessary the concrete should be kept cool using water sprays. The Northwick Park cylinder and cube moulds were not removed after a couple of hours but after 24 hours and the concrete was not kept cool during this time with the result that any build up in temperature would not have been dissipated but would have accumulated within the body of the concrete. It can be argued that the Northwick Park specimens are too small for a temperature build up to occur, but an increase in temperature is the only clear explanation for why a fast rate of conversion should occur at the centres of the moulded specimens. The heat of hydration released within the cast insitu piles is dissipated through the surrounding cold clay soil resulting in a lower conversion rate and higher strength concrete.

During the service life of precast HACC beams, the surface skin effect is associated with drying out of the surface and slow subsequent carbonation. The Northwick Park specimens only remained in air for a very short time (order of a few days) after demoulding and so it seems more likely that whatever initiated the outer skin formation happened prior to demoulding. It is likely that the build up of heat is concentrated at the centre resulting in a faster rate of conversion. In the limited number of cubes examined, the skins tend to be thicker on the surface in contact with the metal mould. This may be due to the higher rate of heat dissipation from the metal surfaces.

The formation of an outer more impermeable skin in a precast HACC specimen increases its sulphate resistance. Further protection can be attributed to surface carbonation because it was found from the examination of the 15-year-old buried cylinders at Northwick Park (Harrison, 1990) that the small amount of carbonation which occured during the time between demoulding and burial had a significant influence on their sulphate resistance. However, the accelerated laboratory tests on HACC have shown that, once sulphates penetrate this protective layer and attack the highly converted porous concrete immediately below, catastrophic expansions will occur. This effect is very unpredictable as it depends not only on the thickness and relative permeability of the outer skin but it will also be variable because of the geometry of the cubes with their vulnerable corners and edges. In general, expansive sulphate attack is more prominent in the extra-high workability mixes with total W/C ratios in excess of 0.54 and is more prevalent in strong sodium sulphate solutions.

The only correlation that can be made between the laboratory-stored cubes and the buried concretes is the performance of the MIX23 cylinders compared with the MIX23 cubes as both types were cast in moulds and therefore converted at comparable rates. In fact no difference was detected as both the cubes (apart from MIX23C) and cylinders showed good sulphate resistance after 15 years. It is impossible to tell whether precast cylinders made with higher W/C ratio MIX25 and MIX27 would have performed so well had they been made and buried at Northwick Park for 15 years.

Temperature is another factor which affects the rate of conversion and therefore affects the sulphate resistance of HACC. Previous work (Lea, 1970) has shown that laboratory-stored specimens kept in water lose more strength than equivalent specimens buried in the ground and the reason is almost certainly connected with the mean temperature increase during storage. Again only the MIX23 specimens can be directly compared in the current work and after 15 years, the mean compressive strength of the water-stored cubes (mean temperature of storage 20°C (Harrison et al, 1981)) was found to be 75% of the mean strength of the fully compacted cylinders (mean temperature of groundwater 13°C (Harrison et al, 1981)). The temperature differential could also account for the MIX23 cube centres being fully converted after 15 years and the cylinders only being highly converted.

The rate of conversion is slower in the MIX24 cubes (MIX25+admixture) and there is no evidence at all of expansive sulphate attack. The admixture had a minimum effect on the workability during mixing since no adjustment to the W/C ratio was made but it appears to have had a marked influence on the rate of conversion of the specimens. As this effect has only been found in precast specimens, then it is probably related to the build up of heat released during the period between casting and demoulding.

7 Conclusions

After 15 years, the buried HACC concretes have shown very good sulphate resistance to the sulphate-rich soils of Northwick Park.

The high heat of hydration produced at the centre of precast specimens made using HACC is sufficient to cause a relatively fast rate of conversion. A slower rate of conversion is found at the surface of these moulded specimens where the temperature increase is not as pronounced and this results in a less converted, less permeable outer skin.

The laboratory-stored cubes show good sulphate-resisting properties as long as the sulphate anions in the storage solutions fail to penetrate the outer protective skin. As soon as this happens, the sulphate anions start to react expansively with the more vulnerable, highly converted cube centres.

More evidence of expansive sulphate attack was found in the laboratory-stored cubes cast with total W/C ratios in excess of 0.54 and in the cubes immersed in strong sodium sulphate solution.

The incorporation of a water-reducing admixture during mixing appears to have had a marked influence on reducing the rate of conversion and improving the sulphate resistance of precast HACC cubes.

After between 10 and 15 years exposure to weak magnesium sulphate solution, the strengths of all the intact cubes has diminished in comparison with the intact cubes stored in the other two, stronger, sulphate solutions and water. Severe surface erosion has also been found in the solution C cubes with mixes containing W/C ratios greater than 0.5.

8 Acknowledgements

The Author would like to thank Bill Harrison, Kelvin Pettifer, Roger Bollinghaus, Mike Bowley and Freda Halliwell for help with the sample analysis and presentation of this work.

The work described has been carried out part of the research programme of the Building Research Establishment of the Department of the Environment and this paper is published by permission of the Director.

9 References

Bate S.C.C. (1984) **High alumina cement concrete in existing building superstructures.** Building Research Establishment Report S040 HMSO.

Building Research Station Information Paper. (1981) **Assessment of chemical attack of high alumina cement concrete.** IP 22/81 HMSO.

Crammond N.J. (1985) Quantitative X ray diffraction analysis of ettringite, thaumasite and gypsum in concretes and mortars. **Cem and Conc Res, 15,** 431.

Harrison W.H. and Teychenne D. C.(1981) **Sulphate resistance of buried concrete; second interim report on long-term investigations at Northwick Park.** Building Research Establishment Report, London S027: HMSO.

Harrison W.H. **Long-term tests on sulphate resistance of concretes at Northwick Park:** third report; to be published as Building Research Establishment Report (1990) HMSO.

Lafarge Fondu International. (1979) **Guide for using cement fondu Lafarge in structural units for civil engineering.** France.

Lea F.M. (1970) **The chemistry of cement and concrete.** Third edition, Arnold, London.

16 BEHAVIOUR OF HIGH ALUMINA CEMENT IN CHLORIDE SOLUTIONS

W. KURDOWSKI, L. TACZUK, B. TRYBALSKA
University of Mining and Metallurgy, Cracow, Poland

Abstract

The durability of HAC in strong chloride solutions, occurring in salt mines, was investigated. These solutions contain large amounts of sodium, magnesium and calcium chlorides.

The paste of HAC undergoes rapid shrinkage in the first month and then the stabilisation of volume change occurs. However, total shrinkage is small and does not exceed 11 mm/m.

Simultaneously the formation of a very dense layer on the surface of the samples is observed. Also very distinct differences in the paste composition of the different layers appear. In the dense layer the recrystallisation of $Al(OH)_3$ gel occurs and a large amount of Friedel salt is formed. Some basic magnesium chloride is also formed. Additionally a much higher content of CAH_{10} appears in the external layer while in the internal layer the concentration of cubic C_3AH_6 is substantially greater.

This reaction layer is of very limited thickness and, because of its very low porosity, evidently hinders the migration of the ions from the liquid phase to the interior of the paste. This external layer also shows much higher strength.

In up to three years the HAC paste did not show any significant change of properties.

1 Introduction

The exploitation of salt in salt mines presents several difficulties caused in part by special composition of mine waters. These waters represent always very strong chloride solutions, especially of sodium, magnesium and sometimes of calcium and potassium. It is therefore difficult to assure the high durability of cement and concrete in such a strongly corrosive environment. Special attention to the choice of cement must be given.

The good resistance of High Alumina Cement paste in various corrosive solutions is well known, but the reasons are not fully understood (2). George suggests that the main reason is the low porosity of the paste, because of the presence of colloidal aluminium hydroxide (2). Midgley believed that the very little attack of aggressive media on HAC pastes may be ascribed in part to a denser, more-resistant surface layer.

Also down to the pH 3.5-4.0 the alumina gel in the aluminous cement paste is not soluble and it is stated that it is for this reason that it is stated to be resistant to this level of acidity. In this aspect the absence of calcium hydroxide in this paste is also of great importance. It is well known that his phase undergoes rather rapid transformations in corrosive solutions.

There are few papers describing the mechanisms of chloride corrosion. The process of corrosion in chloride solutions appears very often in the form of high shrinkage (4). The destructive action of chlorides consists first of the formation of new phases. In the case of HAC paste it may be Friedal salt. The magnesium chloride gives also basic magnesium chlorides and brucite. According to Smolczyk the phase $MgO.Mg(OH)Cl.5H_2O$ is expansive (3).

In particular cases in solutions of moderate magnesium chloride concentration, on the surface of concrete am amorphous layer of brucite can be formed, which protects it from further corrosion (1).

In chloride solutions the lowering of pH of pore solution in concrete is very important. However as it was underlined earlier, the HAC paste has a rather good resistance in case of solutions of moderate acidity.

In this paper the results of experiments with HAC paste immersed in strong chloride solutions are described.

2 Material and Methods

One type of cement, namely HAC "Gorkal 60" from industrial production was chosen. This cement is produced by sintering in rotary kilns a mixture of limestone and aluminium hydroxide. The main cement phase is CA with small amounts of CA_2 and A. The last is added during cement grinding.

HAC cement was used in the form of pure paste with w/c ratio = 0.5. Form the paste standard samples of 4x4x16 cm were obtained. These samples were immersed in three chloride solutions compositions of which are presented in Table 1. These chemical compositions correspond with mine waters found in one of the salt mines. In the investigations the following methods were used:

the volumes changes of the samples were measured with Graf-Kaufmann apparatus,
X ray determination of phase composition of different layers of the samples after two years of immersion,
texture characteristic examined by scanning electron microscopy,
semiquantitive analysis of chemical composition of different layers of grains by EDAX.
porosity measurements with mercury porosimeter,
microhardness measurements to estimate the strength.

Table 1. Chemical composition of Solutions.

Component	Concentration g/1		
	Solution No.		
	1	2	3
NaCl	195	240	14
$CaCl_2$	-	-	164
$MgCl_2$	77	65	272
KCl	98	15	10
KBr	8	-	4
$MgSO_4$	3	-	-

3 Results of Experiments

The samples immersed in chloride solutions showed rapid shrinkage during the first 28 days. After one month the length of samples stabilised (Fig. 1).

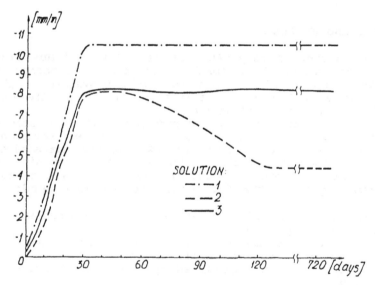

Fig.1. Shrinkage of HAC paste.

The samples immersed in solution 2 showed, in the period between two and four months, a slight expansion. The chemical composition of chloride solutions in the tested range has a rather limited influence on shrinkage. However it must be noted that the highest shrinkage of the samples appeared in solution 1. The temperature of solutions was always 22°C+/-1°C.

Observation of the samples revealed that on their surface a relatively thick coating was formed. The structure of this coating was rather complicated: on the surface a dense layer appears, composed of very fine crystals of brucite, indistinguishable under SEM (Fig.2)

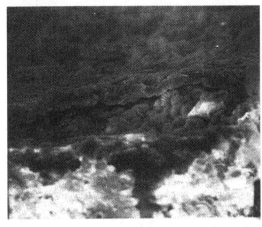

Fig. 2
SEM of HAC paste immersed two years in Solution 1

a) dense layer and hydrargillite

b) region of hexagonal aluminate

and needles. The X-ray examinations show it to be brucite and basic magnesium chloride (needles) (Fig. 3). Under the dense, superficial layer a zone of different crystalline phases is formed, which is rather poorly linked to the surface of the dense layer. Between this zone and the surface of the dense layer several microcracks or even macrocracks appear, parallel to this surface and to the surface of the sample (Fig. 2). Owing to this effect the whole coating can be detached rather easily from the sample.

In some regions under the dense layer of the coating the zone of the hydrargillite crystals is present, which appear rather frequently. In the others, nests of crystals of hexagonal aluminate hydrates are found (Fig.2b).

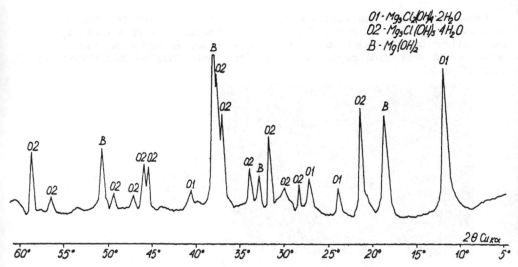

Fig. 3. X-rays pattern of coating

It seems probable that this "crystalline zone" represents the original surface of the sample, modified by the influence of the corrosive solution.

A section of the samples, obtained by cutting with a diamond saw, revealed the existence of one layer, which differs in the shade of colour (darker). The thickness of this layer can be evaluated to be of about 2mm. The qualititive examinations of the concentration of some components by electron microprobe showed a very limited diffusion of magnesium into the volume of the samples (Fig.4).

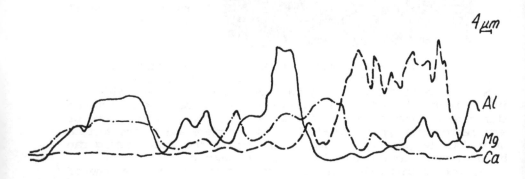

Fig.4. ESCA analysis of coating and dense layer of the sample (from right to left).

The X-ray examination was limited to the samples immersed in solution 1, which ought to be strongly corrosive for HAC paste, because of the high sodium concentration. In this solution the highest shrinkage of the samples was also noted. The examinations were applied to the different layers obtained by successive wearing of the samples.

Very clear differences in the phase composition of different layers were recorded (Fig. 5). The external layer contains a great amount of basic magnesium chloride. The Friedal salt is present in all layers, even in the "unchanged" layer. Much higher content of CAH_{10} appears in the external layer while in the internal layer the concentration of cubic C_3AH_6 is substantially greater. Also a well pronounced recrystallisation of aluminium hydroxide occurs with the formation of hydrargillite. Its content increases in the direction of the sample's core. It is known that hydrargillite can be stabilised with sodium.

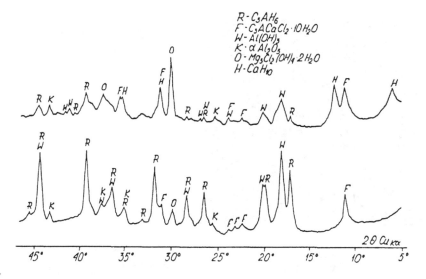

Fig.5. X-rays pattern of HAC paste after two years
    A - external layer
    B - internal layer

The texture of successive layers of the samples is clearly differentiated. The external layer is very compact with no visible crystals (Fig. 6). In the interior of the sample numerous crystals of C_3AH_6 in more porous zones are easily visible (Fig. 6b).

The mercury porosimetry measurements demonstrate a very pronounced porosity decrease of the external zone of the samples in comparison with the internal layer. In the first one the small pores with the diameter under 10mm prevail (Fig. 7), while in the internal zone of the samples the content of the pores with diameter between 29 and 80nm is the highest.

Fig. 6
SEM HAC paste:
a) external layer
b) internal layer

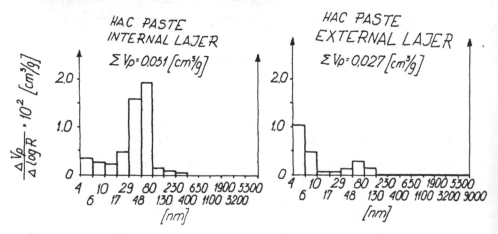

Fig. 7 Porosity of HAC paste after two years

In Table 2 the results of microhardness examinations are presented. The approximate thickness of the layers are also marked. The highest values of strength of external layer are clearly marked.

Table 2. Microhardness of different layers after two years of immersion of the samples.

Layer external		intermediate	internal
150		78	83
	Thickness		
2mm dark		farther unchanged	

4 Discussion of the results

The obtained results give evidence of a good durability of HAC paste in chloride solutions. On the surface of the samples a coating is formed, in which the external layer is very dense. This coating, even in spite of rather poor linkage with the surface layer of the sample, seems to have some protective influence. The coating is formed practically of two phases: brucite and basic magnesium chloride. Especially partly amorphous brucite seems to play an important role in densifying this layer. Also the sample itself underwent some microstructural modifications. Namely the porosity of external layer became much smaller and its phase composition underwent substantial changes. Some new formations appeared, among others Friedal salt, and basic magnesium chloride. However a large quantity of CAH_{10} crystals remained unchanged, while the alumina hydroxide gel showed the transformation towards a crystalline phase. All these processes lead to the densification of the microstructure. These transformations, up to a period of two years, have positive influence on the stability of the paste in corrosive solutions. The dense layer has high strength and also the volume changes of the samples are stabilised.

Further examinations of the samples will continue and with no doubt, will give more complete image of the process. Particularly the differentiation of the aggressiveness of different solutions should be possible.

A very interesting question is the intensity of chloride ion diffusion and consequently the rise of concentration of Friedal salt and its influence on the physical behaviour of HAC paste.

5 References

Calleja, J. (1980) 7th Int. Congress on the Chemistry of Cement, Paris vol.I, page VII-2/1, Paris.
George, C.M. (1983) Chapter 9 in "Structure and Performance of Cements, Ed. by P. Barnes, Applied Science Publishers, London.
Smolczyk, H.G. (1968) 5th Int. Congress on the Chemistry of Cement, vol. III, page 274, Tokyo.
Van Aardt, J.H.P. (1960) 4th Int. Congress on the Chemistry of Cement. vol. II, page 835, Washington.

17 ACIDIC CORROSION OF HIGH ALUMINA CEMENT

J.P. BAYOUX, J.P. LETOURNEUX, S. MARCDARGENT
Lafarge Coppée Recherche, Viviers, France
M. VERSCHAEVE
Lafarge Fondu International, Neuilly-sur-Seine, France

Abstract
A study of the durability of Fondu pastes and mortars in 12 different types of acidic aggressive media enables the definition of a comprehensive model of acidic corrosion based on the decomposition of hydrates followed by the solubilisation of the salts formed.

Some examples are given to quantify the relative importance of the parameters : thermodynamic stability of initial phases, acid strength, concentration of the aggressive acid solution, porosity, neutralizing character of initial phases, solubility of acid salts formed and temperature.

Keywords : High Alumina Cement, Ciment Fondu, Corrosion, Acid, Acid Strength, Solubility, Porosity, Temperature, Aggregate.

1 Introduction

The durability of HAC such as Fondu in acidic media is generally good, compared to Portland cement, because of the absence of free hydrated lime. This particular property enables the use of Fondu cement in aggressive environments such as dairyworks, chemical plants and sewage pipes [1, 2, 3].

However, corrosion can occur, but only significantly in some defined conditions.

The aim of this study is to precise the mechanisms of acidic corrosion, and to define, for given acid types, the acceptable limits of durability.

2 Acidic corrosion model

Acidic corrosion of concrete is generally due to the dissolution of the cement paste and eventually of the aggregates [4]. The mechanism of acidic corrosion of cement paste comprises three main steps : decomposition of hydrates, formation of acid salts of calcium and aluminum, and eventually dissolution of these salts. The chemical reactions involved are written below, for high alumina cement and Portland cement :

- with HAC, the first reaction is the aluminate hydrate decomposition by the acid. In reaction (1), monoacid type AH displace aluminic acid from its calcium salt :

$$x CaO.y Al_2O_3.n H_2O + 2x\ AH \longrightarrow x\ CaA_2 + 2y\ Al(OH)_3 + (x+n-3y)\ H_2O \quad (1)$$

This reaction leads to the formation of calcium salt CaA_2 and aluminum hydroxide is released.
Aluminum hydroxide is amphoteric. So reaction (1) is followed by attack on the aluminum hydroxide, which is only possible if the pH value is below 4 approximately (5) :

$$2y\ Al(OH)_3 + 6y\ AH \longrightarrow 2y\ AlA_3 + 3y\ H_2O \quad (2)$$

An aluminum salt is formed, and water is released. So, for a complete decomposition of aluminate hydrate, $2x+6y$ moles of AH type acid are necessary.
Reaction (2) is also possible for aluminum hydroxide directly formed during the hydration of the cement.

- with a Portland cement, mainly based on calcium silicate, the first reaction is the direct attack of acid on calcium hydroxide; because of the high basic character of calcium hydroxide, this reaction is very easy and explains the poor behaviour of Portland cements in acidic media.
The second reaction is the silicate hydrate decomposition :

$$x CaO.y SiO_2.n H_2O + 2x\ AH \longrightarrow x\ CaA_2 + y\ Si(OH)_4 + (x+n-2y)\ H_2O \quad (3)$$

Calcium salt is formed, with silicon hydroxide releasing. silicon hydroxide is generally not decomposed by acids, except by hydrofluoric acid.

After these reactions of decomposition, the salts formed are dissolved according to their solubilities.

An other possibility is the formation of insoluble complex calcium aluminum hydrates, corresponding to the general formula :

$$Ca_2Al(OH)_6^+, \quad (X^{n-})_{m/n}, \quad \{Al(OH)_4^-\}_{1-m}$$

The best known products of this family are calcium monocarboaluminate ($X^{n-}=CO_2^{2-}$), and monosulphoalumi-nate ($X^n =SO_4^{2-}$), but it is also possible to obtain defined products from most of other anions.

The stability of these complex hydrates depends on the pH value, or more precisely on the acid strength and concentration. This hydrate formation generally slows down the decomposition reaction rates by filling the porosity. However, some of these hydrates can expand during their crystallisa-tion and hence destroy the structure.

This mechanism enables to define the main param-eters of acidic corrosion : thermodynamic stability of initial phases, acid strength, concentration of the aggressive acid solution, porosity, neutraliz-ing character of initial phases, solubility of acid salts formed and temperature. Some examples of effect of these parameters are described in this paper.

3 Experiments

A comparison of the behaviour of HAC formulations in 12 acidic aggressive media has been made, mainly by measuring the weight loss of test samples. The 12 media have been choosen for: their organic or mineral nature, their acidic strength represented

	pK values			Solubilities (g/kg)	
	pK1	pK2	pK3	Ca	Al
Acetic	4,75	–	–	374	v.sl.s
Citric	3,08	4,74	5,4	8,5	–
Lactic	3,86	–	–	31	v.s
Maleic	1,83	6,07	–	29	–
Oxalic	1,23	4,19	–	0,007	i
Succinic	4,16	5,61	–	1,9	–
Hydrochloric	0	–	–	745	699
Nitric	0	–	–	1212	637
Ortho-Phosph.	2,12	7,21	12,67	18	i
Pyro-Phosph.	0,85	1,49	5,77	sl.s	–
Sulphuric	0	1,92	–	2,4	313

TABLE 1 :pK values of Acids studied and solubilities of Calcium and Aluminum Salts

sl.s = slightly soluble, v.sl.s = very slightly soluble, v.s = very soluble, i = insoluble

by the pK values, and the solubility of their calcium and aluminum salts. In table 1 are reported
the values of pK and the solubilities of calcium
and aluminum salts of 11 acids.
Ethylene diamine tetracetic acid (EDTA) has been
used in the form of the sodium salt, because of the
very low solubility of free acid. EDTA is characterised by an large dissolving effect due to the
possible formation of chelates.
    The investigation of these aggressive media has
been made for Fondu neat pastes, mortars hydrated
at 20°C (CAH_{10} mainly formed) and converted at 50°C
in cubic hydrate C_3AH_6 and aluminum hydroxide. Test
samples of 2x2x5 cm were immersed in 50 ml of acid
solution. The solution was renewed twice a day.
    In these conditions, the rate of corrosion is
defined as the weight loss of test samples as a
function of the number of solution renewals.

4 Thermodynamic stability

The Table 2 shows the weight losses measured after
20 solution renewals on neat pastes of FONDU
W/C 0.235, with 12 different acid solutions at 0.1N
concentration.

	F 20°C	F 50°C
Acetic	3,2	1,7
Citric	20,7	5,8
Lactic	10,7	5,1
Maleic	11,2	3,1
Oxalic	-0,6	-0,2
Succinic	0,7	0,9
E.D.T.A.Na	2,9	1,5
Hydrochloric	12,4	4,8
Nitric	14,9	4,5
Ortho-Phosphoric	1,3	0,6
Pyro-Phosphoric	3,6	2,4
Sulphuric	2,8	3,3

TABLE 2 : Weight losses (%) after 20
renewals of 0.1N solutions

The weight losses for F 50°C (Fondu hydrated 24h at
20°C, and converted 6 days at 50°C) are generally
lower than those observed for F 20°C (Fondu hydrated

233

28 days at 20°C).

The open porosity is about 2% for F 20°C and 15% for
F 50°C. Even with this increase of porosity,the high
thermodynamic stability of cubic hydrate C_3AH_6, compared with hexagonal hydrates, explains the
improvement of the durability obtained by conversion.

5 Acid strength

Strong acids are very aggressive, for a given concentration, because they decompose the hydrates
very easily.
It can be seen from Table 2 that hydrochloric
and nitric acids are some of the most aggressive.
However, sulphuric and phosphoric acids, which are
strong acids, are not always very aggressive, and
we can observe that weak organic acids like citric,
lactic or maleic lead to large weight losses.
So it is possible to say that acid strength is
only a secondary parameter for acidic corrosion.

6 Acid concentration

The figure 1 shows the weight losses for Fondu neat
pastes, W/C 0.235, hydrated 28 days at 20°C, after

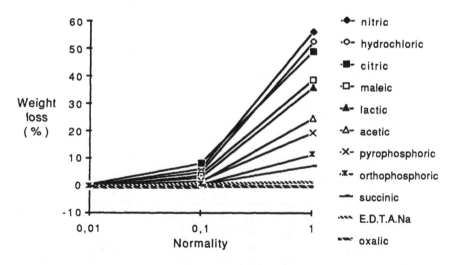

FIGURE 1 :Weight loss (%) of F20°C as a function of
solutions normality, after 8 renewals

8 renewals of the acid solution at 0.01N, 0.1N and N concentrations.

The intensity of aggression is thus in relation to the molar concentration of the acid : low for 0.01N, medium for 0.1N and strong for N solutions.

At low concentrations, it is possible to observe an increase of weight, due to the precipitation of complex calcium aluminates described above.

The pH value is very often used to define the concentration of an acid solution, because it is more difficult to measure concentration level than pH. So it is of a practical interest to define the acceptable limits of pH without exagerated corrosion.

Figure 2 shows clearly that acceptable limits are only significant for a given acid, because of the differences of acid strengths. It is particularly interesting to observe that the pH limits are lower for strong acids, like hydrochloric or nitric, than for weak acids like lactic or acetic.

This is easily explained : weak acid solutions have a higher pH than strong acid solutions, for a given concentration. For example, 0.1N acetic acid has a pH of 2.9 while 0.1N hydrochloric acid pH is only 1. So, for a given pH, strong acids have a lower concentration than weak acids and may be less aggressive.

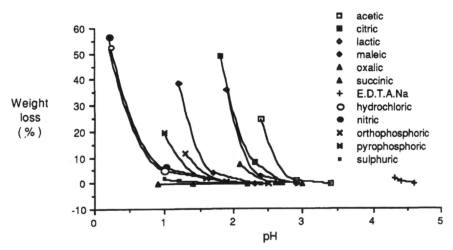

FIGURE 2 :Weight loss (%) of F20°C as a function of solutions pH, after 8 renewals

7 Porosity

The importance of porosity depends on the solubility of salts formed. This parameter is very important if the salts solubility is high : Figure 3
shows the weight loss of ISO Fondu mortars,
hydrated for 6 days at 50°C and immersed in 0.1N
lactic acid solution. The rate of corrosion is significantly increased by the water/cement ratio,
related to the porosity.

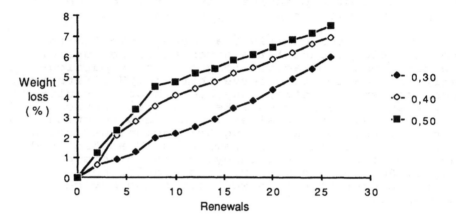

FIGURE 3 :Weight loss (%) of F50°C mortars in 0.1N lactic acid
as a function of renewals number, for W/C 0.30, 0.40
and 0.50

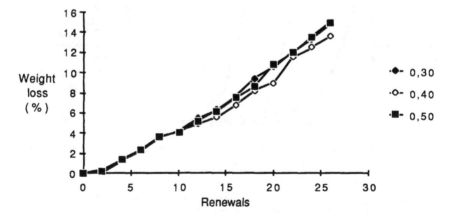

FIGURE 4 :Weight loss (%) of F50°C mortars in 0.1N
sulphuric acid as a function of renewals number,
for W/C 0.30, 0.40 and 0.50

On the contrary, if the salts have a low solubility, like with sulphuric acid (Figure 4), they crystallize in the pores and seal the surface. In this case, the degradation is practically independent from the porosity, for a normal water/cement ratio range.

8 Neutralizing character of initial phases

The neutralizing character of initial phases is very important if the amount of acid is limited, for example in a closed system. Figure 5 shows the weight losses of Fondu mortars made with siliceous or calcareous aggregates, and immersed in pH1 sulphuric acid solution. In this case, calcium carbonate aggregates are better than silica, because they neutralize the acid solution and then decrease the decomposition of cement paste [6].

Moreover, the sample surface is regularly dissolved with reactive aggregates; on the contrary, with unreactive silica, the paste is dissolved around the aggregates, which are quickly extracted, increasing the decomposition rate.

We call these reactive aggregates "sacrificial", because they are attacked by the acid solutions and then protect the interstitial cement paste.

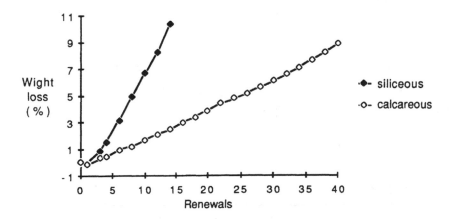

FIGURE 5 :Weight loss (%) of F50°C mortars in pH1 sulphuric acid as a function of renewals number, for siliceous and calcareous aggregates

9 Salts solubilities

The solubility of salts formed is a very important parameter. The comparison of the weight losses in Table 2 or Figure 1, with the salts solubilities in Table 1 enables to explain the low aggressivity of oxalic, succinic, phosphoric and even sulphuric acids, compared to dissolving acids like citric, lactic, maleic, hydrochloric and nitric.

This is a very effective parameter of acidic corrosion, particularly for relatively strong acids.

The calcium salt solubility is the most important, because this is the first salt formed by reaction (1) described above. The aluminum salt is only formed in a second step if the residual pH is below 4.

10 Temperature

The temperature effect has been measured on ISO Fondu mortars, W/C 0.40, immersed in 0.1N solutions of acetic, lactic and sulphuric acids. An increase of temperature from 20 to 40°C does not modify significantly the rates of acetic and lactic corrosion, as shown in Figure 6.
On the contrary, the sulphuric corrosion is increased of about 25% (Figure 7).

This result can be explained by an increase of the calcium sulphate solubility with the tempera-

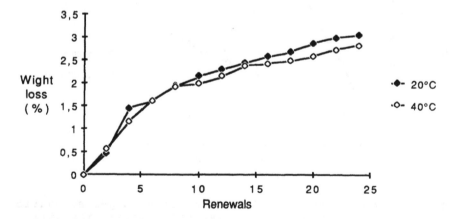

FIGURE 6 :Weight loss (%) of F50°C mortars in 0.1N acetic acid as a function of renewals number, at 20°C and 40°C

238

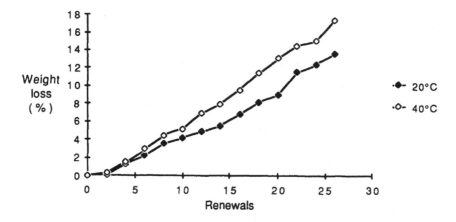

FIGURE 7 :Weight loss (%) of F50°C mortars in 0.1N
        sulphuric acid as a function of renewals number,
        at 20°C and 40°C

ture while the solubilities of calcium acetate and
lactate are sufficiently high that their variation
with the temperature has a negligible effect.

11 Conclusion

The corrosion of high alumina cements, mortars and
concretes in aggressive acidic media is due to a
decomposition of hydrate and aggregate materials,
followed by the dissolution of salts formed.
    The main parameters governing acidic corrosion
are :
- the thermodynamic stability of initial phases,
which explains the relatively good behaviour of
converted Fondu;
- the acid strength, which is only a secondary parameter;
- the acid concentration, which is directly related
to the intensity of corrosion;
- the porosity, which is only important if the
salts formed are soluble;
- the neutralising effect of initial phases, which
reduces the aggressive character of the acidic
solution ;
- the salts solubility, particularly calcium salts,
which is the most important parameter;
- the temperature, which effect is only sensitive
if the salts solubilities are low.

12 References

[1] T.D.ROBSON, "High alumina cements and concretes", Contractors Record Ltd, 1962

[2] A.MATHIEU, P.SOUKATCHOFF, L'eau, l'industrie, les nuisances, 111, 06/1987, 48-51

[3] P.SOUKATCHOFF, Mat. et Constr., 1985, vol.118, n°104, 115-122

[4] H.GRUBE, W.REICHENBERG, Rev. Béton, 11/1987, 446-451

[5] K.H.GAYER, L.C.THOMPSON, O.T.ZAJICEK, Canad.J.Chem, 1958, 36, (9), 1268

[6] See also S.STUTTERHEIM, J.P.LOCHNER, J.F.BURGER, Mag.Concr.Res.,1954, 6, (16), 39

18 THE EFFECT OF LIMESTONE FILLERS ON SULPHATE RESISTANCE OF HIGH ALUMINA CEMENT COMPOSITERS

WOJCIECH G. PIASTA
The Technical University of Kielce, Institute of Civil
Engineering Technology, Kielce, Poland

Abstract

This paper deals with the effect of limestone fillers (powder and porous and dense aggregates) on sulphate resistance of HAC pastes, mortars and concretes. Application of limestone fillers improves very significantly resistance of HAC at normal temperature. It is mainly caused by the formation of more stable calcium carboaluminate hydrate phase. That means that conversion is considerably limited. However, higher temperature (about 30°C or 40°C) seems to be also dangerous for unfavourable phase changes.

Keywords: Sulphate Resistance, Conversion, Paste, Mortar, Concrete, Limestone, Powder, Aggregate, Phase Content, XRD, DTA

1 Introduction

It is generally known that HAC composites, if only correctly made are of high sulphate resistance. Due to low ratio of calcium oxide to acid oxides in HAC as well as formation of AH_3-gel films surrounding main aluminate hydrates of lower basicity CAH_{10} and C_2AH_8, the attack of sulphate ions on HAC is considerably limited (Lea 1971 and Kurdowski et al. 1986). Also physical properties as porosity pore size distribution and permeability, being the results of HAC pastes morphology, protect HAC composites from deep internal approach of sulphate ions and formation of ettringite. However, hexagonal calcium aluminate hydrates are metastable and they change in the conversion reaction in stable cubic phase C_3AH_6, AH_3 and water. The conversion changes not only phase content but also physical features. The conversion products are of substantially higher density, than hexagonal products, and release of free water reduce the volume of solids in the cement paste by about 20% to 25% (Lea 1971, Kurdowski et al 1986 and George C.M. 1980). It is equivalent to the same range increase in porosity of HAC paste that indicates the general decrease in strength and durability. Higher sulphate vulnerability of HAC composites is not only connected with increase in porosity and permeability but also with lower resistance of C_3AH_6 than that of CAH_{10}. Thus, converted HAC cannot be considered as high sulphate resistance cement (Midgley, H.G. 1980), particularly when the conversion had taken place very rapidly in hardened composite, e.g. under higher temperature (about 40°C).

However, the conversion of HAC composites can be limited under normal conditions if powdered limestone is added to HAC mix (Fentiman C.H. 1985). Also limestone fillers as fine and coarse aggregates can protect HAC concretes from detrimental conversion (Cussino L.et al. 1980). The main reason of maintenance HAC with limestone fillers in good state of high strength and durability is their chemical interaction. Calcium monocarboaluminate hydrate $C_3A. CaCO_3.H_{12}$ is the main phase that eliminates considerably the harmful effects of conversion (Fentiman. C.H. 1985 and Cussino et al 1980). This phase and the associated AH_3 are also of relatively high sulphate resistance. Moreover formation of carboaluminate reduces porosity and permeability of HAC composites that is the second but physical way of sulphate resistance improvement (Bachiorrini A.et al 1986)

Thus, it seems that application of limestone fillers contrary to inert ones (e.g. quartz) is effective in protection of HAC from conversion and improvement of HAC resistance to sulphate attack as well (Piasta 1989).

2 Materials and test procedure

The Polish commercial "GORKAL 40" HAC was applied to all composites used in tests. The cement was of the following oxide composition: CaO-36. 1%; Al_2O_3-45. 1%; SiO_2-6.6%; Fe_2O_3-8.4%; Na_2O-0.15%. Two powders of the same fineness 3100 cm^2/g as that of the HAC were ground from pure limestone aggregate and quartz sand. The powders were employed to substitute 25% by mass the cement. These two cements with the additions and the plain cement were used to test sulphate resistance of paste of w/s=0.4 and quartz sand mortars of w/s=0.55. The assessment of sulphate resistance of the paste was carried out at t_1=18°C in two solutions (1.5% Na_2SO_4; 4.5% Na_2SO_4) and tap water by flexural strength and expansion measurements of the 10 by 10 by 60 mm and 16 by 16 by 160 mm bars, respectively. Moreover strength test was conducted for all three pastes at t_2=28°C and t_3=38°C in solution of 4.5% Na_2SO_4 and water. Expansion of mortars stored at t_1=18°C in solution 4.5% Na_2SO_4 and water was tested. In order to find sulphate resistance of HAC concrete the samples of the dimension 40 by 40 by 160 mm sawed out from large members were stored in 4.5% Na_2SO_4 and water and flexural strength was tested. The factors as kind of crushed aggregate (limestone of porosity 5% and 36%, granite) and time of curing period (age: 28 days, 270 days) were taken into account apart from environments and time. The content of cement was 350 kg/m^3 in all concrete mixes. Natural quartz sand was applied as fine aggregate.

After 28 days (or 270 days) of curing at t_1=18°C and RH=70% the first measurements of flexural strength, length of samples and phase content by XRD and DTA were performed. Then the samples were immersed in sulphate solution of comparatively in tap water.

3 The experimental results and their analysis

3.1 Effects of limestone powder

3.1.1 Tests of sulphate expansion

The results (Fig 1) of thirty-month expansion tests in sulphate solution (4.5% Na_2SO_4, 1.5% Na_2SO_4) and water of HAC pastes (w/c=0.40) with 25 per cent additions of $CaCO_3$ and SiO_2 powders and without any additions showed generally statistically significant effect of the kind of environment.

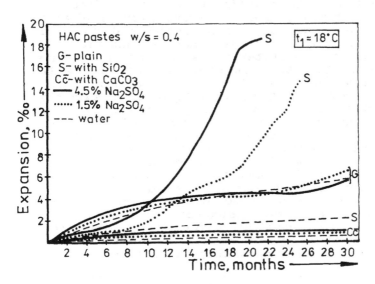

Fig. 1. Expansion of HAC pastes.

This observation indicates, thus, imperfect sulphate resistance of HAC composites population. However, an attention should be paid that this assessment resulting from the two-factor variance analysis is connected mainly with the deteriorating sulphate expansion of HAC paste with addition of quartz powder.

Statistically estimate effects of the kind of addition applied or of the lack suggest the great significance (even at significance level lower than α = 0.001 of the kind of paste i.e. chemical composition of the addition applied. Thus, the expansion test also indicates very distinctly nearly total elimination of phenomenon of swelling of HAC paste with 25 per cent limestone addition in Na_2SO_4 solutions for the period of 30 months. The volumetric stabilization as an effect of calcium carbonate is very considerable in sulphate as well as in water. The sulphate expansion of plain HAC paste is quite substantial, however, it is similar to the swelling deformations in water.

According to statistical estimators however, the sulphate expansion of plain HAC paste is considerably greater than that of the paste with limestone powder. Particularly different effect is noted when quartz powder is applied as the addition to HAC paste. Application of this powder is the reason for the decrease in very high extent of sulphate resistance of HAC paste. The total failure (Fig.1) of the HAC paste with quartz powder was already observed after 20 months of immersion in sulphate solution. Whereas the same quartz filler stabilises quite well (statistically significant) the swelling of HAC paste in water. Thus the application of quartz powder to HAC paste is particularly undesirable when it is to be immersed in Na_2SO_4 solution. The significant negative effect of interaction of these two levels of factors (solution of Na_2O_4 and quartz addition) was confirmed by variance analysis. Such effect was not stated for the limestone addition.

According to the criterior assessment of sulphate expansion at the level 5% it can be stated that only the HAC paste with limestone powder showed the low expansion, i.e. 1.19% after 30 months of attack of Na_2SO_4 solution, suggesting perfect sulphate resistance of HAC composites with limestone filler. The test of the plain HAC paste revealed the sulphate expansion of 5.29% i.e. above the critical magnitude. It suggests relatively good resistance to sulphate waters. Whereas the sulphate expansion of 18.47% of the paste with quartz powder, indicates very low durability in sulphate environments.

The results (Fig 2.) of sulphate expansion test of mortars made of modified HAC's, plain HAC and quartz sand (w/s=0.55) showed statistically different resistance (for the same period of immersion). Due to great difference (for all composites) between swelling in the both environments the effect of kind of environment appears as generally significant.

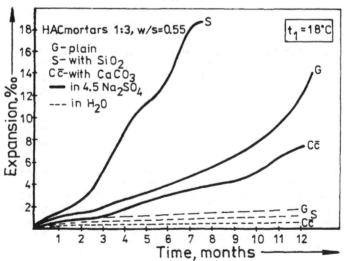

Fig. 2. Expansion of HAC mortars.

244

3.1.2 Test of changes in strength

The further indications of improvement of HAC paste resistance to sulphate solutions and other properties through application of limestone powder are presented by the results of strength tests (Fig. 3). Three-factor variance analysis showed that the effect of limestone powder on HAC paste strength (averaged in the range of time and kind of environment factors) is only positive. Moreover the investigation indicates not only higher strength of HAC paste with limestone but also ascending character of its strength changes in any period of time. Whereas application of quartz powder of HAC paste was the reason of initial improvement of strength. However, statistically significant decrease in strength of this paste was observed soon after. The assessement of effect of interaction between material factor (the levels of material factor: limestone addition; quartz addition; the lack of addition) and environment factor (the levels of environment factor: 4.5% Na_2SO_4; 1.5% Na_2SO_4; H_2O) indicates positive and nearly equal interaction between limestone fillers and any environment,. Whereas the effects, statistically estimated, of interaction betwen quartz filler and all environments are negative in strength tests.

The above mentioned analysis concerns strength expressed in MPa and also the influence of 28 - day strength is included

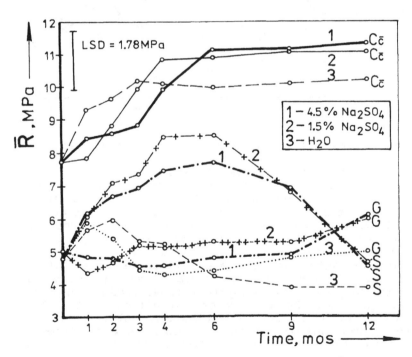

Fig. 3. Flexural strength of HAC pastes.

in it. Thus, the analysis of relative strength, $r=R_r/R_{28}$, was enclosed (fig. 4). Such transformation of strength results is simultaneously direct estimation of durability in relation to initial state. Generally gradual and monotonic increase in strength properties of HAC paste with limestone powder can be stated for every environment. Whereas the changes of relative strength of other HAC pastes indicate their lower sulphate resistance than that of HAC paste with limestone. The final assessment of effects of interaction between time of sulphate attack and kind of HAC paste composition confirms statistically significant improvement of HAC paste durability through application of limestone filler and contrary changes in durability when quartz powder was applied.

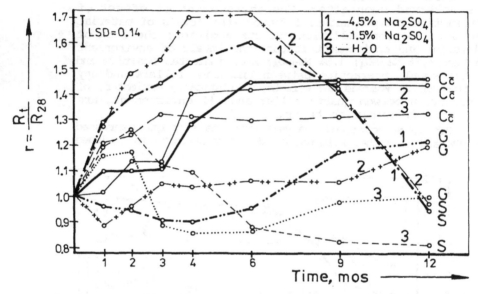

Fig. 4. Relative flexural strength of HAC pastes.

3.3 Phase content analysis at $t_1=18^{\circ}C$

In order to explain exact reasons connected mainly with the changes in phase content of the HAC pastes the investigations of HAC paste morphology have been carried out systematically with the help of XRD and DTA. The first observations were carried out after 28 days of curing when the strength of HAC pastes was measured and then the samples were immersed in solutions and water. It was stated that phase content of the HAC pastes with limestone, quartz and without additions was generally the same. The hexagonal calcium aluminate hydrates CAH_{10}, C_2AH_8 were the main phases. In all pastes the phase CAH_{10} was formed in the highest amount. Also the gel of aluminium hydroxide occured in considerably amount (stated by DTA) in each paste. There was no C_3AH_6 in the phase content of any paste.

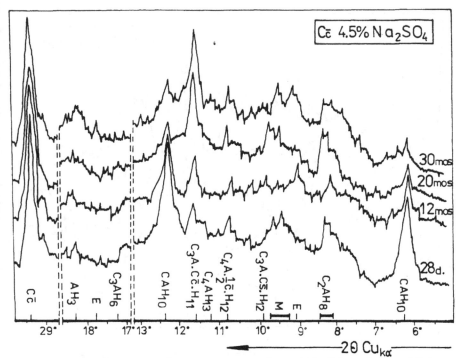

Fig. 5. XRD analysis of HAC paste with limestone (Cc),

After 12 months of immersion in 4.5% Na_2SO_4 solution the next XRD-analysis was carried out (Fig.5,6,7). Then the considerably amount of ettringite was found in the paste with quartz and without additions. Whereas only slight amount of ettringite was stated in the paste with $CaCO_3$. These observations (Fig. 5) confirmed very well the changes in the strength and expansion changes. Moreover the decrease in amount of CAH_{10} and C_2AH_3 connected with the formation of considerably amount of calcium monocarboaluminate hydrate $C_3A.C\bar{c}.H_{11}$. Whereas some amounts of C_3AH_6 were stated for other HAC pastes. The confirmation of internal changes was obtained for the pastes immersed in water i.e. formation of great amount of $C_3A.C\bar{c}.H_{11}$ in the paste with $CaCO_3$ and the beginning of conversion in both other pastes indicated by formation of C_3AH_6. After 20 months immersion it was generally stated the progress of changes in content in all HAC pastes (Fig.5,6.7). Particularly considerable increase in ettringite amount in the paste with quartz explains the total deterioration of this samples. Also the further progressive formation of $C_3A.C\bar{c}.H_{11}$ in the paste with $CaCO_3$ and conversion of hexagonal aluminates in C_3AH_6 and AH_3 in the plain paste are in very good agreement with changes in their strength and expansion.

After 30 months it was stated (Fig.5,6,7,8) for both environments that formation of $C_3A.C\bar{c}.H_{11}$ in the great amount protects very well HAC paste from formation of ettringite under

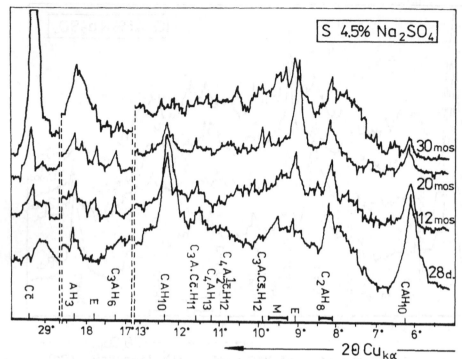

Fig. 6. XRD analysis of HAC paste with quartz (s).

sulphate attack and conversion of hexagonal hydrates into C_3AH_6 and AH_3. Whereas the HAC paste with inert filler lowers its strength and sulphate resistance mainly due to the conversion of CAH_{10} and C_2AH_8 into C_3AH_6 and AH_3 which was very well detected for the HAC pastes with quartz and one without any addition. The finally presented changes in phase content (Fig. 6,8) for the paste with quartz stored in 4,5% Na_2SO_4 solutions are connected with the total earlier deterioration of samples and great surface of chemical attack. So degregation of nearly all calcium aluminate hydrates (also part of ettringite) occured. The main products are: $CaCO_3$, AH_3, ettringite. It indicates that attack of Na^+ ions and free CO_2 was also present in the latest period. For the plain HAC paste only the further progress of the conversion and sulphate attack is observed. It should be noted that the presence of AH_3, mainly in the form of gel, has probably considerably influence on behaviour of HAC composites, particularly under chemical attacks. However, it has not been explained satisfactorily.

3.2 Effect of limestone aggregate

3.2.1 Effects of kind of aggregate

The results of concrete strength tests (Fig. 9) and three-factor analysis of variance indicate the highly significant

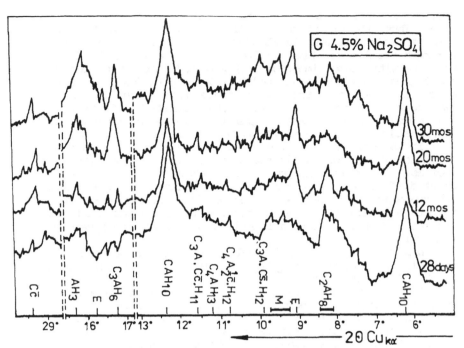

Fig. 7. XRD analysis of plain HAC paste (6).

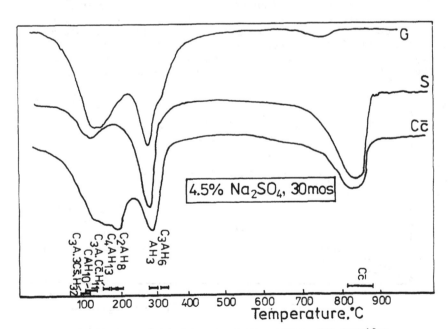

Fig. 8. DTA of HAC paste after 30 months.

effect of the kind of aggregate as well as environment and time. Thus, the significance of the problem was shown from the side of material and durability. The analysis of final strength results (in MPa) confirmed that the strength of HAC concrete with dense limestone (WZ) aggregate was statistically the highest in both environments (4.5% Na_2SO_4, H_2O) among all concretes. Besides, the HAC concrete with porous limestone (WP) aggregate achieved finally the magnitude of strength statistically equal to that of granite concrete in spite of its reasonable lower strength after 28 days of curing.

It is connected not only with considerably increase in strength of the porous limestone concrete but also with the decrease in strength of granite concrete. (GR). In spite of great differences in properties of both limestone aggregates the concretes with them showed very similar increasing strength in both environments. Whereas granite aggregate was inert and did not affect and limit deleterious processes of conversion and sulphate attack.

However, the best statistical assessment can be presented for relative strength ($r=R_t/R_{28}$) Fig.10) excluding the effect of initial strength.

It was stated that there are no statistical differences in relative strength of concretes with porous and dense limestone aggregate and that the relative strength (measure of durability) of granite concrete is significantly lower than those of both limestone concretes. Considerable, statistically equal, positive magnitude of assessments of effect of interaction between each of limestone aggregate and sulphate environment. Was noted statistical results suggest high durability of both concretes in Na_2SO_4 solution. Whereas on the ground of statistically negative estimation of effect of interaction between granite and both environments the lower durability of HAC concrete with granite aggregate must be

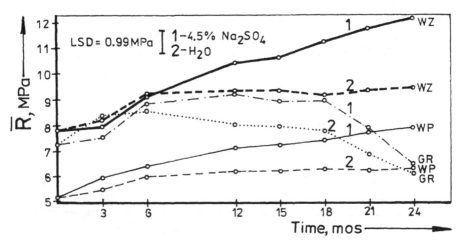

Fig. 9. Flexural strength of HAC concrete.

250

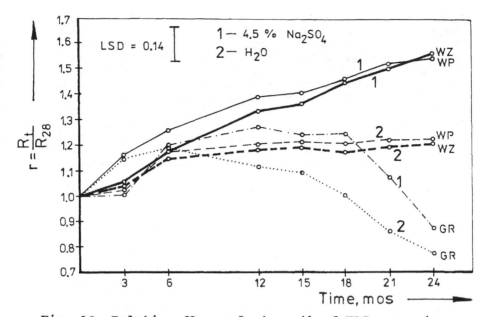

Fig. 10. Relative flexural strength of HAC concretes.

ascertained. Additionally the effect of interaction between granite and the time factor at the level of 24 months (the last one) is particularly negative contrary to the effect of both limestones.

3.2.2 Effect of curing period

Besides, analogous investigation of durability of HAC concrete with porous (WP) and dense (WZ) limestone aggregates cured for 270 days was carried out (Fig.11). The results of slightly

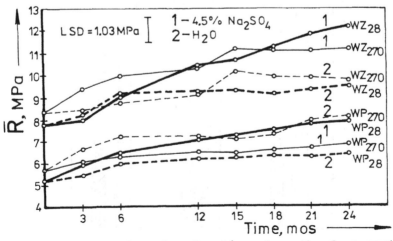

Fig. 11. Effect of curing duration strength of concretes.

higher or the same strength of concretes cured for 270 days
suggest equally high durability like that of HAC concrete cured
for 28 days. Also the influence of two difference physical
forms of limestone aggregates (dense and porous) on sulphate
resistance appeared to be very similar for concretes cured for
270 days and those cured for 28 days. The effect can be
particularly noted for relative strength of concretes. Thus,
it can be stated that in the case of durability of HAC concrete
rather composition of aggregates (limestone -CaCO₃) is of
higher significance than their physical forms. In other words
it can be stated that the age of HAC concrete with limestone
aggregate affects strength and sulphate resistance positively.

3.3 Test of sulphate resistance of HAC pastes at elevated temperatures

Strength test of sulphate resistance of HAC pastes at $t_2=28^{\circ}C$
and $t_3=38^{\circ}C$ was carried out similarly as that at normal
temperature. General resistance of the HAC pastes at t_2 and t_3
to the 4.5% solution of Na_2SO_4 is much lower (Fig.12) than that
at $t_1=18^{\circ}C$. In this case not only HAC paste with quartz powder
but also plain HAC paste and the paste with limestone were very
detrimentally attacked by sulphate ions and then also slightly
by free CO_2 in the presence of alkalis. Also decrease of
strength of all HAC was observed in water at t_2 and t_3 due to
rapid conversion. Interaction of conversion process at
elevated temperature and sulphate attack was the main reason of
deterioration of HAC pastes. However, it was not explained why
the interaction of temperature $t_2=28^{\circ}C$ was more detrimental
than that of $t_3=38^{\circ}C$.

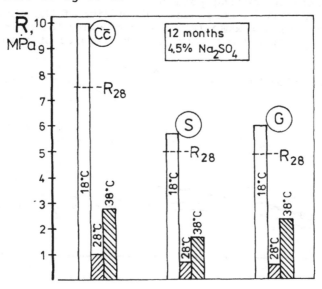

Fig. 12. Effect of temperature on strength of HAC paste.

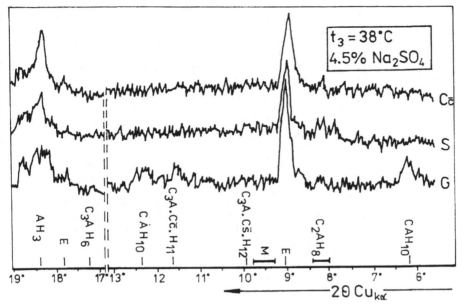

Fig. 13. XRD analysis of HAC pastes at $T_3=38^\circ C$.

4 Discussion

Comparing the results of durability of HAC concretes with those of HAC pastes but with the same kind of fillers, respectively, the great similarity can be noted in their behaviour. The limestone fillers (porous and dense aggregates, fine powder) are the reasons for significant improvement of durability of HAC composites in low and normal temperatures.

The main reason of low durability of HAC composites with inert fillers (quartz powder, granite aggregate) in sulphate solution and water is connected with progressive increase in porosity and permeability. These deleterious changes in physical properties are direct effects of conversion of hexagonal calcium aluminate hydrates into C_3AH_6 and AH_3. Whereas application of limestone fillers eliminates substantially the harmful process of the conversion. As effect of chemical interaction of HAC paste and $CACO_3$ calcium monocarboaluminate hydrate, $C_3A.C\bar{c}.H_{11}$, is formed. This complex hydrate becomes the predominant phase in the composition of HAC composites with limestone fillers due to its permanent formation. It must be noted that formation of monocarboaluminate increases volume of solid phases in HAC paste. It was stated theoretically and then confirmed in the test of porosity of HAC paste with $CaCO_3$ powder by A. Bachiorrini and L. cussino (8). This effect of formation of $C_3A.Cc.H_{11}$ clearly explains the increased durability and strength properties of HAC composite with limestone filler. It can be noted that $C_3A.C\bar{c}.H_{11}$ can be formed in hardened paste for a long time.

An attention should be also paid to the chemical side of protection from sulphate attack with the formation of $C_3A.C\bar{c}.H_{11}$ phase. The formation of monocarboaluminate eliminates substantially calcium aluminate hydrates like C_4AH_{13}, C_3AH_6 which are very susceptible to sulphate attack. But also the reactivity of $C_3A.C\bar{c}.H_{11}$ with sulphate ions is considerably limited due to lower solubility of monocarboaluminate in pore solution of HAC paste than that of ettringite (Lea, 1971)

Whereas the high susceptability of converted HAC composite to sulphate attack is the result of high porosity and content of more reactive phase (C_3AH_6).

The interface of HAC paste-inert aggregate is a region of the highest w/c ratio, highest porosity, lowest strength and additionally often cracked in HAC concrete. Thus, it is most susceptible to conversion as well as to chemical attack. Whereas the interface of HAC paste-limestone aggregate is the region which is relatively dense and strongly interconnected with cement matrix and aggregate. Calcium monocarboaluminate hydrate is the main phase in the interface region (Bachiorrini et al 1986 and Piasta, 1989).

5 Conclusions

The application of limestone aggregate and microfiller to HAC composites improves very significantly their durability in sulphate environments and water in the range of normal cold temperatures.

Due to formation of calcium monocarboaluminate hydrate, $C_3A.C\bar{c}.H_{11}$, at the presence of $CaCO_3$ the application of limestone fillers is of great importance:
- it limits conversion process and magnitude of porosity and permeability what in physical way increases the durability in sulphate solution and generally strength properties,
- it increases sulphate resistance of HAC paste by the elimination of susceptible phases (C_4AH_{13}, C_3AH_6) with formation of $C_3A.C\bar{c}.H_{11}$ of lower reactivity with sulphates.

HAC concretes with limestone aggregates contrary to inert aggregate form the perfect morphology of cement paste-aggregate interface. The porosity and permeability of the interface are being lowered progressively, this is what improves structural properties and durability.

Limestone fillers (fine and coarse 0-20mm) are mainly recommended to apply to HAC concretes of all building purposes including the objects under sulphate attack, however, not in hot climate.

Behaviour of HAC with limestone powder is not totally explained at elevated temperatures. Further test and studies of sulphate resistance of HAC with limestone fillers should be continued at elevate temperatures.

6 REFERENCES

Lea F.M. Chemical Publishing Company, Inc. N.Y. 1971.
Kurdowski et al., VIII Internat. Congr. Chem.Vol.1, pp. 293-318, Rio de Janeiro, 1986.
George C.M. VIII Internat. Congr. Chem. Cem. Vol.1, pp. V-1/25, Paris, 1980.
Midgley, H.G., Mag of Concr. Res. Vol. 27, No. 91. June 1978.
Midgley, H.G., VIII Intern. Cong. Chem. Cem., Paris 1980, pp. V85-V87.
Fentiman C.H., Cement & Concr. Res. Vol. 15, pp. 622-630, 1985.
Cussino L. Et al., VIII Internat. Congr. Chem. Cem. Vol. VIII, pp. V62/67.
Bachiorrini A., et al. VIII Internat. Congr. Chem. Cem. Vol. IV, pp. 383-388, Rio De Janerio, 1986.
Piasta W.G. Resistance of high alumina cement concrete with limestone aggregates to sulphate attack. doctor thesis, University of Poznan, Poland 1989.

6 REFERENCES

PART SIX

BLENDED SYSTEMS

19 HYDRATION OF CALCIUM ALUMINATES IN PRESENCE OF GRANULATED BLASTFURNACE SLAG

A.J. MAJUMDAR, R.N. EDMONDS, B. SINGH
Building Research Establishment, Watford, UK

Abstract
The hydration chemistry of high-alumina cements (HAC) is intimately connected with the chemistry of formation and stability of calcium aluminate hydrates CAH_{10}, C_2AH_8 and C_3AH_6. At higher than ambient temperatures in moist environments C_3AH_6 is the stable phase and chemical processes producing this phase from other calcium aluminate hydrates lead to a reduction in the strength of HAC. The addition of a sufficient quantity of ground granulated blastfurnace slag (ggbs) to HAC has the beneficial effect of counteracting the strength loss under hot and humid conditions.
   A study of the hydration chemistry of CA and two types of HAC, 'Ciment Fondu' and 'Secar 71' in the presence of 50% (by mass) of ggbs has revealed that the beneficial effect of slag addition may be due to the formation of stratlingite (C_2ASH_8). By removing Ca from the solution ggbs also hinders the formation of C_2AH_8 and C_3AH_6. When water is first added to CA or cements in which CA is a major phase, the first hydrate to appear at 20°C or below is an amorphous calcium aluminate phase with C/A ratio close to 1. Amorphous phases persist in HAC hydration for an extended period of time. The heat evolution characteristics of calcium aluminates are modified by the presence of slag and are dependent on the surface area of the cement.
Keywords: High-alumina cement, Calcium aluminate hydrates, Conversion, Granulated blastfurnace slag, Strength, Stratlingite.

1 Introduction

Commercial high-alumina cements (HAC) such as 'Ciment Fondu' manufactured by Lafarge Special Cements show a reduction in compressive strength when kept in a hot and humid atmosphere over a prolonged period of time (Midgley and Midgley 1975). For refractory aluminous cements sold under the trademark 'Secar' (also from Lafarge Special Cements) a possible reduction in strength at relatively low temperatures is of no great significance as the high-temperature strength of these materials depends on the formation of ceramic bonds among their constituent substances. We have claimed recently (Eur. pat. 1989) that the strength loss suffered by HAC under hot and humid conditions can be counteracted by the addition of sufficient quantities of ground granulated blast furnace slag (ggbs) to the cement. If these results are confirmed by others and the mechanical properties of the blended HAC materials remain satisfactory over a long period of time, a case can perhaps be made for an assessment of these materials for structural use. In addition to high early strengths these cements are likely to possess excellent chemical resistance.

It is commonly believed that the reduction in HAC strength under hot and humid conditions is due to chemical processes (called 'conversion') whereby the calcium aluminate hydrates formed initially make way for the more stable hydrate, C_3AH_6[1]. The effect of slag addition on the properties of HAC and the hydration chemistry of calcium aluminates, particularly the conversion reactions, has been under investigation at the Building Research Establishment over the past few years. We have already reported on the hydration chemistry of CA and $C_{12}A_7$ in the presence of ggbs (Edmonds and Majumdar 1989b, Majumdar, Singh and Edmonds 1989) and on Secar 71 + ggbs mixtures (Majumdar, Edmonds and Singh 1990a) and our results on Ciment Fondu+ggbs mixtures are awaiting publication (Majumdar, Singh and Edmonds 1990b). Although our study on these materials is still continuing, for instance the hydration chemistry of CA_2 with or without ggbs, we present here a summary of our findings to date. We also describe the effect of particle size on the hydration of calcium aluminates.

2 Experimental

2.1 Materials
The cements, Ciment Fondu and Secar 71 were obtained from their manufacturer. The ggbs used was 'Cemsave', a product of Frodingham Cement Company. On analysis the slag gave 41% CaO, 33% SiO_2, 13% Al_2O_3 and small amounts of MgO and SO_3. For the present work ggbs was ground to a specific surface area of 4400 cm^2/g. The individual calcium aluminate hydrates CAH_{10}, C_2AH_8, and C_3AH_6 were prepared in the laboratory (Edmonds and Majumdar 1988). The mixtures containing ggbs had 50% by mass of the slag in each case.

2.2 Techniques
The compressive strength of the two cements, Fondu and Secar, and of mixtures of the individual cements with ggbs was measured by using 10 mm cubes prepared with a water to solid ratio (w/s) of 0.40. The cubes were cured in moist air for 24 hours and subsequently transferred to water tanks kept at 20°, 40° and 50°C.

The hydration chemistry of CA, Ciment Fondu and Secar 71 and individual mixtures of these materials with ggbs was studied mainly by conduction calorimetry and quantiative X-ray diffraction. Samples (1–20g) of the anhydrous materials were hydrated under nitrogen using decarbonated distilled water, mostly with a w/s of 0.40. Conduction calorimetry measurements were made using a Wexham Developments JAF

---

1   Cement chemistry notation, $C=CaO$, $A=Al_2O_3$, $S=SiO_2$, $H=H_2O$.

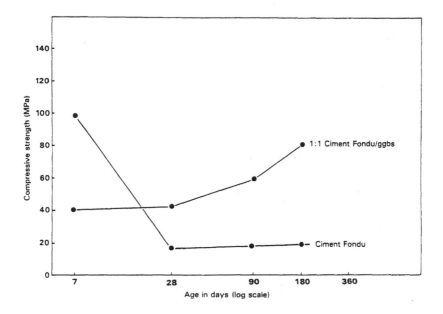

Figure 1. Compressive strength of 10mm cubes made from 1:1 Ciment Fondu
+ ggbs mixture and neat Ciment Fondu with w/s ratio of 0.40.   The
samples were stored under water at 40°C.

conduction calorimeter.   Details of our QXDA measurements are given
elsewhere (Edmonds and Majumdar 1988, 1989a).   Differential Scanning
Calorimetry measurements were made on some hydrated samples using a
Dupont 900 instrument.

3 Strength Properties
The increase in compressive strength of a 1:1 Ciment Fondu/ggbs paste
kept under water at 40°C over a period of six months is shown in
Figure 1.   The compressive strength of the neat cement paste is also
shown.   It is clearly seen that the strength of Ciment Fondu is
reduced very sharply during the first 28 days and the six month value
of ~20MPa is only about a fifth of the 7-day value.   The strength of
the 1:1 Fondu+ggbs mixture on the other hand has increased
progressively from ~40MPa at 7 days to double this value at 180 days.
In as much as the reduction of strength in the neat cement is
ascribable to the 'conversion' processes it is clear that the presence
of slag must modify these processes and apparently makes them
harmless.
    The results from Secar 71 and the 1:1 Secar+ggbs mixture hydrated
at different temperatures as shown in Figure 2 support the above
argument.   After one day the strength of the mixture is nearly half
that of the neat cement indicating little interaction between the
constituents of the mixture at this stage.   But with the passage of

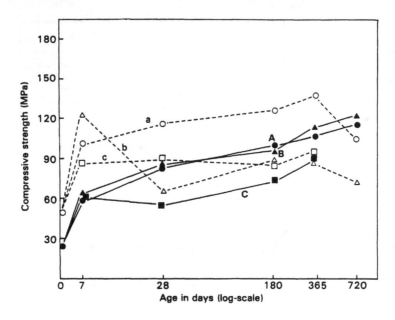

Figure 2. Compressive strength of paste made from Secar 71 cement (---)
and 1:1 Secar 71/ggbs mixture (---)with w/s ratio of 0.40 and stored
under water at (a,A)20°C, (b,B)40°C and (c,C)50°C.

time the mixture gains considerably in strength at all the
temperatures studied. At 20°C the one day value was 24 MPa and the
two year value 116 MPa, the latter being slightly higher than the
strength of neat cement. These results indicate interactions between
the cement and slag. At higher than ambient temperatures the
development of strength in the 1:1 mixture is much faster after six
months and the two year value at 40°C is several times greater than
the corresponding start strengths. Contrasted to the
neat cement paste, the mixture of Secar and ggbs did not show any
reduction in strength with time when cured at 40°C for a period
extending up to two years.

4 Hydration Chemistry
Ciment Fondu is a complex multiphase material and several of its
constituents including the calcium aluminates CA and $C_{12}A_7$ are
hydraulic. Constitutionally, Secar 71 is a simpler material
comprising only two major phases CA and CA_2. Our study on the
hydration of CA_2 is not complete yet and hence we are not able at the
present time to relate the hydration chemistry of the simpler Secar
71 cement (with or without ggbs) to that of its individual components
CA and CA_2 (with or without ggbs). However, for both cements the
influence of CA hydration is crucially important in early life. Some
of our observations on the 1:1 CA+ggbs mixture are compared in this

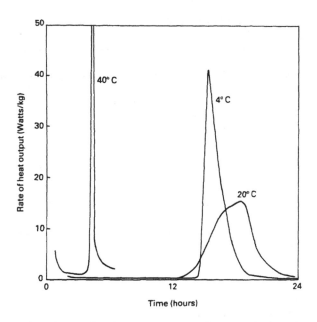

Figure 3. Conduction calorimeter output from CA hydration at various temperatures. The maximum output at 40°C was 820 Watts/kg.

paper with our findings on neat HAC's and the 1:1 HAC+ggbs mixtures.

4.1 CA and 50% CA 50% ggbs (by mass)

The hydration chemistry of CA and the 1:1 CA+ggbs mixture at different temperatures has been discussed in detail in our previous publications (Edmonds and Majumdar 1988, 1989b). Only the important observations are given here using the results obtained with pastes prepared with w/s ratio 0.40.

The conduction calorimeter output from CA hydration at 4°, 20° and 40°C is illustrated in Figure 3. It is clear that the rate of hydration is slower at 20°C than at 4°C. At 4°C, crystalline CAH_{10} first appears at the beginning of the peak on the calorimeter trace whereas at 20°C CAH_{10} forms more slowly appearing only towards the end of the peak in heat ouput. Equally important is the observation that at both 4° and 20°C the first reaction product when water is added to CA is an amorphous calcium aluminate hydrate. Even after 14 days of hydration at 20°C, as much as a quarter of the solid mass may be due to this amorphous hydrate; less at 4°C. A broad DSC peak at about 65°C appearing approximately 12 hours after mixing at 20°C may be due to this amorphous phase. As hydration progresses the signal becomes narrower and stronger and by 14 days is resolved into 2 peaks at 75 and 110°C, the latter being due to crystalline CAH_{10}. A small amount of C_2AH_8 is detected at 20°C after 48 hours but none at 4°C up to 7 days. The final phase composition of hydrated CA is specifically dependent on the amount of water available for hydration. At 20°C and

Table 1. Effect of slag addition (50% by mass) on heat
evolution at 20°C.

Material	Surface Area (cm^2/g)	Heat Evolution		Integrated Heat (kJ/kg)
		Start (hours)	Maximum (hours)	
CA	3900	3.0	8.6	250
Secar	3100	2.3	4.3	185

w/s ratio of 0.40 the proportion of unreacted CA when all the water
has been consumed is ~45% by mass.

In the presence of 50% (by mass) ggbs, the initial product of CA
hydration at 20°C is also the amorphous calcium aluminate hydrate.
CAH_{10} crystals begin to appear at the calorimeter peak but no evidence
is found for the formation of C_2AH_8 over a period of 28 days. This
may be due to the reaction of calcium ions with the slag. After 3
months a small amount of stratlingite (C_2ASH_8) is detected. After six
months, conversion of CAH_{10} has produced C_3AH_6 and there is some
indication that another hydrogarnet phase containing SiO_2 is also
formed.

The effect of slag addition on the heat of hydration of CA is
summarised in Table 1. At 20°C the heat output occurs earlier when
slag is present and the integrated heat is somewhat smaller than that
of CA alone.

The details of the hydration chemistry of CA and 1:1 CA+ggbs
mixture at 40°C are shown in Figures 4 and 5. The illustrations are
self-explanatory. The sharp peak in heat output in both Figures is due
to the formation of C_2AH_8. Conversion of C_2AH_8 to C_3AH_6 begins about
24 hours after mixing in the case of CA attended by an enthalpy change
of about 120kJ/kg. This reaction releases water which allows more CA
to react. The complete conversion of all hydrates to C_3AH_6 and AH_3
should give a mixture of 9% CA, 50% C_3AH_6 and 41% AH_3 in the case of
hydration of CA alone at a w/s ratio of 0.40. This composition is
reached in 28 days.

When slag is present, hydration of the mixture at 40°C also
produces C_2AH_8 but the amount of CA consumed suggests that some of the
available water is used to hydrate the slag. The slag has a clear
effect on the subsequent reactions; it reacts with calcium ions and
thus inhibits the conversion of C_2AH_8 so that after 28 days C_3AH_6
makes up only 5% of the solid mass and most of the C_2AH_8 formed during

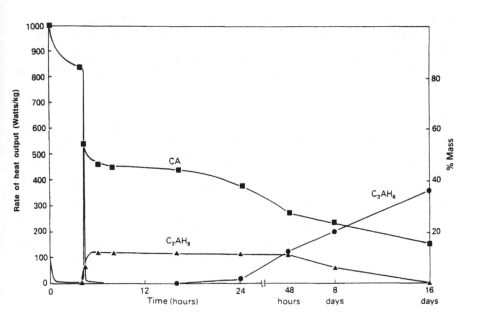

Figure 4. Calorimeter output and composition of CA hydrated at 40°C.

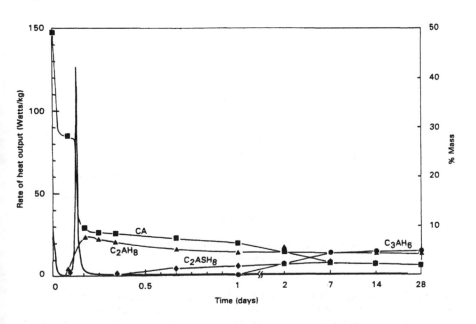

Figure 5. Calorimeter output and composition of CA+ggbs hydrated at 40°C.

initial hydration is still present. The reaction with slag produces C_2ASH_8. Only a small amount of this phase is present after 28 days but by 3 months much more is formed.

As at 20°C, hydration of CA in the presence of slag at 40°C produces less heat than with CA alone, 210 kJ/kg compared to 280 kJ/kg for CA.

4.2 Secar 71 and 50% Secar 50% ggbs (by mass)

The calorimeter output from the hydration of Secar at 4°, 20° and 40°C are shown in Figure 6. Compared to CA the hydration at 4°C begins much earlier with the cement. At 4°C the only crystalline product of hydration is CAH_{10}, which by 90 days may be as much as 50% of the final product at w/s ratio of 0.40. At 20°C a small amount of C_2AH_8 and C_3AH_6 is measured after 28 days; no corresponding conversion of hexagonal hydrates to C_3AH_6 is detected in pure CA. At 40°C the initial products of hydration are C_2AH_8 and alumina gel. C_3AH_6 forms soon afterwards and is the main crystalline hydration product after 2 days and continues to grow at the expense of C_2AH_8, reaching more than 20% after 3 months. Only a small amount of C_2AH_8 is detected at that time.

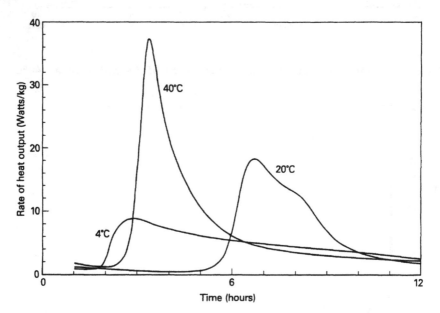

Figure 6. Calorimeter output from the hydration of Secar at various temperatures.

The conduction calorimeter output from the hydration of 50:50 Secar/ggbs mixture at 20°C with w/s ratio of 0.40 is similar to that of neat cement. The main calorimeter peak at about 4 hours (Table 1) represents the formation of CAH_{10}. After 3 months the hydrated material contains 39% CAH_{10} and small residues of CA and CA_2. This composition is consistent with the use of most of the available water in hydrating the calcium aluminates. Unlike in pure Secar C_2AH_8 and C_3AH_6 are not detected in the hydration products from the Secar/ggbs mixture up to 90 days.

At 40°C only a small amount of C_2AH_8 is detected during the calorimeter peak. As with pure Secar much of the hydrated material remains in the amorphous form. The conversion of C_2AH_8 to C_3AH_6 is much slower in the Secar/slag mixture; it begins only after about 7 days and after 3 months C_3AH_6 makes up less than 10% of the material. At the same time, reaction with the slag produces C_2ASH_8. As these reactions proceed C_2AH_8 disappears. It does so more quickly than in pure Secar and is all gone by 28 days. At this stage the total amount of crystalline material makes up only a small proportion of the total mass and incorporates only a fraction of the water originally added. The remainder of the water may be combined in amorphous hydrates formed by activation of the slag. This amorphous material containing silica may give rise to crystalline C_2ASH_8 at a later stage. As mentioned previously reactions involving slag inhibit the formation of C_3AH_6 and may thus lead to a different microstructure from that normally formed in Secar after conversion. A DSC peak at around 150°C obtained with a 90 day old sample may be due to C_2ASH_8.

The effect of slag addition on the heat output of Secar is illustrated by the figures in Table 1. At 20°C, the hydration reaction appears to start sooner when the slag is present and the total heat produced is much reduced (see Table 2). At 40°C, the heat produced by the Secar/ggbs mixture amounts to 205 kJ/kg of dry solid.

4.3 Ciment Fondu and 50% Ciment Fondu 50% ggbs (by mass)

A detailed account of our work on the hydration chemistry of Ciment Fondu and Fondu + ggbs mixture will appear in print shortly (Majumdar, Singh and Edmonds 1990b) and only a few points are discussed here. In Figure 7 the conduction calorimeter output from the hydration of the 1:1 Fondu + ggbs mixture at 4°, 20° and 40°C are shown. The w/s ratio used in these pastes is 0.30. A similar pattern is observed with Fondu alone, although the calorimeter signals from neat cement are sharper and narrower than those of the slag mixture. The heat output from the cement + ggbs mixture during the first 24 hours of hydration are 170, 210 and 240 kJ/kg of dry solid at 4°, 20° and 40°C respectively. The corresponding figures for the neat cement are 200, 250 and 320 kJ/kg. The slag mixture thus generates slightly less heat than the pure HAC, but in both systems the extent of hydration is determined mainly by the availability of water rather than by the amount of the cement clinker.

The hydration of Ciment Fondu at different temperatures gave results similar to those found by others (Midgley and Midgley 1975). Hydration of Fondu/ggbs mixture produces C_2ASH_8 at both 20° and 40°C. At 20°C, CAH_{10} forms initially but is later replaced by C_2ASH_8. At 40°C, the normal hydration and conversion reactions give C_3AH_6 the proportion of which reaches a maximum at 28 days. Considerable amounts of C_2ASH_8 are produced afterwards and after one year this

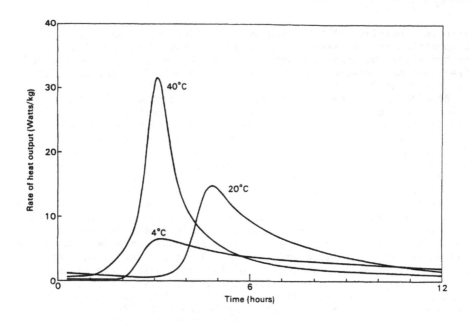

Figure 7. Conduction calorimeter outputs from the hydration of 1:1
Ciment Fondu/ggbs mixture at 4°, 20° and 40°C.

phase is the predominant crystalline hydrate at both 20° and 40°C.
Compared to pure calcium aluminates and Secar, the addition of ggbs
produces more C_2ASH_8 in Ciment Fondu at equivalent stages of
hydration. The formation of C_2ASH_8 is dependent on the activation of
the glassy slag to release silica, so it seems that slag activation is
enhanced by some of the minor phases present in Ciment Fondu, possibly
the alkali metal oxides.

4.4 Effect of particle size
The effect of particle size on the heat evolution characteristics of
calcium aluminates and cements derived from them at 20°C is
illustrated in Table 2 by reference to CA and Secar. It is clear that
in both cases hydration reactions begin earlier and the maximum in
heat output is achieved faster with coarser powders. We are unable to
offer a satisfactory explanation for this interesting result at
present but we think that the formation of calcium aluminate hydrate
gel at the first stage of hydration and subsequent crystallisation of
CAH_{10} may be associated with this observation. The total heat
produced by the hydration does not seem to depend strongly on the
fineness of the powder as seen in the results given in Table 2.
The calorimeter traces at 40°C resulting from the hydration of
Secar cement powders having the particle sizes given in Table 2 are
shown in Figure 8. It is interesting to note that at this temperature
hydration reactions seem to commence earlier with fine powders as is

Table 2. Effect of particle size on heat evolution at 20°C.

Material	Surface Area (cm^2/g)	Heat Evolution		Integrated Heat (kJ/kg)	
		Start (hours)	Maximum (hours)		
CA	3900	9.0	13.5	275 ⎫	up to
	4750	12.0	16.0	280 ⎬	40
	5600	11.0	18.0	285 ⎭	hours
Secar	3100	5.0	6.7	230 ⎫	up to
	6500	7.0	8.8	265 ⎬	24
	7900	15.0	17.9	230 ⎭	hours

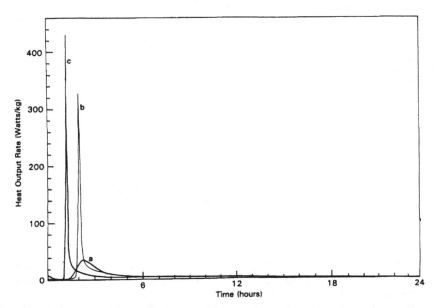

Figure 8. Heat output of Secar cement having different surface areas (a) 3100 cm^2/g (b) 6500 cm^2/g (c) 7900 cm^2/g and hydrated at 40°C.

to be expected. It is worth pointing out that at 40°C, C_2AH_8 and not CAH_{10} is the first crystalline calcium aluminate hydrate formed.

Bushnell-Watson and Sharp (1986) and others have observed that the setting times of Secar HACs coincide with the onset of heat evolution and these times are temperature dependent. We have shown here that the onset of heat evolution also depends on the particle size. It is, therefore, imperative to study the effect of particle size on the setting behaviour of HACs at different temperatures.

Finally it should be pointed out that the encouraging results that have been described in this paper arising from the interaction of HAC and ggbs were obtained when the cement pastes were cured under water. There is insufficient information on the properties of these cements in drier environments. Further research in this area and on the microstructure of these blended HACs is planned for the future.

5 Conclusions

At ambient temperatures the first product of hydration of calcium aluminates and probably of cements in which they are principal constituents is an amorphous calcium aluminate hydrate phase. This amorphous phase may persist for long periods of time and may have a major influence on the final constituion of hydrated cements.

The rate of hydration of some calcium aluminates and high-alumina cements derived from them is dependent on the particle size of the powder. At 20°C coarser powders hydrate faster.

The addition of a sufficient quantity of granulated blast furnace slag to high alumina cement reverses the reduction of strength with time observed for the neat cement when cured under water at 40°C. One of the reasons for this phenomenon is believed to be the formation of stratlingite (C_2ASH_8) in preference to C_3AH_6.

The silica from the slag also reacts with calcium in solution inhibiting the formation of C_2AH_8 initially and C_3AH_6 after conversion.

In the presence of granulated slag the amount of C_2ASH_8 formed is much greater in Ciment Fondu than in Secar. The amount of C_2ASH_8 increases with an increase in the temperature and the amount of water available for hydration.

Acknowledgements
Dr R N Edmonds is now on the staff of Eton College. This paper is published by kind permission of Director of Building Research Establishment.

References
Bushnell-Watson, S.M. and Sharp, J.H. (1986) **The effect of temperature upon the setting behaviour of refractory calcium aluminate cements.** Cem. Concr. Res., 16, 875-884.
Edmonds, R.N. and Majumdar, A.J. (1988) **The hydration of monocalcium aluminate at different temperatures.** Cem. Concr. Res., 18, 311-320.
Edmonds, R.N. and Majumdar, A.J. (1989a) **The hydration of Secar 71 aluminous cement at differnt temperatures.** ibid., 19, 289-294.
Edmonds, R.N. and Majumdar, A.J. (1989b) **The hydration of mixtures of**

monocalcium aluminate and blastfurnace slag. ibid., 19, 779–782.

European Patent Application 0312323 (1989).

Majumdar, A.J., Singh, B. and Edmonds R.N. (1989) **Hydration of mixtures of $C_{12}A_7$ and granulated blastfurnace slag.** ibid., 19, 848–856.

Majumdar, A.J., Edmonds, R.N. and Singh, B. (1990a) **Hydration of Secar 71 aluminous cement in presence of granulated blastfurnace slag.** ibid., 20, 7–14.

Majumdar, A.J., Singh, B. and Edmonds, R.N. (1990b) **Hydration of mixture of 'Ciment Fondu' aluminous cement and granulated blastfurnace slag** (to appeaer in Cem. Concr. Res.)

Midgley, H.G. and Midgley, A. (1975) **The conversion of high alumina cement.** Mag. Concr. Res. 27, 59–75.

20 THE EFFECT OF CURING CONDITIONS ON THE HYDRATION AND STRENGTH DEVELOPMENT IN FONDU : SLAG

C.H. FENTIMAN, S. RASHID
Lafarge Special Cements, West Thurrock, Grays, Essex, UK
J.P. BAYOUX, A. BONIN, M. TESTUD
Lafarge Coppée Recherche, Viviers, France

Abstract
The hydration and strength development of 50:50 mixes containing calcium aluminate cement (Ciment Fondu) and ground granulated blast Furnace slag (ggbfs), has been studied. A range of curing temperatures from 5 to 60°C were looked at with neat paste samples being used for XRD and DTA analysis whereas concrete samples were used for the measurement of strength. These studies showed that the early hydration of Ciment Fondu resulted in the formation of hydrates that would normally be expected from the pure cement. There is evidence that all the hydrates other than C_3AH_6 subsequently react, incorporating silica, to form stratlinghite, C_2ASH_8. On curing at higher temperatures C_3AH_6 was found to be the principal hydrate and strength development was low. The probable reasons for these low strengths are discussed.
Keywords: Calcium Aluminate Cement, Ciment Fondu, Blast Furnace Slag, Stratlinghite, X-Ray Diffraction, Differential Thermal Analysis, Hydration.

1   Introduction

It has been demonstrated that blends of calcium aluminate cement with ground granulated blast furnace slag exhibit modified hydration chemistry as compared with calcium aluminate cement alone. The principal effect is the formation of the phase known as stratlinghlite (C_2ASH_8) which compensates for the effects of the "conversion" reaction associated with calcium aluminate cements.

$$3 \ CAH_{10} \longrightarrow C_3AH_6 + 2AH_3 + 18H_2O$$

The mechanism for this alteration of the system was not clear and early thoughts centered around the formation of C_2ASH_8 and changes to the microstructure of the hydrated products (Edmonds and Majumdar 1989) C_2ASH_8 is widely considered to be a stable phase, but early work (Fentiman 1985) on the use of calcareous fillers showed that the

formation of stable hydrates alone does not prevent the
conversion/strength loss, if metastable hydrates are
formed at the same time and are able to transform into
C_3AH_6. It was therefore decided to investigate the
formation of hydrates at different temperatures and
different water/powder ratios to see if metastable
hydrates such as CAH_{10} form. If this is so what are the
subsequent reactions and what are the effects of these on
strength development ?

2   Experimental procedure

The calcium aluminate cement used in this research was
Ciment Fondu supplied by Lafarge and the Ground Granulated
Blast Furnace Slag "Cemsave", was supplied by Civil and
Marine Limited, UK. The chemistry of the components is
given in table 1, below.

Table 1,

Compound	Calcium Aluminate Cement	GGBS
CaO	38%	41.5
SiO_2	4%	34.55
Al_2O_3	39%	11.00
Fe_2O_3 + FeO	16.5%	0.00
MgO	1%	9.10
TiO_2	2%	0.33
Na_2O	0.2%	0.27
K_2O		0.36

Mineralogically the calcium aluminate cement comprises
principally of monocalcium aluminate CA and Ferrite phase
with minor amounts of $C_{12}A_7$. The GGBS is principally a
glass but also contains some crystalline materials such
as mervenite.
   All the present work was carried out on a 50:50 blend
of the GGBS (fineness approx 4600 cm^2/grm) and calcium
aluminate cement (fineness approx 3000 cm^2/grm), hereafter
known as Fondu : Slag. The experimental programme was
divided into two parts;

i)   Mineralogical analysis by XRD and DTA of neat paste
      sample which were all prepared in the same way and
      comprised by the Fondu : Slag blend, mixed at a
      water/ powder ratio was 0.5.

ii)   Strength development of concrete mixes consisted of
        1 part Fondu : Slag, 2 parts zone 2 sand, 2 parts
        10mm rounded gravel and mixed at a water/powder
        ratio of 0.5.

The principal aim was to evaluate the effect of curing
temperature on the mineralogical and strength developments
of Fondu : Slag.
    A range of temperatures from 5°C to 80°C were
investigated in 5°C increments.   In most cases the testing
was continued up to 28 days.   The samples were stored from
the time of mixing in water baths set up at the
appropriate temperature to within ± 2°C.
    The neat pastes were placed in the water baths prior to
setting and the utmost care was taken to ensure that the
surface of the pastes were not disturbed by immersion.
    The concretes were also placed in the appropriate water
baths immediately after mixing and were demoulded at 24
hours.

3   X-ray diffraction

X-ray measurements were made on a Philips PW 1840 powder
diffractometer using copper K α radiation operating at
35 Kv and 26 mA.
    In this research the hydration reaction was not stopped
prior to measurement because it was possible to analyse
the samples immediately after crushing and grinding.
    Due to the high background noise it was difficult to
estimate when the phases first appeared.   This introduced
a high probability of experimental error.   A standard
sample was checked at the beginning of each scan but
because of the poorly defined peaks correction was only
made if absolutely necessary.   In order to reduce the
experimental error the height of the characteristic peaks
were measured for each phase and are given by Table 2.

Table 2.

Phase	2θ Angle
CA	30.1°
CAH_{10} mean of	12.4°, 6.2°
C_2AH_8	8.4°
C_3AH_6 mean of	17.3°, 31.9°, 39.2°
C_2ASH_8 mean of	7.0°, 14.1°, 21.3°

For C_2ASH_8, C_3AH_6 it was possible to measure the mean of 3
peaks, however for the remainder only one or two peaks
could be used.

RELATIVE
PEAK HEIGHT

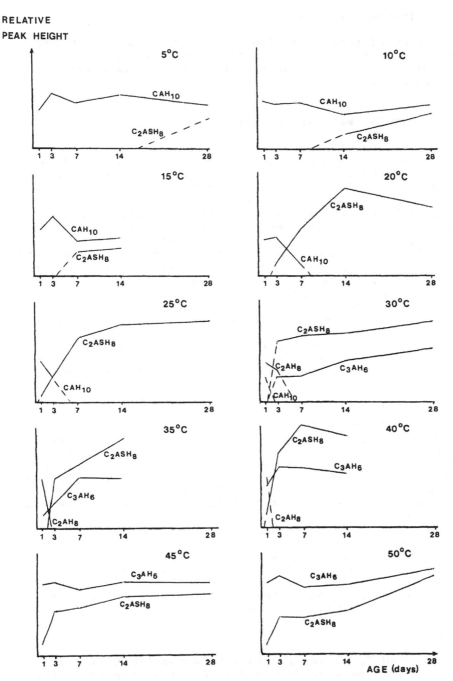

Figure 1.   Effect of curing temperature on relative XRD
              peak heights of Fondu : Slag paste.

4   Thermal analysis

Differential thermal analysis was made on a Dupont 990
instrument with a sample of about 40mg and a heating rate
of 20°C per minute.   The reference sample used was
calcined alumina.

5   Results and discussions

These experiments were the starting point for a better
understanding of the Fondu : Slag cement system.   It was
well known that the hydrates present in calcium aluminate
cement depend largely on temperature.   The aim was
therefore to establish the type of hydrates formed at
different temperatures and how these evolved with time.
The x-ray results obtained are presented in Figure 1 for
test age up to 28 days.   When interpreting the data from
this figure it should be born in mind to view it as only
qualitatively, however the trends observed appear to be
significant enough to establish the temperature ranges
which favour the formation of each phase.
    CA peaks were absent from the traces, this is not
surprising in view of the fact that the water/cement ratio
in respect of Ciment Fondu was 1.00.   Additional
experiments at a water/powder ratio of 0.3 showed the
presence of some residual CA.   However the following data
corresponds to a water/powder ratio of 0.5 throughout.

CAH_{10}

At temperatures of 5°C to 25°C CAH_{10} was the only phase
detected by XRD in 24 hours.   At 10°C or less the CAH_{10}
peaks remained approximately at a constant level up to 28
days although a decrease may well have taken place.   At
15°C and 20°C there is much more evidence of this decrease
and at 20°C CAH_{10} was not detected after the 28 days.   At
25°C and 30°C the rate of disappearance of CAH_{10} was much
more rapid.
    As expected from the hydration chemistry of Calcium
Aluminate Cements, CAH_{10} becomes less stable with an
increase in temperature and at 30°C is gone within 3 days
up to 25°C.   This reduction in CAH_{10} is undoubtedly
related to a reaction leading to the formation of C_2ASH_8.
The lack of C_3AH_6 shows that this is not a conversion
reaction of the classic type.

C_2AH_8

As expected C_2AH_8 was formed within the first 24 hours at
temperatures between 30°C and 40°C but disappears within 7
days.   As with the CAH_{10} the rate of decrease seems to be
related to temperature.

C_2ASH_8

C_2ASH_8 was observed at all temperatures and its formation appeared to show an inverse relationship to the formation of CAH_{10} and C_2AH_8. At 5°C it first appears at 28 days, whereas at temperatures above 25°C it is already present at 24 hours. The maximum "amounts" of C_2ASH_8 appear to be at temperatures of around 35°C and 40°C. At the high temperature, up to 50°C, C_2ASH_8 is present at 24 hours.

C_3AH_6

C_3AH_6 was recorded at temperatures above 30°C. At 30°C C_3AH_6 was not present at 24 hours but was present after 3 days and increased slowly. At temperatures of 35°C and above C_3AH_6 was present at 24 hours in increasing amounts but quickly reaching a stable amount. Above 45°C temperature did not seem to increase the quantity of C_3AH_6 formed.

6 General discussion

At a water/powder ratio of 0.5, up to 28 days, at temperatures from 5°C to 25°C only two phases are observed, that is CAH_{10} and C_2ASH_8. The general picture is one of CAH_{10} decreasing as C_2ASH_8 forms at a rate which appears to be dependant on temperature. The amorphous calcium aluminate hydrate phase (Payne and Sharp) is known to be closely linked with CAH_{10} and conversion. As a result of this it was felt that its role in the formation of C_2ASH_8 was important and had to be understood. Therefore DTA traces were obtained for neat paste samples cured at 20°C, Figure 2. The DTA traces showed the presence of CAH_{10} and the gel phase only up to 7 days followed by the formation of C_2ASH_8 and C_3AH_6. It can be seen from Figure 2 that as C_2ASH_8 forms the other phases show a marked decrease, this implies that the formation of C_2ASH_8 passes through an intermediary stage, such as CAH_{10}.

The x-ray diffraction data showed the presence of C_2AH_8 at temperatures between 30°C and 40°C, but the question of the involvement of this hydrate, in the formation of C_2ASH_8, is clouded by the presence of C_3AH_6.

At temperature of 45°C and above only C_3AH_6 and C_2ASH_8 are observed by XRD and are confirmed by DTA which also shows the presence of AH_3. The DTA data for samples cured at 50°C show that C_3AH_6 is present throughout whereas C_2ASH_8 forms more gradually. In a separate experiment curing at 60°C and 80°C produced only C_3AH_6 and no C_2ASH_8 was observed.

Unlike the metastable phases CAH_{10} and C_2AH_8. C_3AH_6 appears to be immune from reacting with silica to form C_2ASH_8.

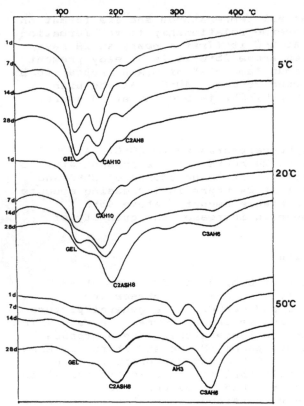

Figure 2.

DTA of Fondu : Slag pastes cured at different temperatures.

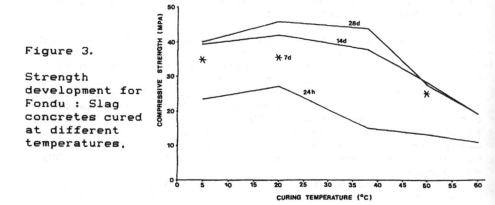

Figure 3.

Strength development for Fondu : Slag concretes cured at different temperatures.

278

The strength development across the temperature range shows some marked differences, Figure 3, with lower strength development at the temperature extremes. At 5°C quite good 24 hours strength are observed but only moderate gain thereafter. This can be attributed to higher strengths being obtained through the formation of CAH_{10} and gel phase. The lower rate of subsequent strength gain compared with the following probably reflects the slow formation of C_2ASH_8. At 20°C the formation of CAH_{10} was high in the first 24 hours, resulting in similar 24 hours strengths as found at 5°C, but much more rapid formation of C_2ASH_8 and CAH_{10} (Midgley and Rao 1978) led to slightly better subsequent strength developments. At 40°C, C_2AH_8 and C_3AH_6 are present at 24 hours, and strengths are relatively low but the subsequent increase in strength may be related to the formation of C_2ASH_8. At higher temperature still development of strength is much lower. This seems to relate to an increase in C_3AH_6 hence a more porous microstructure, a lesser formation of C_2ASH_8, and with no CAH_{10}.

It seems very likely that the 24 hours strengths are related to the first formed hydrates from the hydration of CA alone.

Curing at low temperatures leads to the formation of CAH_{10}. In order to hydrate all the CA to form CAH_{10} one requires a water/CA ratio of 1.14. This is in fact slightly greater than the water/calcium aluminate cement ratio of 1 in the Fondu : Slag system, i.e. all the water is used up in hydration.

Hydration to form C_3AH_6 requires a water/CA ratio of slightly less than 0.5, whereas in the present system there is much more water available for hydration. The excess being unused in the hydration of CA and the net result is greater porosity.

Since C_3AH_6 does not lead to the formation of C_2ASH_8 there is little potential for a subsequent increase of strength.

Various reactions can be postulated leading to the formation of C_2ASH_8. Of these some appear more likely than others but it seems clear from the results that there is more than one path leading to its formation. In this work the formation of C_2ASH_8, in each case, appears to have gone, through an intermediary stage. However there is no direct evidence that C_3AH_6 is involved in these reactions.

The following schematic equations are assumed to occur through solution.

1. $2CAH_{10} \xrightarrow{\text{(S)}} C_2ASH_8 + AH_3 + 9H_2O$

2. $CAH_{10} + CSH \longrightarrow C_2ASH_8 + 3H_2O$

3. CAH_{10} $\xrightarrow{\text{(C)(S)}}$ C_2ASH_8 + $2H_2O$

4. C_2AH_8 $\xrightarrow{\text{(S)}}$ C_2ASH_8

5. C_2AH_8 + $2CSH$ + AH_3 + $3H_2O$ $\longrightarrow$ $2C_2ASH_8$

6. C_2AH_8 + AH_3 $\xrightarrow{\text{(C)(S)}}$ $2C_2ASH_8$

Reactions of types 2 and 5 between the hexagonal hydrates and CSH have already been postulated in the literature (Rao and Viswanath 1980).

Reactions of types 1, 3, 4 and 6 may also occur but as yet still need to be proven.

7 Conclusion

This study carried out on the hydration mechanism shows the hydrates to be dependant on temperature. At low curing temperatures the hydrates formed initally are;

$$CAH_{10}, \text{ gel phase, } C_2AH_8 \text{ and } AH_3$$

These hydrates undergo a reaction with silica to form C_2ASH_8. However the temperatures at which each hydrate forms corresponds to that found in the pure Ciment Fondu system. At higher temperatures C_3AH_6 increasingly replaces the existing metastable hydrates and appears not to undergo the reaction to form C_2ASH_8. As a result of the lessening of metastable hydrates the amount of C_2ASH_8 formed decreases.

At very low temperatures e.g. 5°C the transformation from CAH_{10} to C_2ASH_8 is very slow, but at the higher temperatures the rate of this transformation increases rapidly. At intermediate temperatures (eg 30°C), where the metastable hydrate is C_2AH_8, the formation of C_2ASH_8 has already begun by 24 hours.

The strength development at different temperatures seems to relate to the mineralogical transformation taking place. At low temperatures, fairly good 24 hours strengths were obtained and subsequent strength development followed the formation of C_2ASH_8. The best overall strength development was obtained at 20°C with the formation of CAH_{10} followed by the rapid formation of C_2ASH_8.

At 40°C C_2ASH_8 was still formed in significant quantities and the strength development was almost as high as that found at 20°C. However, at temperatures above 40°C there was a rapid reduction in strength development. This is associated with C_3AH_6 forming instead of metastable hydrates and a lesser formation of C_2ASH_8. The reason for the decline in strength at high temperatures is almost certainly due to the low water requirement needed to hydrate CA to form C_3AH_6 as compared with CAH_{10}.

Therefore in order for the claimed improvement of strength development to be achieved, up to 28 days, the temperature during curing of concrete made with Fondu:Slag cement must be kept low.

Long term behaviour of this system will be further investigated to give a more comprehensive picture.

8   References

R.N. EDMONDS and A.J. MAJUMDAR." The Hydration of Mixtures of Monocalcium Aluminate and Blast Furnace Slag" Cement and Concrete Research. Vol. 19, pp 779-782 1989.

C.H. FENTIMAN."Hydration of Carbo Aluminous Cement at Different Temperatures" Cement and Concrete Reseach, Vol. 15, pp. 622-630, 1985.

H.G. MIDGLEY and P.B. RAO."Formation of Stratlingite $2CaO.SiO_2.Al_2O_3.8H_2O$ in Relation to the Hydration of High Alumina Cement" Cement and Concrete Research. Vol. 8, pp. 169-172, 1978.

D.R. PAYNE and J.H. SHARP "The Nature of the Gel Phase in Calcium Aluminate Cements" to be publised.

P.B. RAO and V.N. VISWANATHAN, 7th Intl. Congress on Chemistry of Cement (paris), 5, 51 (1980).

21 THE MICROSTRUCTURE OF BLASTFURNACE SLAG/HIGH ALUMINA CEMENT PASTES

I.G. RICHARDSON, G.W. GROVES
Department of Materials, University of Oxford, Oxford, UK

Abstract
The microstructure of two year old ground granulated blast-furnace slag (ggbfs)/HAC 3:1 blends has been studied by a combination of analytical (transmission) electron microscopy (AEM) and electron microprobe analysis (EMPA) mapping techniques. Phase identification was supplemented by differential scanning calorimetry (DSC) and X-ray diffraction (XRD). These revealed that all crystalline components of the HAC had been consumed by this age. The major hydration products were gehlenite hydrate and two distinct, mainly amorphous gels: an outer gel product, occupying originally water filled space; and an inner gel product, occupying space released by slag reaction.
Keywords: Blast-furnace Slag, High Alumina Cement, Microstructure, Microchemistry, Gehlenite Hydrate, Transmission Electron Microscopy, Electron Microprobe Analysis.

Introduction

Ground granulated blast-furnace slag (ggbfs), a by-product from the production of pig iron, is most commonly used by the construction industry as a partial replacement for Ordinary Portland Cement (OPC). However, recent research (Edmonds and Majumdar (1989)) has indicated that it may also be of value blended with high alumina cement (HAC). The hydration of the ggbfs within this system produces a gehlenite hydrate phase (C_2ASH_8) which has the effect of restricting the undesirable conversion of CAH_{10} and C_2AH_8 to the stable cubic hydrate, C_3AH_6. This paper considers the microstructure and microchemistry of a mature high-slag ggbfs/HAC blend as revealed by electron microscopy and microanalysis.

Experimental

The materials used in this work were a Frodingham cement company 'Cemsave' ggbfs and SECAR 71 high alumina cement. An approximate phase composition for the HAC and oxide compositions for both the slag and HAC are shown in table 1.

Samples of 3:1 ggbfs/HAC were prepared by mixing the required amounts of solids and de-ionised water at a water:solids ratio of 0.4. The resulting slurry was placed in 1cm diameter cylindrical

Table 1. Oxide compositions of ggbfs and HAC (% by weight), and phase composition of HAC.

	CaO	SiO$_2$	Al$_2$O$_3$	Fe$_2$O$_3$	SO$_3$	MgO	K$_2$O	Na$_2$O	TiO$_2$	Mn$_2$O$_3$
ggbfs	41.7	37.2	11.0	0.38	3.68	7.74	0.55	0.64	0.68	0.73
HAC	28.7	0.35	70.5	0.10	–	–	–	–	0.05	–

	C$_{12}$A$_7$		CA		CA$_2$		Al$_2$O$_3$	
HAC	6		44		30		20	

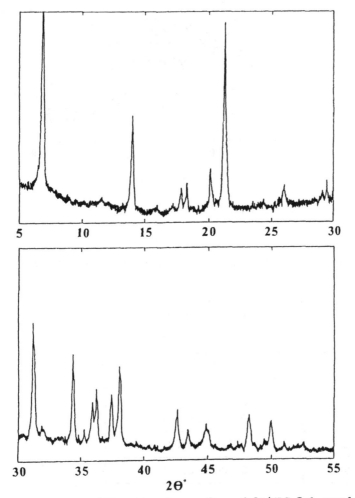

Fig.1. Powder X-ray diffraction trace for ggbfs/HAC 3:1 sample.

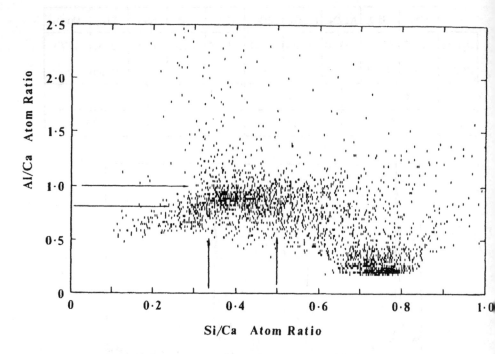

Fig.2. Plot of Al:Ca against Si:Ca atom ratios.

Table 2. Atom ratios for gehlenite hydrate determined by EMPA.
n = 524.

Si:Ca	0.409 ± 0.044
Al:Ca	0.887 ± 0.047
S:Ca	0.027 ± 0.018
Mg:Ca	0.057 ± 0.048
K:Ca	0.007 ± 0.001
Ti:Ca	0.015 ± 0.008
Fe:Ca	0.006 ± 0.001

plastic tubes which were then sealed in bags before placing in a 20°C
cure bath. After two years curing the samples were removed,
sectioned, and washed in propan-2-ol to stop the hydration.

Powder X-ray diffraction (XRD) patterns were obtained using a
Philips PW1050 goniometer employing Cu Kα radiation. Thermal
analysis was carried out using a Mettler DSC20, differential scanning
calorimeter at a heating rate of 10°C min^{-1} in an atmosphere of air.
A JEOL 2000FX electron microscope with a Tracor Northern TN5500 X-ray
microanalysis facility was used for analytical TEM. Details of the
sample preparation techniques for TEM have been described by Groves et
al. (1986). Analytical TEM is ideal for the examination of phases
which exist on a sub-micron scale, and for determining major element
relationships within these phases. As TEM does not lend itself to the
observation of areas of a sufficient size to be considered
representative of a bulk hardened cement paste, analysis on a large
scale, and the determination of minor element/major element
relationships, was performed on a CAMECA CAMEBAX electron microprobe
equipped with two WDS and one EDS detectors. This was used to
generate a 56x56μm (1μm increment in both x and y) compositional map
of a selected area of a bulk, flat and polished cement section. Ten
elements were analysed simultaneously comprising, Na, Mg, Al, Si, S,
K, Ca, Ti, Mn and Fe. The correction procedure is discussed in the
next section.

Results and Discussion

Hydration Phases
Figure 1 shows an X-ray diffraction trace for the ggbfs/HAC 3:1
sample. The peaks can all be indexed as reflections of gehlenite
hydrate, C_2ASH_8, as listed by Taylor (1964). This indicates that all
the crystalline components of the HAC fraction have been consumed by
this age. Differential scanning calorimetry (DSC) confirmed gehlenite
hydrate as the major hydration product, but also revealed the presence
of an amorphous gel phase, and possibly a small amount of crystalline
AH_3 and a poorly crystalline hydrogarnet. However, gehlenite hydrate
was the only crystalline phase present in significant quantities.

Morphology and composition of the gehlenite hydrate phase
The extensive occurrence of the gehlenite hydrate is illustrated by
the electron microprobe analysis (EMPA) mapping data. Figure 2 shows
a plot of Al:Ca against Si:Ca atom ratios. The region corresponding
to what is considered to be essentially pure gehlenite hydrate is
indicated. There are 524 points included within this area from a
total of 3136. Many other points around this area correspond to
gehlenite hydrate-rich phase mixtures. Pure gehlenite hydrate
(C_2ASH_8) has atomic ratios of Al:Ca = 1 and Si:Ca = 0.5. The mean
values for the area on figure 2 are slightly low; Al:Ca = 0.89 and
Si:Ca = 0.41. The low Al can probably be explained by considering the
correction procedures adopted. This point is outlined below but will
be considered more thoroughly in a future publication. Details of the
correction procedures and available correction programs for EMPA can be
found in Reimer (1985). The minimum useful analysis volume in the EMPA

Fig.3(a). TEM micrograph showing a plate of gehlenite hydrate and an amorphous gel.

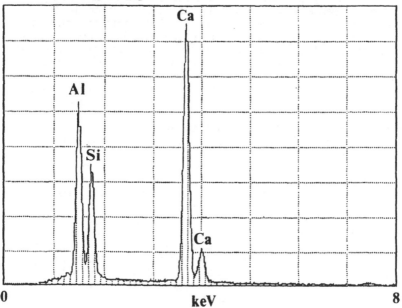

Fig.3(b). EDXA spectrum for the gehlenite hydrate in Fig.3(a).

mapping technique used is ~1μm³. In order to estimate the relative proportions of the elements present, the number of characteristic X-ray counts for each element in the specimen must be compared under identical conditions with those arriving from standards of known composition. In this case ten elements were analysed, utilizing nine different standards. As the specimen differs from each standard in its density and average atomic weight, several matrix correction procedures need to be applied to obtain more accurate compositions. However, as the correction factors depend upon the unknown concentrations themselves, they have to be applied as an iterative method which may become prohibitive when a large number of points are involved. This was the case in the present study where only partial corrections were used in order to reduce the required computing time. However, the feasibility of applying more comprehensive corrections is currently being investigated as the partial correction has probably caused the lower than expected value for Al by underestimating the ZAF corrections for Al. It must also be noted that the multi-phase mixtures of phases of widely varying densities and compositions which commonly occur in hardened cement pastes on a sub-micron scale may affect the reliability of corrected data. This is because the correction procedures assume the X-ray generation volume to be fully dense and homogeneous.

Matrix effects probably do not account for the reduced Si concentration, which may be real. This was confirmed by the analysis of an ion-beam thinned section of the sample in the TEM, where matrix corrections are not required if the specimen is thin enough, and a simple K-ratio method will suffice (Cliff and Lorimer (1975)). Figure 3(a) is a TEM micrograph showing a plate of gehlenite hydrate surrounded by an amorphous gel. Deposits of gehlenite hydrate were commonly observed to cover tens of square μm in area. Figure 3(b) is the EDXA spectrum corresponding to this feature. A number of such analyses produced a mean Al:Ca atom ratio of 0.97 ± 0.04 and a Si:Ca ratio of 0.43 ± 0.03. This implys that the gehlenite hydrate is slightly deficient in Si, but confirms that the low Al count in the mapping data was probably caused by matrix effects. Other minor constituents present in the gehlenite hydrate, as revealed by the EMPA, are given in table 2.

The form of gehlenite hydrate present in this sample was electron-beam sensitive, and damaged easily in the microscope; selected area electron diffraction patterns could not be obtained before damage occurred. This is in contrast to a preparation of synthetic gehlenite hydrate crystals which could be examined for long periods at magnifications up to ~50K without noticeable loss of crystallinity (Rodger et al., unpublished). These occurred as thin hexagonal plate crystals of varying size. The unstable, large-plate morphology of the ggbfs/HAC gehlenite hydrate is very similar to that reported by Richardson et al. (1990) for the AFm present in mature ggbfs/OPC blends.

Morphology and compositions of the amorphous gel phases
In addition to gehlenite hydrate, the major hydration products were two distinct, mainly amorphous gels : an outer gel product, occupying originally water filled space; and an inner gel product, occupying

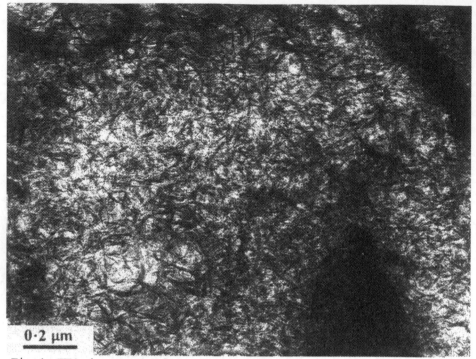

Fig.4. TEM micrograph of outer gel product.

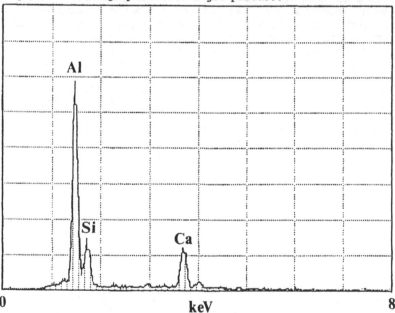

Fig.5. EDXA spectrum for outer gel product.

Table 3. Atom ratios for EMPA points with Al:Ca >1.5, corresponding to mixtures of outer gel product, Al_2O_3, and gehlenite hydrate.

	$\bar{x} \pm \sigma_{n-1}$		High Value	Low Value
Si:Ca	0.363 ±	0.183	1.190	0.077
Al:Ca	6.124	9.346	81.912	1.504
S:Ca	0.049	0.046	0.353	0.000
Mg:Ca	0.114	0.206	1.676	0.000
K:Ca	0.020	0.015	0.118	0.006
Ti:Ca	0.032	0.041	0.412	0.001
Fe:Ca	0.015	0.016	0.147	0.005

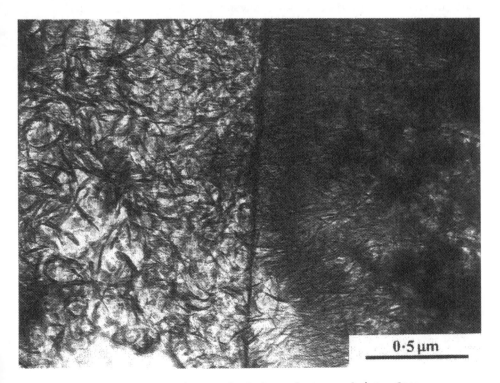

0·5 μm

Fig.6. TEM micrograph showing typical inner/outer gel interface.

space released by slag reaction. No inner product corresponding to HAC phases could be readily identified.

The outer gel product is observed by TEM to have a relatively fine morphology, figure 4, not unlike that found in other high slag content systems, and similar to that found in unblended hydrated HAC (Poon and Groves (1988)). The outer gel product had a Ca:Si atom ratio of 1.1 ± 0.23, approximately that of the ggbfs itself (and typical of outer product C–S–H gel produced by non-calcium bearing activators), but had a very high Al:Ca ratio of 7.70 ± 1.23. This value is at the Al-rich end of the compositional range reported by Poon and Groves (1988) for the gel phase in hydrated HAC. Figure 5 shows an example EDXA spectrum of outer gel product. The Ca:Si ratio, and the absence of crystalline reflections in the electron diffraction pattern suggests that it may be an intimate mix of C–S–H and alumina gels. No points on the microprobe map corresponded with this composition, but many could be attributed to mixtures of outer product and gehlenite hydrate. Mean atom ratios for these points are listed in table 3.

A dense well-defined inner/outer gel interface was a notable feature of this system, an example of which is shown in figure 6. This is in contrast to ggbfs/OPC blends, where the interface is generally less abrupt and usually only denser than the corresponding inner and outer products when the precipitation of Mg-rich phases within the inner product is well advanced. The ggbfs inner gel itself appears to have a rib-like morphology, figure 7, a variation of which is also, although less commonly, observed in ggbfs/OPC blends. The ggbfs inner gel has a lower Ca:Si and higher Al:Ca atom ratio than in a corresponding ggbfs/OPC blend, as shown in table 4. However, as is characteristic of ggbfs hydration, the Mg, Ti and Mn remained within these regions, yielding a Mg:Si ratio comparable with other high ggbfs systems, table 4. This Mg-rich character of the inner product is illustrated in figure 8, an EDXA spectrum for a region of inner gel, and in figure 9, a plot of Mg:Ca against Al:Ca atom ratios for the mapping data. No discrete Mg–Al hydroxide type phases were observed in the ggbfs/HAC blend. The occurrence of these in the slag inner product of ggbfs/OPC blends is responsible for the variation in the Mg:Si ratio reported in table 4.

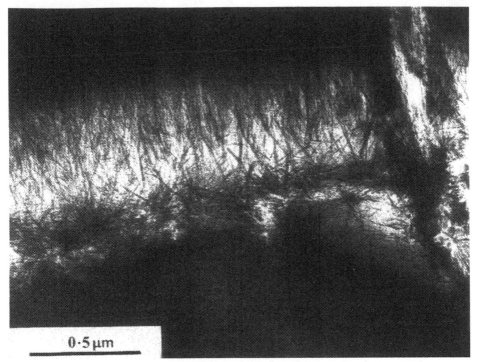

Fig.7. TEM micrograph of inner gel product.

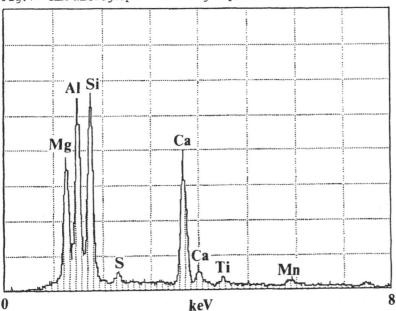

Fig.8. EDXA spectrum for inner gel product.

Table 4. Atom ratios determined by AEM for slag inner product in ggbfs/HAC 3:1 (w/s = 0.4, 20°C, 2yrs) and ggbfs/OPC 3:1 (w/s = 0.4, 20°C, 14 months) blends.

	n = 13	n = 20
	ggbfs/HAC	ggbfs/OPC
Ca:Si	0.69 ± 0.11	1.40 ± 0.11
Al:Ca	1.32 ± 0.57	0.23 ± 0.13
Mg:Al	0.93 ± 0.10	2.33 ± 0.87
Mg:Si	0.93 ± 0.14	0.85 ± 0.56

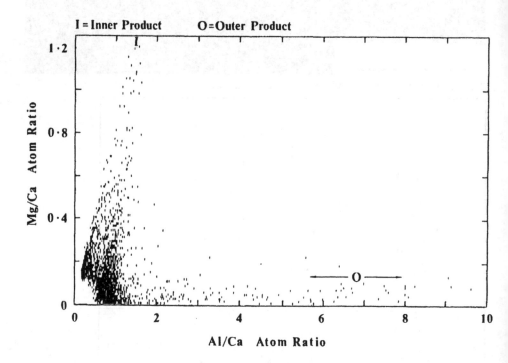

Fig.9. Plot of Mg:CA against Al:Ca atom ratios.

References

Cliff, G. and Lorimer, G.W. (1975) The quantitative analysis of thin specimens. **J. Micr.**, 103, 203–206.

Edmonds, R.N. and Majumdar, A.J. (1989) The hydration of mixtures of monocalcium aluminate and blast-furnace slag. **Cem. Concr. Res.**, 19, 779–782.

Groves, G.W., LeSueur, P.J. and Sinclair, W. (1986) Transmission electron microscopy of tricalcium silicate paste. **J. Amer. Ceram. Soc.**, 69, 353–356.

Poon, C.S. and Groves, G.W. (1988) TEM observations of a high alumina cement paste. **J. Mater. Sci. Lett.**, 7, 243–244.

Reimer, L., (1985) **Scanning Electron Microscopy : Physics of image formation and microanalysis.** Springer-Verlag, Berlin.

Richardson, I.G., Rodger, S.A. and Groves, G.W. Hardened cement pastes and some effects of leaching. **Mater. Res. Soc. Proc.**, 176, (to be published 1990).

Rodger, S.A., Groves, G.W. and Richardson, I.G., unpublished work.

Taylor, H.F.W., (1964) in Appendix 1, **The Chemistry of Cements** (ed. H.F.W.Taylor), Academic Press, London.

22 EFFECT OF MICROSILICA ON CONVERSION OF HIGH ALUMINA CEMENT

S. BENTSEN, A. SELTVEIT
Norwegian Institute of Technology, Trondheim, Norway
B. SANDBERG
Elkem Materials a/s, Kristiansand, Norway

Abstract
The mineral conversion in different cement pastes has been investigated as a function of time and temperature. The investigations have been done by means of x-ray diffraction analyses. The cement pastes consisted of "Secar 80" and "Secar 51" with varying amounts of "Elkem Microsilica" and water/powder ratio of 0.5.

Examinations of the hydration pattern showed that the following conversions took place at the respective temperatures:

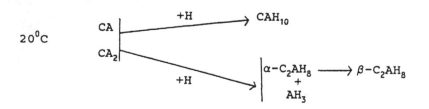

Examinations of the temperature dependence of the mineral conversions showed that the dominant phases for the given temperature ranges were:

$$
\begin{array}{lll}
0 & - & 20^\circ C & CAH_{10} \\
20 & - & 70^\circ C & C_2ASH_8 \\
70 & - & 100^\circ C & C_3AH_6
\end{array}
$$

α-and β-C_2ASH_8 was found be important intermediate products. Traces of the phases α-C_4AH_{13}, $C_4A\overline{C}H_{11}$, and $CaCO_3$ were found.   Unreacted microsilica was shown by comparing x-ray diagrams for pure microsilica and hardened calcium aluminate cement with microsilica.

1   Introduction

The mineral conversion during hydration of calcium aluminate cements with addition of microsilica has previously been examined by Siri Bentsen (1984).   During this work emphasis was placed on determining which mineral phases had been formed after three months, as after such a long hardening time it could be assumed that the system was close to equilibrium.   For a selection of these samples it was also tried to investigate the time dependency for the mineral conversion during the first week of the hydration (hardening).   The cement pastes were for this reason analysed after 4 hours, 12 hours, 1 day and 1 week.   However, it appeared that it was necessary with even shorter time intervals between the analyses to obtain a realistic picture of the time dependency.   This was not particularly the case for the hardening times between 1 day and 1 week.   In the present work it is consequently aimed to repeat a part of the tests with shorter time intervals between the analyses.
    In previous work (Bentsen 1984) only three different hardening temperatures were investigated.   The long-term experiments showed that the hardening temperatures were determinant for the mineral phases that were formed.   It was consequently necessary to carry out a more thorough investigation of the mineral phase formations temperature dependence.   The hardening times chosen were 1 day and 1 week.   1 day was chosen because it is relatively common hardening time in practical utilization of refractory cement.   1 week was chosen because it is supposed that the system has got relatively close to equilibrium and showed that very limited conversion took place between 1 week and three months.   The temperature range between 0 and 100°C was chosen for pure practical reasons.

2   Experimental

2.1   Raw materials
The following raw materials have been used:

Calcium aluminate cements   "Secar 80"
                                   "Secar 51"

Microsilica   "Elkem Microsilica 970"

Deflocculant   Sodium-hexa-metaphosphate

Distilled water

2.2   Sample preparation and hardening condition

All the samples have been prepared in the same way.
Calcium aluminate cement and microsilica were first mixed
dry.   The total weight of the powder was 10.0g. To the
dry powder was then added 5.0g distilled water such that
the water/powder ratio became 0.5.   In the cases where
deflocculant was used, this was first dissolved in water.
The amount of deflocculant used corresponds to 0.12g per
10g sample.

   The samples were mixed and cured in 25mm plastic
beakers (height 40mm, lower diameter 28mm, upper diameter
32mm).   During curing these beakers were placed floating
in water inside a desiccator or in a glass bottle sealed
with a glass lid and rubber seal.   In this way the
atmosphere above the specimen is approximately saturated
with water vapour, one has a relatively even temperature
and the heat of reaction is quickly conducted away from
the sample or specimen.

   The various curing conditions are as shown in table 1.
The table also shows the maximum temperature, average
temperature and temperature stability during curing.   In
table 2 it is given an overview of what compositions,
curing temperatures and curing times.   In table 3 there is
a corresponding review of compositions and curing
conditions for the investigation of the mineral
conversions temperature dependence.

2.3   X-ray diffraction analyses

The x-ray diffraction analyses were carried out with the
following equipment:

Philips x-ray generator   PW 1730/10
Philips x-ray tube   PW 1043/01
Philips goniometer   PW 1050/25
Philips electronic calculation and
control unit   PW 1360
Philips recorder   PM 8203

The recording were done with Cu $K\alpha$-radiation and nickel
filter.   The voltage was 40 kV and current 20 mA.   The
specimens were taken out by "drilling" with a spatula.

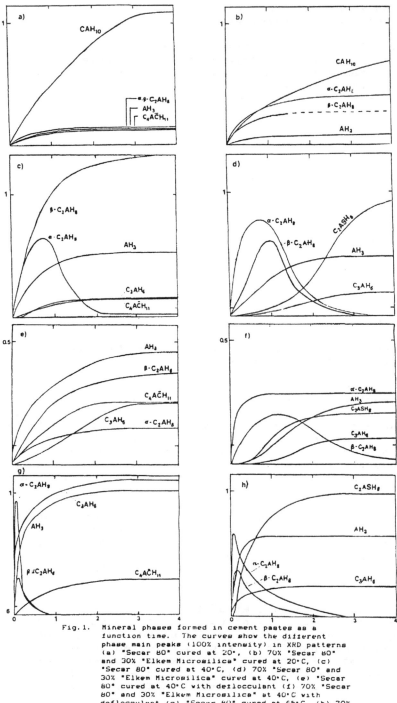

Fig. 1. Mineral phases formed in cement pastes as a function time. The curves show the different phase main peaks (100% intensity) in XRD patterns (a) "Secar 80" cured at 20°, (b) 70% "Secar 80" and 30% "Elkem Microsilica" cured at 20°C, (c) "Secar 80" cured at 40°C, (d) 70% "Secar 80" and 30% "Elkem Microsilica" cured at 40°C, (e) "Secar 80" cured at 40°C with deflocculant (f) 70% "Secar 80" and 30% "Elkem Microsilica" at 40°C with deflocculant (g) "Secar 80" cured at 65°C, (h) 70% "Secar 80" and 30% "Elkem Microsilica" cured at 65°C.

Specimens were crushed down to a fine powder and placed in
sample holders of plexiglass.

Table 1.   Review of test temperatures
           (t = average temperature, t_{max} = maximum
           temperature measured in the water surrounding
           the specimen)

t (°C)		tmax (°C)
	±	
5.0	0.5	5.5
17.0	1	17.5
19.0	1	19.0
20.0	0.5	20.05
21.0	2	-
30.0	0.2	-
40.0	0.2	-
50.0	0.2	-
60.0	0.2	-
65.0	0.2	-
70	5	76
90	5	93
100	1	100

3   Results and discussion from the examination of the curing as a function of time

3.1   Introduction
The highest x-ray peak for the various mineral phases
have been plotted as a function of the curing time.   To
compensate for a variable net voltage, and because there
is less cement in microsilica-containing paste, the peak
height has been calculated with alumina Alcoa T-60 20% by
weight as a reference.   The intensity for the alumina peak
that in accordance to literature gives 100% reflection was
set to 1.

The figure 1 thus shows the time dependency for the
formation of the various hydrate phases in cement paste
that were cured at 20, 40 and 65°C.   The cement pastes
consisted of "Secar 80" with and without addition of
"Elkem Microsilica".   In figures 2 to 4 shows that the
remainder of original phases CA and CA_2, with same type of
plot.   The uncertainty for the relative peak is ± 0.1.

3.2   The curing pattern at 20°C
The results from the investigations of the curing pattern
for <u>pure calcium aluminate cement</u> shows good agreement
with previously published results referred to by Bentsen
1984.   The most important phase that is formed is CAH_{10}.
This phase grows close to linear in three days before it
starts to stabilise.   In addition to CAH_{10} traces of α-

and β-C_2AH_8, $C_4A\bar{C}H_{11}$ and AH_3 also formed.

The formation of CAH_{10} happens at the same time as the consumption of CA (cfr. figure 2), but the consumption of CA is stagnating after only 2 days. If one compares the consumption of CA_2 with the formation of $C_4A\bar{C}H_{11}$ and α-, β-C_2AH_8, it appears that they have approximately the same pattern, but with the opposite sign. This is in agreement with Borbov's (1974) theory that CA_2 cannot convert to CAH_{10}.

Table 2.   Review of the test conditions by examination of
           the curing products for different cement pastes.

Composition (%)			Curing Temp. (°C)	Curing Time (t = hours, d = days)						
Secar 80	Micro silica	Defloc-culant								
100			20	4t	12t	1d	2d	3d	7d	
70	30		20	4t	12t	1d	2d	3d	7d	
100			40	4t	12t	1d	2d	3d	4d	7d
70	30		40	$\begin{cases} 4t \\ 6d \end{cases}$	12t 7d	1d	1.5d	2d	3d	5d
100		0.12g	40	4t	1d	2d	3d	4d	7d	
70	30	0.12g	40	4t	2d	2d	3d	4d	5d	7d
100			65	2t	4t	6t	8t	12t	1d	2d 3d
70	30		65	2t	4t	6t	8t	12t	1d	2d 3d

Table 3.   Summary of test conditions by investigation of
           the mineral phase formation dependence on
           temperature.

Composition (%)		Curing Temperature (°C)
Cement	Microsilica	
"Secar 80" 100		After 1 day: 5, 10, 17, 21, 30, 40, 50, 60, 70, 90 and 100.
70	30	
50	50	
"Secar 51" 100		After 1 week: 5, 10, 19, 21, 30, 40, 50, 60, 70 and 90.
70	30	
50	50	

When microsilica is added to the cement, the curing
pattern is notably changing character. The main mineral
phase is still CAH_{10}, however it is formed far less in
this system compared with pure cement. Rather, it is
formed more of both α- and β-C_2AH_8 (β-C_2AH_8 is dotted
because α- and β-peaks are so close to each other in the
x-ray diagram. When one of them is large, it is difficult
to determine intensity on the other one).

The reaction rate decreases noticeably when microsilica
is present in the cement. This is especially easy to see
in figure 2.

Also in this cement paste the AH_3 is formed as a by-
product, but $C_4A\bar{C}H_{11}$ is not found.

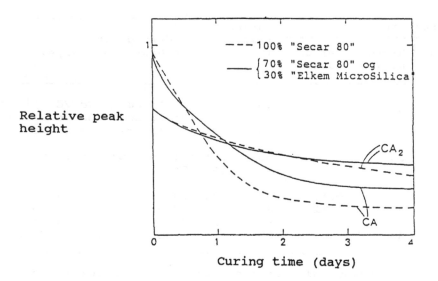

Fig 2.   Consumption CA and CA_2 as a function of time in
          cement paste of "Secar 80" with and without
          "Elkem Microsilica". The curves shows the
          different mineral phase's main peaks (100%
          intensity) in the x-ray diffraction diagram.
          Curing temperature 20° C.

3.3 Curing patterns at 40°C

By curing of <u>pure calcium aluminate cement</u> at 40°C, the C_2AH_8 is the dominating phase the 4 first days. See figure 1c. Turriziani (1984) states that of the various possible modifications for C_2AH_8 *, α_1-and β-C_2AH_8 that exists in connection with aqueous solutions, and that α_1-C_2AH_8 is most stable. The results from the test we have done here shows the opposite, namely that it is β-C_2AH_8 that is the most stable phase in pure calcium aluminate cement paste. The curves in figure 1c indicates that α-C_2AH_8 is an intermediate product during the formation of β-C_2AH_8. Apart from this, the results are in good agreement with the literature. AH_3 is formed as a by-product, and it is also formed smaller quantities of C_3AH_6 and $C_4A\bar{C}H_{11}$.

In <u>microsilica-containing cement paste</u> the hydration is somewhat slower than in pure cement paste, but nevertheless faster than at 20°C (cfr. figure 3). Otherwise the mineral conversion during the first day is about the same as in pure "Secar 80" (figure 1d). After 1 day of curing, however the β-C_2AH_8 reacts with the microsilica and forms C_2ASH_8. This phase stabilizes after 3 to 4 days, at the same time as C_2AH_8 completely disappears. C_3AH_6 and AH_3 show the same pattern in both pastes.

3.4 The influence of deflocculant on the curing pattern at 40°C

When deflocculant is added to the cement paste and this is cured at 40°C, the reaction rate in the cement paste is lowered both with and without addition of microsilica (figure 4), but still such that the microsilica-containing cement paste reacts slower than the pure one. The use of deflocculant is also resulting in all phases in the paste showing low intensity in the x-ray diagram. The figures 1e and 1f, that illustrate how the reaction products have been formed and are consequently drawn in a double scale. The amorphous nature of the hydration product is also confirmed by thermogravimetry.

In the <u>pure cement paste with deflocculant</u> about equal amounts of C_3AH_6, $C_4A\bar{C}H_{11}$ and AH_3 are formed as in the pure cement paste without dispersing agent, however far less of both α- and β-C_2AH_8 are detected in the x-ray diffraction diagram. The formation of C_2AH_8 from CA and CA_2 should give AH_3 as a by-product. As it now appears that substantial amount AH_3 has been formed, this would indicate that the conversion has taken place. As other reaction products are not found by x-ray diffraction and analyses, this would indicate that they are x-ray amorphous.

---

* α_1 is here used to be an agreement with Turriziani, but if otherwise in the report only referred to as α.

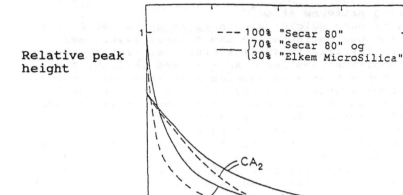

Relative peak height

Curing time (days)

Fig 3. Consumption of CA and CA₂ as a function of time in cement paste of "Secar 80" with and without "Elkem Microsilica". The curves shows the different mineral phases main peaks (100% intensity) in the x-ray diffraction diagram. Curing temperature 40°C.

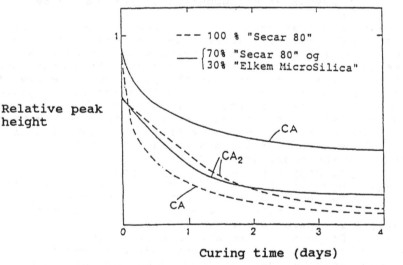

Relative peak height

Curing time (days)

Fig 4. Consumption of CA and CA₂ as a function of time in cement paste of "Secar 80" with and with "Elkem Microsilica" with deflocculant. The curves shows the different mineral phases main peaks (100% intensity) in the x-ray diffraction diagram. Curing temperature 40°C.

In the cement paste with microsilica and deflocculant
however the intensity of the AH_3 peaks are halved compared
to cement with microsilica, without deflocculant. This
could be attributed to the far lower consumption of CA and
CA_2. However the AH_3 peak is nevertheless high compared
to the height of the other main peaks, which means that
here also a relatively high degree of conversion must have
taken place than it may seem in the x-ray diagrams.

Another interesting phenomena that also takes place
where deflocculant has been used, is that neither in pure
cement nor in cement with microsilica, is the $\alpha-C_2AH_8$
reacting further to $\alpha-C_2AH_8$ as it does in all the other
tests. Rather it is stabilized at a certain level after
one day. In cement with microsilica, $\beta-C_2AH_8$ reacts with
microsilica and forms C_2ASH_8, but this is not the case
with $\alpha-C_2AH_8$.

None of the test series at 40°C as curing temperature,
gives CAH_{10} as an intermediate product. If the reaction
is taking place via this phase, it consequently must
further react rapidly to C_2AH_8.

3.5   The curing pattern at 65°C

The curing pattern for pure cement at 65°C shows some
simularity with the curing pattern at 40°C, but the
conversion is taking place must faster (see figure 1g and
5). $\beta-C_2AH_8$ is not stable as it is at 40°C, but reacts
quickly further to C_3AH_6. As a consequence of this, some
AH_3 is also formed, such that C_3AH_6 and AH_3 are becoming
the main mineral phases. They are stabilized after one to
two days of curing. $\alpha-$ and $\beta-C_2AH_8$ which are
intermediate products have their peaks after 2-4 hours of
curing. At this temperature it is formed relatively large
amounts of $C_4A\bar{C}H_{11}$, are also found.

In cement with microsilica the curing pattern is even
more similar to the corresponding series at 40°C, but also
here the conversion is faster. About the same amounts
C_3AH_6, AH_3 and C_2ASH_8 are formed, but they stabilized
already after one to two days. $\alpha-$ and $\beta-C_2AH_8$ have their
peaks after 2-4 hours of curing (see figure 1h).

303

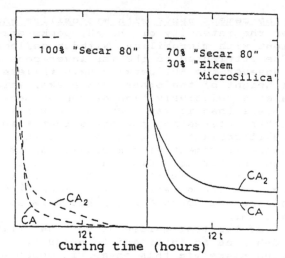

Relative peak height

100% "Secar 80"

70% "Secar 80"
30% "Elkem MicroSilica"

CA₂

CA

CA₂

CA

12 t   12 t

Curing time (hours)

Fig 5.   Consumption of CA and CA_2 as a function of time
         in cement paste of "Secar 80" with and without
         "Elkem Microsilica".   The curves shows the
         different mineral phases main peaks (100%
         intensity) in the x-ray diffraction diagram.
         Curing temperature 65° C.

3.6   Conclusion

All of the test series showed that the microsilica reduced
the conversion rate during curing.   In pure cement the
dominating phases after 4 days is CAH_{10} after curing at
20° C, β-C_2AH_8 after curing at 40° C and C_3AH_6 and AH_3 after
curing at 65° C.   For cement with microsilica, the
respective dominating phases were; CAH_{10} C_2AH_8 and C_2ASH_8
respectively.

In pure cement the maximum content of α-C_2AH_8 is
obtained after about 1 day's curing 40° C and maximum
content of α- and β-C_2AH_8 after 2-4 hours curing at 65° C.
For cement with microsilica the α- and β-C_2AH_8 have their
main peak after about 1 day of curing at 40° C and after 2-
4 hours of curing at 65° C.

The most probable reaction mechanism for formation of
C_2ASH_8 is as follows:

$$\left.\begin{array}{c} CA \\ CA_2 \end{array}\right\} \xrightarrow{\;+\;H\;} \begin{array}{c} \alpha\text{-}C_2AH_8 \xrightarrow{\hspace{2cm}} \beta\text{-}C_2AH_8 \xrightarrow{\quad +S\quad} C_2ASH_8 \\ + \\ AH_3 \end{array}$$

The conversion rate increases with temperature.   The
deflocculant that has been used reduces the conversion
rate and causes some of the reaction products to get low
crystallinity and consequently give lower x-ray peaks in
the x-ray diffraction diagram.

4 Results and discussions from examinations of the mineral conversion temperature dependence

4.1 Introduction

The results from the temperature tests have been plotted in the same way as the time tests, but here the peaks from the x-ray diagram are plotted as a function of the temperature. For cement pastes that contains "Secar 80", the alumina peak is also here used as a reference both for a d-spacing determination and intensity determination. In cement pastes that contains "Secar 51", however the alumina peaks are too small to be used as references on the intensity, but they are nevertheless used for determination of d-spacing values. For the intensity it is used the same scale as for "Secar 80"-curves, but they have been recalculated for microsilica-containing cement paste to compensate for the lower cement content. This is done by taking the peak for the phases in the paste that contained 30% microsilica and multiplied with 100%/70% and the peak height for the phases in the paste that contains 50% microsilica was multiplied by 100%/50%.

The curves that show the temperature dependency of the mineral conversions is given in figures 6-14. Figures 6-12 show the most dominating phases. Figures 13 and 14 show what mineral phases are present in smaller amounts. Figures 15-20 show how much CA, CA_2 and $C_{12}A_7$ that is present in an unreacted form in the pastes after 1 day and 1 week.

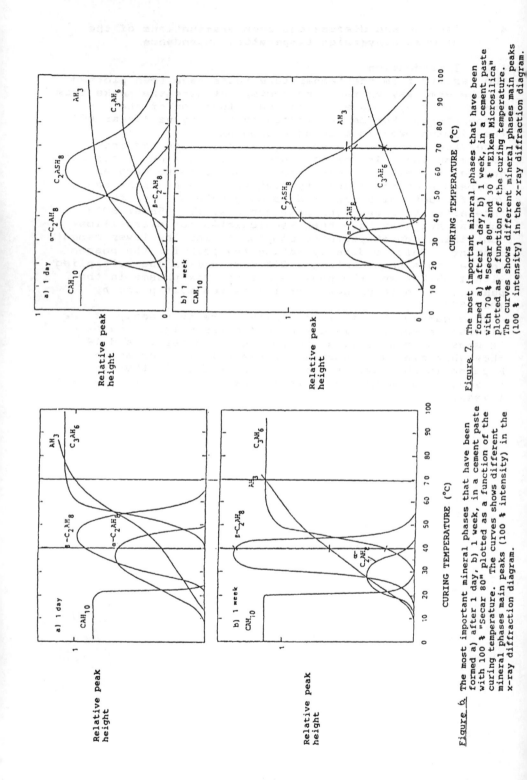

Figure 6. The most important mineral phases that have been formed a) after 1 day, b) 1 week, in a cement paste with 100 % "Secar 80" plotted as a function of the curing temperature. The curves shows different mineral phases main peaks (100 % intensity) in the x-ray diffraction diagram.

Figure 7. The most important mineral phases that have been formed a) after 1 day, b) 1 week, in a cement paste with 70 % "Secar 80" and 30 % "Elkem Microsilica" plotted as a function of the curing temperature. The curves shows different mineral phases main peaks (100 % intensity) in the x-ray diffraction diagram.

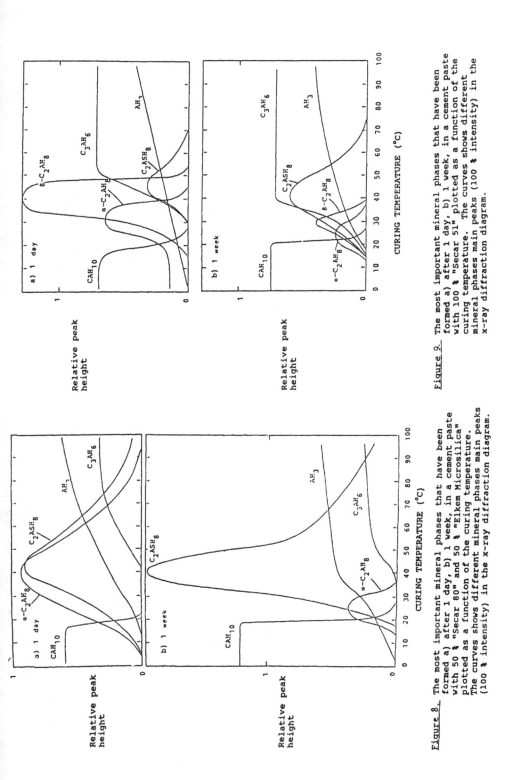

Figure 8. The most important mineral phases that have been formed a) after 1 day, b) 1 week, in a cement paste with 50 % "Secar 80" and 50 % "Elkem Microsilica" plotted as a function of the curing temperature. The curves shows different mineral phases main peaks (100 % intensity) in the x-ray diffraction diagram.

Figure 9. The most important mineral phases that have been formed a) after 1 day, b) 1 week, in a cement paste with 100 % "Secar 51" plotted as a function of the curing temperature. The curves shows different mineral phases main peaks (100 % intensity) in the x-ray diffraction diagram.

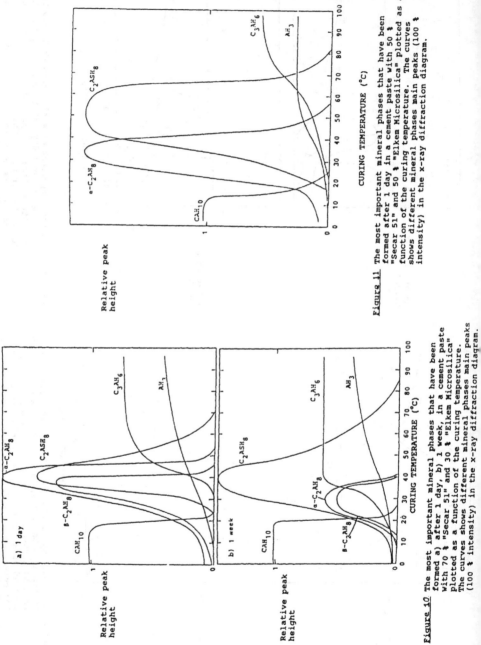

Figure 11 The most important mineral phases that have been formed after 1 day in a cement paste with 50 % "Secar 51" and 50 % "Elkem Microsilica" plotted as a function of the curing temperature. The curves shows different mineral phases main peaks (100 % intensity) in the x-ray diffraction diagram.

Figure 10 The most important mineral phases that have been formed a) after 1 day, b) 1 week, in a cement paste with 70 % "Secar 51" and 30 % "Elkem Microsilica" plotted as a function of the curing temperature. The curves shows different mineral phases main peaks (100 % intensity) in the x-ray diffraction diagram.

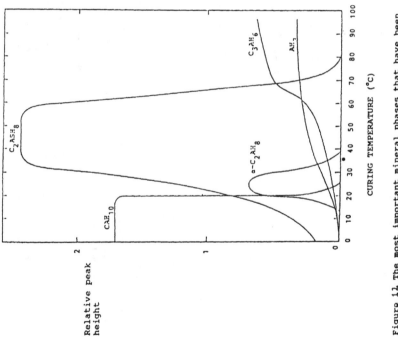

Figure 11 The most important mineral phases that have been
formed after 1 day in a cement paste with 50 %
"Secar 51" and 50 % "Elkem Microsilica" plotted as a
function of the curing temperature. The curves
shows different mineral phases main peaks (100 %
intensity) in the x-ray diffraction diagram.

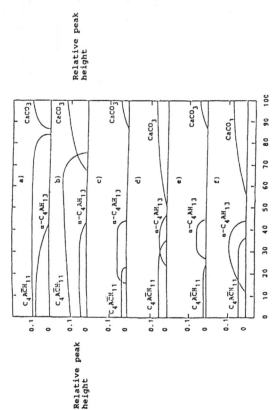

Figure 13 Minor mineral phases after curing, plotted as a function of the curing temperature. The curves shows the different mineral main peaks (100 % intensity) in the x-ray diffraction diagram.

The cement pastes consist of:

a) 100% "Secar 80" cured in 1 day,
b) 100% " cured in 1 week,
c) 70% and 30% "Elkem Microsilica" cured in 1 day,
d) 70% and 30% " " cured in 1 week,
e) 50% and 50% " " cured in 1 day,
f) 50% and 50% " cured in 1 week.

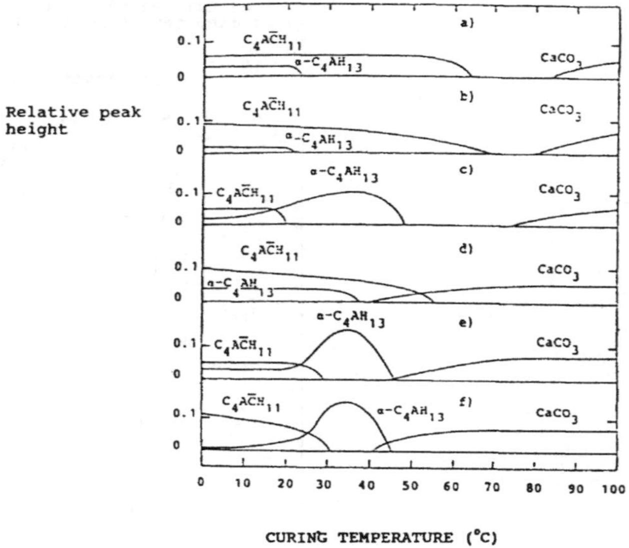

Figure 14 Minor mineral phases after curing, plotted as a function of the curing temperature. The curves shows the different mineral main peaks (100 % intensity) in the x-ray diffraction diagram.

The cement pastes consist of:

a) 100% "Secar 51" cured in 1 day,
b) 100% " cured in 1 week,
c) 70% and 30% "Elkem Microsilica" cured in 1 day,
d) 70% and 30% " " cured in 1 week,
e) 50% and 50% " " cured in 1 day,
f) 50% and 50% " cured in 1 week.

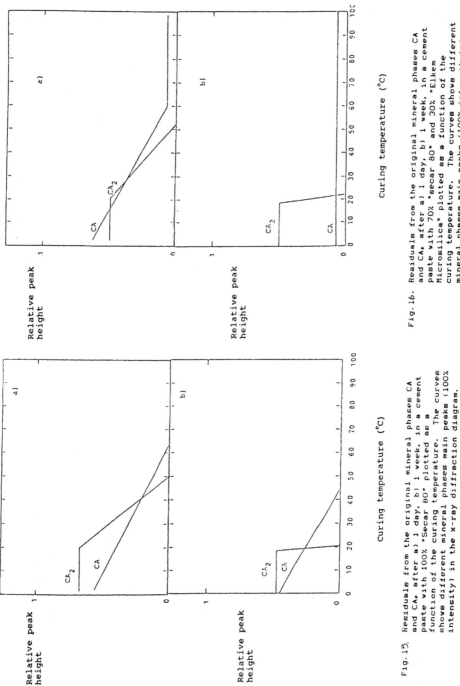

Fig. 15. Residuals from the original mineral phases CA and CA_2 after a) 1 day, b) 1 week, in a cement paste with 100% "Secar 80" plotted as a function of the curing temperature. The curves shows different mineral phases main peaks (100% intensity) in the x-ray diffraction diagram.

Fig. 16. Residuals from the original mineral phases CA and CA_2 after a) 1 day, b) 1 week, in a cement paste with 70% "secar 80" and 30% "Elkem Microsilica" plotted as a function of the curing temperature. The curves shows different mineral phases main peaks (100% intensity) in the x-ray diffraction diagram.

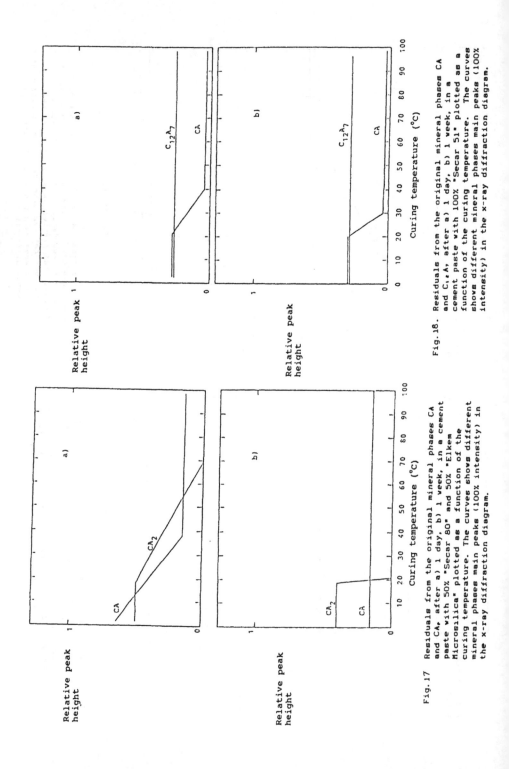

Fig.17 Residuals from the original mineral phases CA
        and CA₂ after a) 1 day, b) 1 week, in a cement
        paste with 50% "Secar 80" and 50% "Elkem
        Microsilica" plotted as a function of the
        curing temperature. The curves shows different
        mineral phases main peaks (100% intensity) in
        the x-ray diffraction diagram.

Fig.18. Residuals from the original mineral phases CA
        and C₁₂A₇ after a) 1 day, b) 1 week, in a
        cement paste with 100% "Secar 51" plotted as a
        function of the curing temperature. The curves
        shows different mineral phases main peaks (100%
        intensity) in the x-ray diffraction diagram.

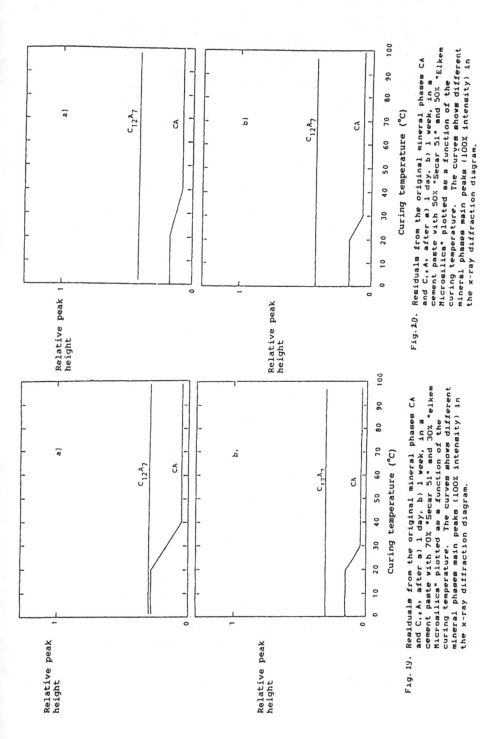

Fig.19. Residuals from the original mineral phases CA and C₁₂A₇, after a) 1 day, b) 1 week, in a cement paste with 70% "Secar 51" and 30% "elkem Microsilica" plotted as a function of the curing temperature. The curves shows different mineral phases main peaks (100% intensity) in the x-ray diffraction diagram.

Fig.20. Residuals from the original mineral phases CA and C₁₂A₇, after a) 1 day, b) 1 week, in a cement paste with 50% "Secar 51" and 50% "Elkem Microsilica" plotted as a function of the curing temperature. The curves shows different mineral phases main peaks (100% intensity) in the x-ray diffraction diagram.

4.2   Consumption of the original phases CA, CA_2 and $C_{12}A_7$

In the two cements that have been investigated the
following calcium aluminate phases are present in the
start: $C_{12}A_7$, CA and CA_2. In "Secar 80" it is CA and
CA_2, in "Secar 51" it is CA and $C_{12}A_7$. According to
Parker and Sharp (1982) $C_{12}A_7$ is the most reactive,
thereafter it is CA and finally CA_2. The latter have also
been reported to be inert. If one now adds CA to CA_2 this
will increase the reaction rates in the mixtures.
Likewise, if $C_{12}A_7$ is added to a CA-cement, the reaction
rate will increase.

The residual quantities of CA and CA_2 in a "Secar 80"
cement paste with and without microsilica is given in
figure 16-18. For a pure cement, the CA
shows a linearly decreasing curve with reaches zero at
about 65°C after 1 day and at 45°C after 1 week.

With the addition of microsilica however, CA will not
go to zero, but there is a residual part left. This
residual part is constant for all temperatures. There is
more CA left when the ratio cement/silica is 1 (i.e. 50%
"Secar 80" and 50% "Elkem Microsilica") than when the
ratio is 2.3 (70 and 30%, respectively). This should
indicate that the silica reduces the reactivity of CA. It
should here be noted that previous investigations (Bentsen
1984) have shown that silica addition in cement pastes
causes a reduction in pH of the paste. These two
phenomena may have a connection.

When it comes to CA_2 it does not seem like this is
dissolved at all for temperatures below 20°C. As it will
be shown later it is in this range that CAH_{10} has it
stability. If these results are comprehended it should
confirm Borbov's (1974) theory that CA_2 does not react to
form CAH_{10}. The constant range of CA_2 below 20°C varies
somewhat in value for the different series, but this is
within the uncertainty that have been given for the
curves.

For curing temperatures above 20°C, the remaining
CA_2 is decreasing linearly and faster than the CA
remainder in all except in one case, and it is reduced to
zero in all the compositions. This should also indicate
that the solubility of CA_2 is larger than the solubility
of CA for temperatures above 20°C.

Figures 19 to 21 the corresponding curves for mixes
with "Secar 51" is given. For these pastes it looks like
there has been an exchange of roles, where $C_{12}A_7$ react
analogues with CA in "Secar 80" and CA analogues with CA_2
in "Secar 80". What is left of $C_{12}A_7$ after 1 day and 1
week is constant for all temperatures, but this value
increase with the microsilica content in the paste.
(This statement has not been controlled statistically).

4.3 CAH_{10}
From figures 7-13 it appears that for the curing times and

curing temperatures which have been investigated here, the stability area for CAH_{10} is 0-20°C for all the test series. For "Secar 80" paste it has been formed considerably more CAH_{10} after 1 week than what was formed after 1 day. This is exactly what one could expect from the results from the investigation of the curing pattern as a function of time.

For 100% "Secar 51" and the composition 70% "Secar 51" and 30% Elkem Microsilica any such increase of the CAH_{10} amount has not been found between 1 day and 1 week. Unfortunately time dependencies studies have not been done that can explain this, but follow the theory that a mixture of $C_{12}A_7$ and CA would be more reactive than a mixture of CA and CA_2 this is not unexpected. The conversion has probably been almost complete after only 1 day. This theory does not need to be invalid even if it now appears that for a composition 50% "Secar 51" and 50% Elkem Microsilica has formed more CAH_{10} between 1 day and 1 week. It has been shown that $C_{12}A_7$ has an accelerating effect on the reactivity of CA and it has previously been shown that microsilica has a retarding effect. Even if this retarding effect has not reduced the reactivity measurable in paste with 30% microsilica, it might well influence the paste with 50% microsilica.

For "Secar 51" the amount of CAH_{10} that is formed increase with the microsilica content, In spite of there being less cement in the sample it is about the same amount of CAH_{10}. This means that the degree of conversion with respect to CAH_{10} is higher in cement with microsilica than in cement without microsilica.

4.4 α- and β- C_2AH_8

Because the formation of α-C_2AH_8 and β-C_2AH_8 are closely related they will here be treated together. In 3.3 and 3.4 it was shown that during curing at 40 and 65°C, the formation of α-C_2AH_8 occurred first. Later this converted to β-C_2AH_8. In pure "Secar 80" at 40°C the β-C_2AH_8 was stable the first 4 days during curing, but it is earlier shown (Bentsen 1984) that after a longer time will react further and from C_3AH_6.

In all temperatures studies α-C_2AH_8 together with β-C_2AH_8 and in some cases C_2ASH_8 are the dominating phases between 20 and 50°C after 1 day of curing. After 1 weeks' curing the amount of α-C_2AH_8 has been reduced considerably and it will probably disappear totally.

In pure "Secar 80" which only contains traces of SiO_2, β-C_2AH_8 is the most important phase between 30 and 55°C already after 1 day. After 1 week it has increased considerably in the temperature range 30-45°C, but at the same time the width of the temperature range it dominates has been reduced to the benefit of C_3AH_6. By addition of microsilica β-C_2AH_8 reacts relatively quickly to C_2ASH_8, however there are still traces of it after 1 day.

315

"Secar 51" contains according to the producer 6.7% SiO_2, which are present predominately in the phases C_2AS and C_2S. Like microsilica, silica from these phases will react with C_2AH_8, but this will not have any significance after 1 days' curing where β-C_2AH_8 still is the most important phase between 20 and 50°C. However, after 1 week C_2ASH_8 will take over. With 30% microsilica in the paste it will after 1 days' curing be about equal quantities of both α-, β-C_2AH_8 and C_2ASH_8 in the same temperature range. After 1 week the main content of α- and β-C_2AH_8 have reacted further to C_2ASH_8. With 50% microsilica in the cement, the production of β-C_2AH_8 is appearing so slow compared to the formation of C_2ASH_8 that it is not noticeable as an intermediary product after 1 day.

4.5 C_2ASH_8

The most important alteration in the mineral conversions by addition of microsilica is represented by formation of C_2ASH_8. In contrast to α- and β-C_2AH_8, C_2ASH_8 is stable also after very long curing time and is therefore supposed to have a positive effect on mechanical strength after curing. Even by heating it will not react further before the water disappears at 220°C. This has been previously shown and discussed by Bensten 1984.

For the sake of simplicity the results have been presented in a table that shows the relative peak height for C_2ASH_8, what temperatures that gives maximum peak height and in which area C_2ASH_8 is the dominating phase. See table 4. It appears from the table that C_2ASH_8 increases with increasing microsilica content in the paste and that more C_2ASH_8 is formed in the "Secar 51" paste than in "Secar 80" based pastes. The maximum intensity varies from 35-60°C, but most of the peaks are around 45°C. The dominating area for C_2ASH_8 increase from 40-70°C after 1 day to 20-70°C after 1 week. The lower limit for the formation of C_2ASH_8 will hardly go below 20°C the upper limit at 70°C might possibly increase (cfr figures 6-12 but especially 12).

4.6 C_3AH_6

In pure cement C_3AH_6 is the dominating phase for curing temperatures above 40°C. This applies both to "Secar 80" and "Secar 51" and after both 1 day and 1 week. By addition of microsilica C_2ASH_8 take over much of the dominance in the range 40-70°C. However, the C_3AH_6 quantity is also influenced outside this area. With respect to temperatures below 100°C it is not reaching the level it has in pure cement. Consequently the microsilica inhibits the conversion to C_3AH_6 in spite of no formation of C_2ASH_8 or other silicon containing cement phases that can be measured by x-ray diffraction analyses.

In some cases it may seem like the microsilica influences the structure of C_3AH_6. When the C_3AH_6 peaks have low intensity double peaks or shift in d-value in the x-ray patterns have been observed. The shift is however, not more than -0.02 in d-value. In the x-ray examinations that have been carried out the uncertainty in d-value ± 0.01 at 30° 2 θ and ± 0.08 at 10° 2 θ. The shift is in other words small compared with the uncertainty so that no firm conclusions can be drawn from this. Even if this phenomenon can be discovered at low intensities only, it is also thinkable that it may occur where C_3AH_6 has high intensity, but that the shift is hidden because of the peak height.

Table 4.

The table shows C_2ASH_8 presence after 1 day and 1 weeks' curing of "Secar 80" and "Secar 51" with and without addition of "Elkem Microsilica". (R.P.H. = Relative peak height of C_2ASH_8 at the maximum, T_{max} = the temperature where C_2ASH_8 gave the highest peak, D.O. = Dominating area for C_2ASH_8.)

Composition of the cement	Cured in 1 day			Cured in 1 week		
	R.P.H.	T_{max} (°C)	D.O. (°C)	R.P.H.	T_{max} (°C)	D.O. (°C)
100% "Secar 80"	0	-	-	0	-	-
70% "Secar 80" 30% microsilica	0.8	60	50-70	1.0	50	30-70
50% "Secar 80" 50% microsilica	0.9	43	45-65	1.9	41	25-70
100% "Secar 51"	0.3	45	-	0.6	45	-
70% "Secar 51" 30% microsilica	1.4	40	42-55	1.5	40	20-55
50% "Secar 51" 50% microsilica	2.0	52	40-70	2.3	35-55	20-70

4.7   AH$_3$

Parker and Sharp (1982) mention AH$_3$ as a by-product from the formation of the different phases of the hydrates. CA$_2$ is the only of the original phases that gives AH$_3$ as a by-product at the formation of CAH$_{10}$. In the tests we have done we have not detected any AH$_3$ before C$_2$AH$_8$ is formed in any of the test series. This reinforces the above theory.

The ratio between the number of mol AH$_3$ that it is formed per mol C$_2$AH$_8$ and the number of mol AH$_3$ that is formed per number C$_3$AH$_6$ is 1:2 for CA, 3:5 for CA$_2$ and 2:9 for C$_{12}$A$_7$. In other words it is formed considerably more AH$_3$ by the formation of C$_2$AH$_8$. If one studies the figures 6-12, it is seen that AH$_3$ increase slowly during the formation of C$_2$AH$_8$ and that the rate of formation increase when C$_3$AH$_6$ is formed. The formation of AH$_3$ also follow the formation of C$_2$AH$_8$ and C$_3$AH$_6$ as is expected from the literature.

4.9   Minority phases

It is also demonstrated minor quantities of the phases α-C$_4$AH$_{13}$, C$_4$A$\bar{C}$H$_{11}$, and CaCO$_3$. For the sake of good order this have been drawn in to separate figures - figures 13 and 14. C$_4$A$\bar{C}$H$_{11}$ and α-C$_4$AH$_{13}$ exists by and large for the temperatures below 50°C with exception of the pure cement pastes where the range for C$_4$A$\bar{C}$H$_{11}$ is stretched all the way up to 80°C. According to Firens, Verhaegen and Verhaegen (1974) the α-C$_4$AH$_{13}$ is an intermediate product during the formation of C$_3$AH$_6$ and C$_4$A$\bar{C}$H$_{11}$. The formation of C$_4$A$\bar{C}$H$_{11}$ occurs either by direct carbonization of α-C$_4$AH$_{13}$ or via C$_3$AH$_6$. CaCO$_3$ exists mainly for temperatures above 40-50°C.

4.10   Conclusion

In pure "Secar 80" and pure "Secar 51" the dominating phases after 1 day are:

```
 0  -   20°C:  CAH₁₀
20  -   50°C:  α-/ β-C₂AH₈
50  - 100°C:  C₃AH₆  and AH3
```

After 1 week:

```
 0  -   20°C:  CAH₁₀
20  -   40°C:  (α-)/ β-C₂AH₈
40  - 100°C:  C₃AH₆  and AH₃
```

In cement pastes that contain "Elkem Microsilica" the dominating phases after 1 day were:

0	-	20°C:	CAH_{10}
20	-	50°C:	$\alpha-/C_2AH_8$
40	-	70°C:	C_2ASH_8
70	-	100°C:	C_3AH_6 and AH_3

After 1 week:

0	-	20°C:	CAH_{10}
20	-	70°C:	C_2ASH_8
70	-	100°C:	C_3AH_6 and AH_3

Minor quantities of the phases $\alpha-C_4AH_{13}$, $C_4A\bar{C}H_{11}$ and $CaCO_3$ were demonstrated in all the pastes.

By comparison of the x-ray diffraction patterns for pure microsilica and silica containing cement paste, unreacted microsilica was shown in the cement pastes after 1 weeks of curing.

5 References

Bensten, S.: "Mineral conversion during curing of calcium aluminate cement with addition of amorphous microsilica." (In Norwegian: "Mineralomsetning under herding av kalsiumaluminatsement tilsatt amorft silicastøv."), main paper by Institutt for silikat- og høytemperaturkjemi, Norges Tekniske Høyskole, 1984.

Bensten, S.: "pH in calcium aluminate cement with varying microsilica content." (In Norwegian: pH i kalsiumaluminatsement med varierande microsilicainnhold."), Institutt for silica for silica- og høytemperaturkjemi, Norges Tekniske Høyskole, 1984.

Borbov, B.S., Zaldat, G.I. Zalizovski, E.V., Izv. Akad Nauk. SSSR Neorg. Mater., 10, Page 2187, 1974, referred from Parker and Sharp.

Firens, P. Verhaegen A., Verhaegen J.P.: "Etude de la formation de l'hydrocarboaluminate de cacium.", Cem Concr. Res. Pergamon Press Inc., vol. 4, page 695-707, 1974.

Parker, K.M., Sharp, J.H. "Refractory Calcium Aluminate Cements.", J. Brit. Cer. Soc, 81, page 35-42, 1982.

Turrisiani, R. in "The chemistry of cement and concrete.". ed. H.W.F. Taylor, Acedemic Press, London, Vol. 1, page 243, 1964.

23 STUDY OF THE HYDRATION PROPERTIES OF ALUMINOUS CEMENT AND CALCIUM SULPHATE MIXES

J.P. BAYOUX, A. BONIN, S. MARCDARGENT
Lafarge Coppée Recherche, Viviers, France
M. VERSCHAEVE
Lafarge Fondu International, Neuilly-sur-Seine, France

Abstract
Hydration of Aluminous Cement/Calcium sulphate mixes has been studied either on neat paste mortars or diluted suspensions with the three different varieties of calcium sulphate : gypsum, hemihydrate, anhydrite.
We have shown that the properties, type of hydrates formed, expansion / shrinkage, mechanical strengths depend greatly on the nature of the calcium sulphate.   In every case anhydrite is favorable.
An explanation of the mechanisms has been proposed. It is based on the relative dissolution kinetics of the various phases available in the mix.
The very important role played by addition of lime has also been investigated.
Keywords : HAC, Ciment Fondu, Calcium Sulphate, Gypsum, Hemihydrate, Anhydrite, Lime, Mechanical Strength, Expansion, Shrinkage, Mineralogy.

1   Introduction

Mixes of Aluminous Cement and Calcium Sulfates are very often used nowadays in the formulation of products for the building industry and in various other types of application like for instance supporting the coal mine vault, transfering and distributing the load on to a metal structure.

The properties we have to achieve are different depending on the type of application. This requires a good knowledge of the mechanisms involved in every case in order to be able to propose the best solution. The aim of this work is to show how, depending on the nature and amount of calcium sulphate we can forecast the characteristics of the binary mixes. The complementary association of lime has also been studied. In every case the mechanisms are studied and explanations are given.

2   Experimental conditions

2.1   Raw materials
The aluminous cement we have chosen is Ciment Fondu Lafarge

(F) ground at 3000 cm^2/g. The calcium sulphates are Gypsum (G) Hemihydrate (SH) both coming from natural products and anhydrite (A) which is of synthetic origin.

All three calcium sulphates have a purity above 97 % and a fineness very similar to that of Fondu.

The Lime we have used is industrial hydrated lime (CH) free from calcite.

2.2 Mixes

The mixes that we have prepared are (F + G), (F + G + CH), (F + SH), (F + A), (F + A + CH). Each mix was studied with various proportions of calcium sulphate always expressed by their equivalence in hemihydrate (equ. SH). In every table the proportions % equu. SH are related to the amount of aluminous cement.

2.3 Tests

1 In neat paste Water/Solid = 0,30
- Setting time with the 300 g Vicat needle
- Strengths on 2 X 2 X 10 cm sample cured 24 hours in the humid cabinet then under water until testing at various dates up to three months
- Mineralogy of the solid phase at different dates after drying by solvent water exchange (acetone - ether)

2 In Reference Mortars [1] at water / solid = 0,4
- Measure of length variations :
     * $\Delta l/l$ "direct" when curing 14 days either in water or in air at 50 % relative humidity.
     * $\Delta l/l$ "delayed" after curing 14 days in a first media (air or water) and transfer into a second one (water or air)

3. In diluted suspensions (W/S = 20) stirred for 24 hours protected from CO_2
This technique is very convenient to analyse simultaneously the solid and liquid phases. These characterisations enable a better understanding of the hydration phenomena.

3   Results

3.1   Setting time
The following table 1 summarises the results.

Table 1 : Setting time (min) Initial / Final of Fondu/Sulphate
mixes

% equ. SH	F + G	F + SH	F + A
0	210/240	210/240	210/240
5	150/245	-	-
10	150/250	49/80	285/310
20	145/228	40/60	360/390
30	140/235	42/58	-/520

We can classify the calcium sulphates into two categories :

    (a) Both hydrated varieties (G) and (SH) are
accelerators, (SH) being the most effective. The shortest
setting times are obtained towards a 10 % equ. SH addition
of calcium sulphate.

    (b) The anhydrous variety (A) on the contrary retards
the set. The effect increasing with the amount of calcium
sulphate added.

3.2 Mechanical strengths
Graphs 1 below show the variations of strength measured on
neat pastes with the various calcium sulphates.

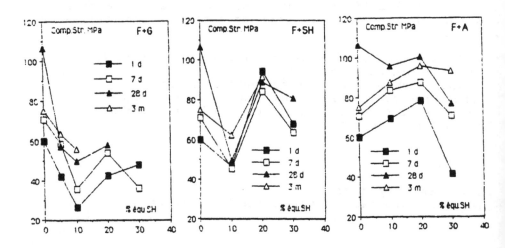

Figure 1 - Variations of Fondu/Sulphate mixes strengths

The mixes containing 30 % equ SH always show a considerable swelling and, moreover, many cracks appear. Because of this, it is meaningless to discuss the low or loss of strength measured.

In the case of additions below 20 % it is possible, as for the setting time, to distinguish two groups.

(a) Both hydrated varieties (G) and (SH), for which we have measured a severe strength loss with the addition of increasing amounts of calcium sulphates up to 10 % equ SH. This strength decrease appears relatively less at longer ages because of the conversion of the neat aluminous cement. The weakest strengths are achieved with (G).

From 10 % to 20 % equ SH additions there is an increase of the strength. The largest increase is for hemihydrate additions (SH) compared to gypsum (G). For (SH) its transitory hydration into (G) certainly favours good strengths at least at shorter ages.

(b) The anhydrous variety (A) for which we have measured a steady increase of the strength when the equ SH increases from 0 to 20 %.

3.3   Mineralogy

The variations of the mineralogy for the solids are given in table 2 for every mix. Neat Ciment Fondu behaves normally. There is a large amount of hexagonal CAH_{10} associated at longer ages with another hexagonal phase C_2AH_8. At 7 days gibbsite (AH_3) and the cubic hydrate C_3AH_6 appear increasing tremendously at 3 months to the detriment of the hexagonal phase C_2AH_8. We have not hydrated the whole amount of CA available in the ciment even after 3 months.

In the mixes with calcium sulphates after hydration we have found calcium sulfoaluminates, Ettringite and Monosulfoaluminate (MS). The higher the équ. SH % the larger the amount of Ettringite or monosulfo aluminate. The amounts depend on the nature of the calcium sulphate involved in the reaction.

It is possible to distinguish the same two groups as in 3.2 :

(a) Both hydrated varieties (G) and (SH) for which we have scarcely seen CAH_{10} and C_2AH_8. At low équ. SH very small amounts of C4AH13 can be revealed and monosulfo shows. At higher équ. SH Ettringite is the major hydrate that can be found.

(b) With the anhydrous variety (A), in the range from 0 to 10 % equ. SH additions, we have found as much CAH_{10} as in the neat Ciment Fondu with additionaly a fair amount of C_4AH_{13}.

At a 20 % equ. SH addition, Ettringite forms to the detriment of CAH_{10}. When increasing the addition up to 30 % equ. SH Ettringite is dominant, considérable quantities of gibbsite (AH_3) with it. Monosulfoaluminate is also present but it is of minor importance.

			XRD Peak heights (mm)									
		Date	CA	SH	G	A	M.S.	Ettring	CAH10	C2AH8	C4H13	C3AH6	AH3
		24 H	68	0	0	0	0	0	16	0	0	0	trace
F		7d	57	0	0	0	0	0	16	7	0	7	4
REFERENCE		28d	40	0	0	0	0	0	20	8	0	15	6
		3m	30	0	0	0	0	0	16	6	0	32	12
		24 H	60	0	5	0	8	7	0	0	11	0	0
	5 %	7d	57	0	0	0	13	0	0	0	7	0	0
	equ. SH	28d	53	0	0	0	trace	4	0	0	6	11	trace
		3m	35	0	0	0	0	0	0	0	11	21	13
		24 H	77	0	0	0	7	34	0	0	6	0	0
	10 %	7d	62	0	0	0	6	30	0	0	trace	0	0
	equ. SH	28d	66	0	12	0	3	70	0	0	0	trace	0
		3m	50	0	8	0	3	75	0	0	4	?	0
F + G		24 H	55	0	14	0	0	75	0	0	0	0	0
	20 %	7d	57	0	27	0	0	81	0	0	0	0	0
	equ. SH	28d	50	0	0	0	0	70	0	0	12	0	0
		3m											
		24 H	49	0	77	0	0	88	0	0	0	0	0
	30 %	7d	38	0	60	0	0	8	0	0	0	0	0
	equ. SH	28d	40	0	43	0	0	87	0	0	0	0	0
		3m	14	0	20	0	0	119	0	0	0	0	0
		24 H	74	0	0	0	10	42	0	0	trace	0	0
	10 %	7d	72	0	trace	0	8	21	0	0	7	0	0
	SH	28d	68	0	0	0	16	28	0	0	6	8	0
		3m	59	0	0	0	19	21	0	0	9	12	0
F + SH		24 H	59	0	14	0	0	77	0	6	0	0	0
	20 %	7d	54	0	11		0	79	0	4	0	0	0
	SH	28d	52	0	trace	0	0	61	0	7	0	0	0
		3m	52	0	0		0	70	0	6	0	0	0
		24 H	57	0	41	0	0	74	0	0	0	0	0
	30 %	7d	34	0	27	0	0	88	0	0	0	0	0
	SH	28d	31	0	18	0	0	101	0	0	0	0	0
		3m	27	0	0	0	0	137	0	0	0	0	0
		24 H	58	0	0	60	0	0	17	0	13	0	0
	10 %	7d	41	0	0	47	0	0	18	0	12	0	0
	equ. SH	28d	38	0	0	40	0	0	16	0	9	0	0
		3m	38	0	0	37	trace	0	19	0	8	0	0
F + A		24 H	36	0	0	110	0	8	10	0	17	0	0
	20 %	7d	35	0	0	108	0	9	14	0	13	0	0
	equ. SH	28d	35	0	0	109	9	13	12	0	13	0	0
		3m											
		24 H	30	0	0	155	10	43	0	0	0	0	0
	30 %	7d	31	0	0	139	10	47	0	0	0	0	0
	equ. SH	28d	30	0	0	145	9	48	0	0	0	0	0
		3m	25	0	0	134	12	49	0	0	0	0	0

Table 2   Solid phases mineralogy

3.4 Length variations

All the results are given on table 3 and graph 4 being associated.

Table 3 : Length variations $\Delta l/l$ at 14 days ($\mu m/m$) for Fondu/Sulphates mixes

| | Direct | | Delayed | |
	water	air	air $\longrightarrow$ water	water $\longrightarrow$ air
F	- 30	- 1150	650	- 600
F + G 25 % equ. SH	850	- 830	800	- 820
F + SH 15 % equ. SH	970	- 400	550	- 920
20 % equ. SH	1980	- 210	1020	- 690
F + A 25 % equ. SH	210	- 710	750	- 480

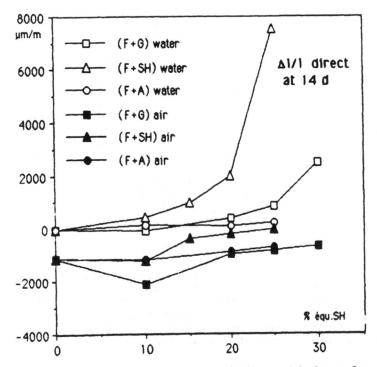

Figure 4 - length variations $\Delta l/l$ at 14 days for Fondu/Sulphates mixes

These show that :
    - Ciment Fondu on its own when cured in water has a
very minor shrinkage (- 30 µm/m)whereas in the air it is
very large (- 1150 µm/m). Delayed variations as defined in
2.3.2 are reversible. The sample swells when transfered
from air to water and shrinks by the same amount when
transferred back to air. Concerning the mixes it is clear
that the expected effect of Ettringite formation on the
reduction of shrinkage in air is limited. The best
reduction has been achieved in the case of F + SH mixes
with a equ. SH % which tends to a limit, beyond which the
sample starts cracking. With (F) + (A) mixes the swelling
under water is by far the smallest among all mixes.
Concerning this characteristic (A) again differs from other
hydrated calcium sulphate varieties (G) or (SH).

3.5   Hydration in diluted suspensions

Mixes (F) + (G), (F) + (SH), (F) + (A) have all been
studied all with 20 % equ. SH additions of the various
forms of calcium sulphate. Figure 5 and table 6 show the
variations of the mineralogy of the solid phase on one hand
and the ion composition on the other. In this paper we
have not presented the case of (F) + (SH) since after the
very quick hydration of hemihydrate into gypsum we return
to the (F) + (G) mix.

    (a) With gypsum (G) the solution is very quickly rich
in sulphate and calcium ions. The amount of aluminium ions
remains low till the massive precipitation of Ettringite.
Beyond this massive precipitation the situation is
different. All the gypsum having reacted with the
aluminous cement the solution becomes nearly free of
sulphate ions. The Calcium / Aluminium molar ratios in the
solution are high and vary greatly from 4,5 to 2 with a
peak at 12 after 9 hours.
    At the same time the analysis performed on the solid
shows that the main hydrate is Ettringite which increases
very quickly. We have no monosulfoaluminate because of the
high level 20 % equ SH present in the mix.

    (b) With the anhydrous calcium sulphate (A) the ions
available in the solution are different. At the beginning
it is aluminium and calcium ions which are at highest
concentrations whereas sulphate ions are depressed. This
situation remains even after the massive precipitation of
Ettringite. The Calcium / Aluminium ions concentration
ratio in the solution are low and vary slightly from 1.6 to
1.5 with a depression of 1,1 at 5 hours.

Table 6 : Analysis of F + G and F + A solutions

F + G at 20 % equ. SH   Analysis of the solution

date	pH	SiO2 mM/l	CaO mM/l	Al2O3 mM/l	Fe2O3	SO3 mM/l	CaO/Al2O3 mM/l
3 min	11,05	0	19,73	3,77	0	14,55	5,23
10 min	10,78	0	19,41	4,45	0	13,66	4,36
45 min	10,68	0	18,70	3,94	0	13,79	4,75
1 h 30 min	10,75	0	18,34	3,82	0	14,03	4,8
3 h 30 min	10,75	0	17,38	2,68	0	14,55	6,49
6 h 00 min	10,45	0	16,02	1,48	0	14,49	10,82
9 h 10 min	10,46	0	14,79	1,22	0	14,59	12,12
10 h 20 min	10,57	0	11,71	1,52	0	11,28	7,70
11 h 40 min	11,47	0	5,32	4,01	0	0,90	1,33
12 h 15 min	12,17	0	7,77	6,17	0	0,00	1,26
13 h 00 min	12,45	0	10,29	7,78	0	0,04	1,32
13 h 30 min	12,37	0	9,68	7,33	0	0,00	1,32
24 h 00 min	12,55	0	7,05	4,81	0	0,00	1,47
30 h 00 min	-	0	6,73	4,31	0	0,00	1,56

F + A at 20 % equ. SH   Analysis of the solution

date	pH	SiO2 mM/l	CaO mM/l	Al2O3 mM/l	Fe2O3	SO3 mM/l	CaO/Al2O3 mM/l
3 min	11,45	0	10,09	6,36	0	2,89	1,59
10 min	11,60	0	13,07	9,11	0	2,66	1,43
30 min	11,50	0	12,61	10,05	0	1,58	1,25
1 h 30 min	11,57	0	13,30	11,21	0	0,76	1,19
2 h 55 min		0	13,30	11,93	0	0,33	1,11
3 h 00 min	11,85	0	14,30	12,44	0	0,26	1,15
4 h 00 min		0	14,52	13,00	0	0,19	1,12
5 h 00 min	12,75	0	15,11	13,63	0	0,11	1,11
6 h 00 min	12,97	0	16,80	14,83	0	0,10	1,13
7 h 00 min	11,69	0	18,13	15,69	0	0,09	1,16
8 h 00 min	11,86	0	18,48	16,23	0	0,08	1,14
9 h 00 min	11,90	0	17,86	14,90	0	0,06	1,20
10 h 10 min	12,93	0	9,29	7,31	0	0,48	1,27
10 h 50 min	12,71	0	8,29	6,56	0	0,70	1,26
12 h 40 min	12,43	0	5,77	4,49	0	0,79	1,29
14 h 20 min	13,00	0	6,93	5,70	0	0,19	1,22
22 h 50 min	12,25	0	7,36	4,88	0	0,00	1,51
24 h 00 min	13,00	0	7,16	4,75	0	0,16	1,51

Again it is clear that both hydrated calcium sulphates greatly differ from the anhydrous one.

The analysis of the solid phase show that there is a
transitory formation of Monosulfoaluminate before the
precipitation of Ettringite occurs. Later on the
Ettringite still increases slightly at the expense of
Monosulfoaluminate. We have not measured any CAH_{10} or
C_4AH_{13} as we did in the neat paste this being a result of
the great dilution in the case of the suspensions.

Furthermore, whatever the calcium sulphate, the massive
precipitation of hydrate always coincides with variations
of the pH and depressions of the ionic concentrations in
the solution.

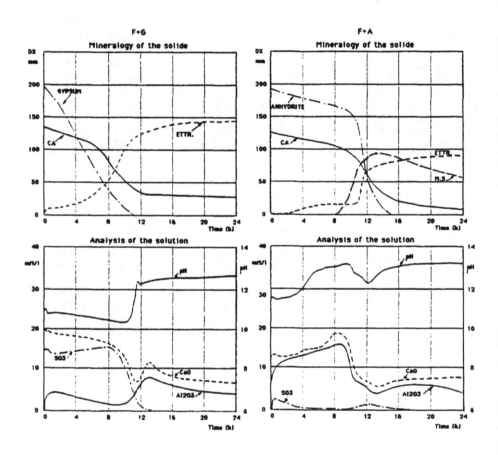

Figure 5 : Comparison of Mineralogy and Analysis of
solution for F + G and F + A mixes.

3.6 Influence of additions of hydrated lime to Aluminous Calcium sulphate mixes

When hydrating monocalcium aluminate CA in presence of calcium sulphate Ettringite is formed as shown by the first equation below.

This equation also shows that AH_3 is formed at the same time. This AH_3 is very poorly cristalline which explains why it has never been identified by the x ray diffraction technique.

In order to take advantage of all the aluminium available in the CA it is necessary, as explained by the second equation below, to add lime. As a consequence this should lead to the formation of Ettringite on its own with a possible increase of the strength of the hydrated mixes.

$$3 \ CA + 3 \ C\bar{S} + 32H_2O \longrightarrow C_3A, \ 3C\bar{S}, \ 32H_2O + 2AH_3$$
$$3 \ CA + 6 \ CH + 9 \ C\bar{S} + 90H_2O \longrightarrow 3(C_3A, \ 3C\bar{S}, \ 32H_2O)$$

Such additions of lime have been tried on F + G mixes containing 20 % eq SH. Two different ways of adding the lime have been tried. Lime has been added either as a solution (200 mM/l) or as a solid. The percentages of Hydrated lime introduced are 3 % of the total weight of the mix in the case of the solution and 5 % and 10 % in the case of solid additions.

Graphs 6 show the results. It appears that :

(a) When the hydrated lime is added as a solution (3 % of Ca (OH)$_2$ calculated on the total mix weight) the composition of the solution when the hydration proceeds is very similar to that of the mix without any lime added to it. However this addition of lime speeds up the hydration reaction without changing the nature of the hydrate which is formed which remains Ettringite.

(b) When adding 5 % of hydrated lime in the solid state which represents slightly more than we have added in form of a solution, the formation of Ettringite is accelerated but to a lesser extent than with 3 % in solution.

(c) When adding 10 % of solid hydrated lime the reaction is much slower than that of the mix without lime. The amount of calcium ions in the solution is very high whereas sulphate ions'concentrations are not modified and aluminium ions'concentration totally depressed. The aluminium ions' concentration remains very small till the precipitation of the hydrates. This time the reaction forms Ettringite, C_4AH_{13} at short ages and monosulfoaluminate at longer ages. Similar tests have been carried on a (F) + (A) mix containing 20 % equ. SH. The results are the same, that is acceleration of the formation of hydrates when hydrated lime is added in small amounts

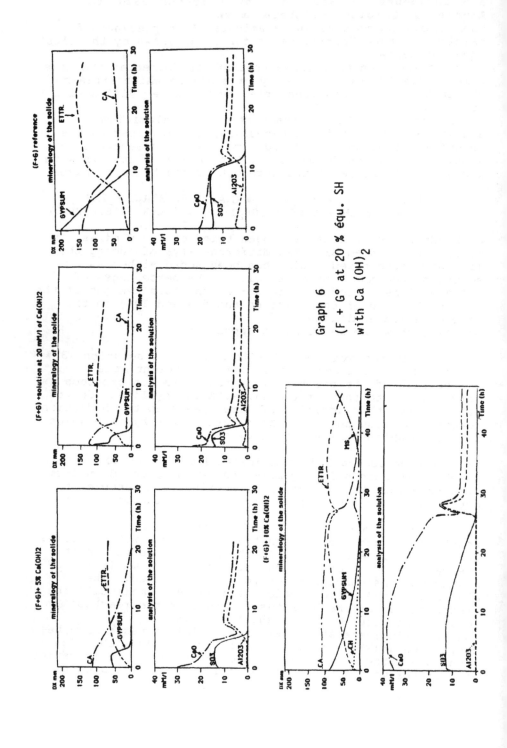

Graph 6
(F + G° at 20 % équ. SH
with Ca (OH)$_2$

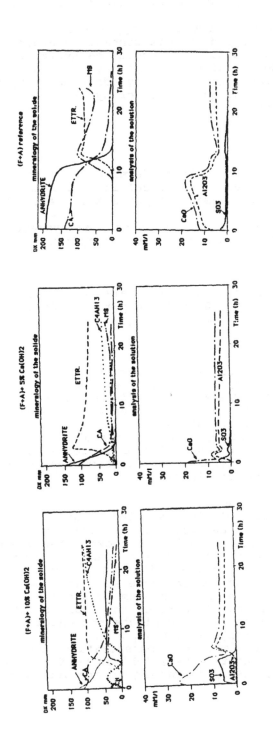

Graph 6 (following)

(F + A) at 20 % équ. SH
with Ca(OH)$_2$

and retardation with large amount of solid hydrated lime added.

Comparing Anhydrite and Gypsum it appears that it requires more lime to retard the mix containing anhydrite. As an example with 10 % solid lime added the hydration is still accelerated whereas with the gypsum the hydration was slower than that of the mix without any addition. Anhydrite mixes tolerate significant quantities of hydrated lime better than gypsum does.

4   Discussion

This work has shown a systematic difference between Anhydrite and the Hydrated forms of calcium sulphate. Anhydrite's behaviour is in many respects much better than Gypsum or Hemihydrate.

Considering only Gypsum and Anhydrite (Hemihydrate can be assimilated to Gypsum as soon as it has hydrated) it has been proven that :

Solubility
At saturation the ionic concentrations are very similar with Anhydrite and Gypsum. At room temperature Anhydrite tends to dissolve a little more than Gypsum, above 42°C it is the opposite.

Dissolution kinetics
Measures of the dissolution kinetics for gypsum Anhydrite and ciment Fondu give respectively 4,65 mM/l/min - 0,59 mM/l/min and 1,44 mM/l/min. It is important to note that ciment Fondu is intermediate between Gypsum and Anhydrite and very close to the latter.

Because of these differences in the dissolution kinetics the composition of the solution in terms of ion concentrations will be determined by the component which has the highest dissolution kinetic.

* With Gypsum
As shown on graph 5 the solution contains large amounts of calcium and sulphate ions and very little aluminium ions. Calcimum Aluminate going into solution slower than gypsum there are a few Aluminium ions and as a consequence the CaO/Al_2O_3 ratio is high. The precipitation of the hydrates will be very much dependant on the rate at which the solution is being fed with the Aluminium ions coming from the monocalcium aluminate. Because of the high CaO/Al_2O_3 ratio the hydrates which will precipitate are calcium monosulfoaluminate, Ettringite or C_4AH_{13}.

In neat paste, at low sulphate concentrations because of the small amount of Ettringite which can be formed and because of the lack of CAH_{10} the strengths are very poor.

As soon as the sulphate percentage increases there is more Ettringite forming and the mechanical strengths increase. Unfortunatly when the sulphate amount gets too high the swelling of the sample due to the very large amount of Ettringite being formed leads to crack formation damaging the strength of the sample.

* With Anhydrite

The solution contains large amounts of calcium (graph 5) and Aluminium ions coming from the monocalcium aluminate which goes into solution quicker than the Anhydrite. The solution is poor in sulphate ions. The CaO/Al_2O_3 is low and close to that of monocalcium aluminate. This time it is the Anhydrite going slowly into solution which will regulate the kinetics of precipitation of the hydrates. Because of the low CaO/Al_2O_3 ratio and in the case of small quantities of Anhydrite being added, CAH_{10} can be precipitated. This explains why in these conditions the mechanical strengths remain very good. When a large amount of Anhydrite is added we still have crack formation damaging the mechanical strength.

As a summary it can be said that, it is the compound having the higher dissolution kinetic which will determine the nature and the amount of the ions in the solution at the very beginning of the reaction with water.

These concentrations determine the nature of the hydrates which will precipitate according to the CaO/Al_2O_3 ratio. The kinetic is dependant on the rate at which the solution is being fed with ions required for the precipitation of the hydrates. The dissolution kinetic can be a limiting factor. Nevertheless we can also assume that mechanical blocking of the reaction may occur.

This blocking can result from the precipitation of impervious illcristallized hydrates covering the anhydrous phases such as, for example, alumina gel. This will cease as soon as these hydrates cristallize. These assumptions are to be compared to the blocking of tricalcium aluminate in the case of Portland cements where the poorly crystallised Ettringite covering the C_3A prevents its hydration [2].

Concerning the addition of hydrated Lime to ciment Fondu / Calcium Sulphate mixes the situation is not absolutly clear. Generally speaking, when adding lime to the mixes we impose a high calcium ion concentration of the solution resulting (because of the ionic equilibrium) in a global reduction of the dissolution of both calcium aluminate and calcium sulphate.

This reduction of the dissolution kinetics should disappear as soon as the lime is combined.

With small additions of lime, the excess calcium is combined rather quickly and we return to the situation without lime addition. Furthermore the fast combination of the lime with the alumina gel explains why, in this case, the hydration reaction of the mixes is accelerated [3]. (The alumina gel is not favorable to the mobility of the ions). With large additions of lime it takes a rather long time to get back to the situation without lime addition. Various publications [4] explain that the lime has a blocking effect of the Ettringite seeds by adsorption on their surface preventing growth.

5   Conclusion

Hydration of aluminous cement - calcium sulphates mixes with formation of calcium sulpho aluminates is always possible whatever the calcium sulphate being used.
   Nevertheless, depending on the nature of the Calcium Sulphates, both hydrated forms on one hand and anhydrite on the other, the resulting hydrates differ and consequently also the mechanical strengths.
   Whatever the characteristics, it is with anhydrite that we can obtain the best results particularly mechanical strength which remains at a high level up to 20 % equ. SH additions.

- Hemihydrate is the best calcium sulphate to reduce hydration shrinkage of Ciment Fondu but it implies its use in large percentages which are fairly close to the point beyond which cracks appear in the structure.

- Typical different behaviours of calcium sulphates can be explained by their different dissolution kinetics.

- Additions of lime aiming at combining the maximum of the aluminous cement into Ettringite leads to different reactions. A small addition plays the role of an accelerator whereas a large one, still smaller than that corresponding to the stoechiometry of the reaction, leads to a dramatic reduction of hydration kinetic.

6   References

[1] References mortars according to AFNORPIS-403
[2] B. COTTIN (1979) Hydratation et expansion des cements
    An. Cim. Fr 4 pp 139-144
[3] A. BONIN, B. CARIOU (1980) On the calcium aluminate
    gypsum lime system. 7th International Congress on the
    Chemistry of Cement Paris.
[4] P.K. MEHTA (1973) Effect of lime on hydration of pastes
    containing Gypsum and Calcium Aluminates or Calcium
    Sulfoaluminates - Journal of the American Ceramic
    Society Vol 56 n° 6 June.

24 ETTRINGITE-BASED CEMENTS

S.A. BROOKS
Pozament Ltd, Burton-on-Trent, UK
J.H. SHARP
School of Materials, University of Sheffield, Sheffield, UK

Abstract
Ettringite is frequently regarded as an undesirable
substance to constitute a major phase in a hardened cement
paste, due to its role in sulphate attack. For many years,
however, ettringite has been used as a structural component
of a diverse range of cementitious systems. One such system
involves the use of ettringite-based cements in the mining
industry. These materials must set rapidly and develop
early strength, yet be capable of being pumped long
distances and be stable when left for long periods, often in
excess of 24 hours, without setting in the pipelines. The
development of suitable low cost systems is reviewed and
their reliance on ettringite discussed. Systems in current
use based on the reaction of high alumina cement with
calcium sulphate in the presence of lime to form ettringite
are emphasised. The importance of the composition is
discussed, indicating the fine balance required between the
components of the system for optimum performance at a
reasonable cost.
Keywords: Calcium Aluminates, Ettringite, Expansive
Cements, High Alumina Cement, Mine Cements, Rapid Setting
Cement, Sulphate.

1 Introduction

Ettringite is well known as a minor, although important,
hydration product of Portland cement. Gypsum or some other
form of calcium sulphate is added to Portland cement clinker
to retard the rate of hydration of C_3A, with the consequent formation of ettringite, $C_6A\bar{S}_3H_{32}$. When the calcium
sulphate has all been consumed, the ettringite is trans-
formed into monosulphate, $C_4A\bar{S}H_{12}$. Monosulphate is
vulnerable to sulphate attack and its conversion back into
ettringite is accompanied by a large increase in volume,
which can lead to catastrophic failure of the concrete. This
is because the density of ettringite is very low (1730
kg/m^3), hence it occupies a large amount of space for a
given mass of material. As a result ettringite is

frequently regarded as an undesirable substance to have as a major constituent of a hardened cement paste. In spite of this ettringite has been used as a structural component of various cementitious systems for many years.

The way in which the formation of the calcium sulpho-aluminate is utilised within such cementitious systems varies with the application. Ettringite formation was the first reaction to be used in expansive cements and today the powerful expansive force associated with its production is used in expansive cements, in the reaction of some additives used to impart expansion to Portland cement mixes and even for non-explosive rock bursting.

One of the earliest forms of expansive cement was developed by Mikhailov (1960, 1973), whose self-stressing cement was composed of 18% high alumina cement (HAC), 15% gypsum and 67% ordinary Portland cement (OPC). Such expansive cements have been used in the USSR since 1942, because they do not experience the problems of shrinkage during setting and hardening which other cements exhibit. They were used to rebuild war-damaged structures, in subway construction to replace lead for caulking joints of cast iron tubing to guarantee water tightness, and in coal pit bracing.

Mikhailov (1960, 1973) also described water-proof cements in the form of quick-setting and rapid-hardening hydraulic binders, which reached high strength in several hours. These were obtained by dry grinding mixtures of HAC, gypsum and hydrocalumite, C_4AH_{13}. Budnikov (1960) discussed the role of gypsum and anhydrite in such cements and demonstrated the potential for cements based on ettringite as efficient binding agents.

Ettringite formation can also give the advantage of high early strength with or without the expansive reaction. Galer and Webb (1983) described a non-expansive rapid hardening ettringite composition for the production of cement boards, while Long et al.(1987) described a range of rapid hardening cements which are inherently expansive or shrinkage compensating, produced from a special cement combining $C_4A_3\bar{S}$ with Portland cements.

If high strength is not required ettringite is very suitable for the production of high yield, low solids systems, such as the pump packing materials to be described below. Other systems use ettringite to produce very light-weight foams, some with density as low as 150-200 kg/m^3 at 1.2:1 water:solids (w:s) ratio, a yield of less than 100 kg/m^3. The use of an even higher w:s ratio of 5:1 can give rise to a very high yield, low cost, low strength infill material.

Another important property of ettringite-based cements can be their resistance to sulphates and acid ground waters. This was exploited in supersulphated cement, as described in British Standard BS 4248:1974. Dense concretes with a water:cement (w:c) ratio of 0.45 or less based on this

Table 1. Composition of expansive cements

Type	SiO$_2$	Al$_2$O$_3$	CaO	SO$_3$	Source of alumina	Relevant mine cements
K	21.6	4.8	62.4	6.2	$C_4A_3\bar{S}$	type IV
M	18.4	7.7	61.1	4.7	$CA + C_{12}A_7$	types I,II,III
S	18.3	8.8	63.7	5.3	C_3A	none

cement have a resistance to weak solutions of mineral acids down to pH 3.5. Supersulphated cement (SSC) was invented by Kuhl in Germany in 1908 as a means of combating sulphate attack. Such cements were extensively used until the wide-scale introduction of sulphate-resisting Portland cement and blended cements containing ground granulated blast-furnace slag as a partial replacement for Portland cement. SSC is comprised of 80-85% granulated blast-furnace slag activated by calcium hydroxide in the presence of calcium sulphate. The lime is usually produced in situ by the hydration of added Portland cement clinker. Midgley and Pettifer (1971) showed that the increase in strength up to 7 days was due to the rapid formation of ettringite. Subsequently calcium silicate hydrate gel was also formed and contributed to the ultimate strength of the cement. The ettringite was formed as long laths (120 µm long, 0.5 µm in diameter).

This morphology of ettringite suggests other potential uses. For example, a US patent (4,140,540) describes a method of producing ettringite fibres up to 150 µm in length, to be used as complete or partial replacement for asbestos fibres, while a Russian patent (2,165,089/33) describes a method of improving strength by incorporating needles of ettringite claimed to be up to 2 mm long.

Three reactions are used in practice as the basis of the expansive cements, called types K, S and M (Kesler et al., 1970; Mehta, 1973). In each case the formation of ettringite is the source of the expansive energy (Kalousek, 1973). The differences between these cements, as can be seen from the data given in Table 1, relate to the source of alumina, but their practical application is based on control of the rate and extent of formation of ettringite, according to the equations:

$$C_4A_3\bar{S} + 8\ C\bar{S}H_2 + 6\ C + 80\ H\ =\ 3\ C_6A\bar{S}_3H_{32} \quad \ldots\ldots(1)$$

$$CA + 3\ C\bar{S}H_2 + 2\ C + 26\ H\ =\ C_6A\bar{S}_3H_{32} \quad \ldots\ldots(2)$$

$$C_3A + 3\ C\bar{S}H_2 + 26\ H\ =\ C_6A\bar{S}_3H_{32} \quad \ldots\ldots(3)$$

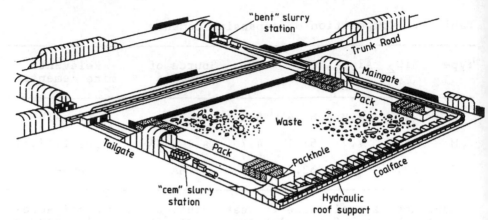

Fig. 1. Use of a pumped pack system in a British coalmine

The composition of several of the more important mine
cements to be discussed below is similar to that of type M
expansive cement, except that the w:s ratio is much higher
(up to 2.5, cf. 0.5). The chemistry of the formation of
ettringite is, however, the same.

It is evident that there have been many and varied
cements based on production of ettringite. Some such
cements have been used (and accepted) for load-bearing
construction work. With the current emphasis on the use of
ettringite-forming cements in mining applications, it is
opportune to recall this long history of ettringite-based
cements in alternative applications.

2 Mine cements

2.1 Background
Longwall advancing is the main method used for the
underground mining of coal seams in European coalfields
(Whittaker et al., 1980). Parallel service tunnels, often
described as roadways, are driven to the coal face, to which
they are connected (Fig. 1). The coal and debris are
removed along these tunnels, which are also necessary to
allow transport of men and materials to the coal face as
well as allowing ventilation of the mine.

To maintain contact with the coalface as it advances, the
roadways have to be progressiveley extended. The roof above
the extracted coal (i.e. in the space between the roadways)
is allowed to collapse as the hydraulic powered supports ad-
vance. The roadways are primarily supported by steel arches
but to prevent these from being crushed, additional support
is required alongside the roadways on the waste side.

A number of different ways of supporting the roadways had
been used in the United Kingdom up to about 1970, including
wooden chocks or concrete blocks which were built into

columns behind the steel arches (Whittaker and Titley, 1971). A technique used in Germany at that time involved the use of anhydrite, which was either blown or pumped to the location requiring support through pipes and which would harden on contact with water (Whittaker et al., 1980). Attempts to use anhydrite packing in the U.K. failed because of the extremely high costs of the equipment and pipes, the general lack of large volumes of distributed compressed air in Britsh mines and the different nature of British coal mining strata. The strata immediately adjacent to seams exploited in the U.K. are generally weaker than the German equivalent and the coal is harder. Anhydrite is of a high compressive strength and has no yield before or after failure. The effect is to promote puncture of the roof and floor and associated floor heave.

A British technique used to produce a "monolithic" support without using compressed air was to generate a filler slurry of bentonite and coal fines with water. This was pumped by means of a large reciprocating piston pump via 100 mm steel pipes to a shuttered packhole behind the face where a rapid setting cement slurry was prepared and injected to harden the mass. Various patents were taken out at this early stage (e.g. UK patents, 1,362,954 and 1,387,075).

An improvement to this method known as the Warbret system involved adding set accelerators, e.g. triethanolamine and alkali metal carbonates suspended in a prehydrated bentonite-coal waste slurry (Kellet and Mills, 1980 a), which resulted in a longer pump life than the rapid setting cement slurries. Even at this early stage it was apparent that the cost and the pump life were critically important factors in the development of an acceptable system.

2.2 Ettringite Mine Cements

To take advantage of ettringite formation from CA (by reaction 2) as well as from C_3A (by reaction 3), Kellet and Mills (1980 b) developed a system, which we will call type I mine cement, that incorporated some HAC and calcium sulphate in the cement slurry in addition to OPC and some minor components. A typical composition is shown in Table 2. All the components of the "cem" slurry were mixed together, while the second slurry contained a mixture of bentonite and solids from the mine.

It is well established that additions of HAC to OPC and vice-versa act as accelerating agents, indeed many such mixtures flash set on addition of water. However, the combination of HAC, calcium sulphate and lime (including that liberated by the hydration of OPC) in this cement, leads to the rapid formation of ettringite, when combined with water, resulting in rapid strength development at the relatively high w:s ratio used.

Although this type I cement was in many respects a considerable improvement over earlier products, it still suffered from its relatively short pump life, which resulted

Table 2. Chemical composition of some mine cements

	Type I Kellet & Mills (1980 a)	Type II Kellet & Mills (1980 b)	Type III Beale & Viles (1982)
	"cem slurry"	"cem slurry"	"cem slurry"
OPC	54	66.65	-
HAC	25	18.4	100
			"bent slurry"
β-CaSO$_4$	18.8	14.7	72.0
CaSO$_4$.2H$_2$O	-	-	5.0
citric acid	0.2	0.25	-
Na$_2$CO$_3$	2	-	-
	"bent slurry"	"bent slurry"	
Na$_2$CO$_3$	-	50	-
Li$_2$CO$_3$	-	-	0.4
CaO	-	-	5.0
bentonite	+	50	17.6
mine solids	+	-	-

in frequent stoppages during application (Nixon and Mills, 1981; Bexon, 1986), and the high capital cost of the equipment and pipes required to pump the coal fines, which could otherwise be sold.

A breakthrough, described by Mills (1988) as the "quantum leap" in the development of mine cements, was to combine the inventive steps of the two patents described above. The resultant product (Kellet and Mills, 1980 b) is based on the chemical reaction leading to the formation of ettringite to provide rapid setting and early strength development. The accelerator, sodium carbonate, was transferred from the cementitious slurry to the bentonite slurry. Neither mine waste solids nor fine coal was incorporated, and these aggregate materials were effectively replaced by water, leading to considerable cost savings and allowing more coal to be sold. The composition of this type II cement is given in Table 2 and it can be seen to be essentially similar to that of type I cement being composed of two slurries, known as a "cem" slurry and a "bent" slurry. The slurries were pumped through pipes and mixed at the desired location in the ratio of 9 parts of "cem" slurry to 1 part of "bent" slurry, giving an ultimate strength in excess of 5MPa.

The transfer of the alkali carbonate accelerator from the "cem" slurry to the "bent" slurry resulted in significant

Table 3. Comparison of mine cements

Principal source of	Function	Supplied by ....in cement		
		Type II	Type III	Type IV
lime	necessary for	OPC	lime	hydralime
sulphate	ettringite	β-CaSO$_4$	β-CaSO$_4$ & CaSO$_4$.2H$_2$O	β-CaSO$_4$
aluminate	formation	HAC & OPC	HAC	C$_4$A$_3\overline{S}$
suspending agent	to maintain suspension until setting occurs	bentonite	bentonite	bentonite
accelerating agent	to accelerate strength development	Na$_2$CO$_3$	Li$_2$CO$_3$	alkali metal salt

improvements in the pump life (to about 20 minutes) and a corresponding reduction in the number of blocked hoses, allowing the cement slurry to be pumped up to 500 m (Mills, 1988). Once these type II mine cements became established, it was apparent that modifications to allow longer pumping distances would be advantageous.

Whereas the type I and II cements developed by Kellet and Mills incorporated HAC as an important component leading to the formation of ettringite, the type III mine cement, the subject of a patent by Beale and Viles (1982), was based on the formation of ettringite from HAC as the largest single component in the mix. The ratio of HAC:total C$\overline{S}$ was maintained at a similar value to that used in the earlier cements, but OPC was omitted, and the other phases were transferred to the "bent" slurry. Sodium carbonate was also omitted, but a small addition of lithium carbonate, a known accelerator for HAC (Robson, 1962), was added to the "bent" slurry (Table 2). The two slurries were mixed in a 1:1 ratio at the packhole and the "cem" slurry pump station was moved next to the "bent" slurry station much further from the coal face (Fig. 1). The high contents of HAC and calcium sulphate lead to the formation of greater quantities of ettringite via equation (2), at the expense of the additional hydrates formed from OPC, which did not contribute much to the strength developed by type I and II cements in the first few hours.

The function of the various components of these mine cements is explained in Table 3. Since 1982 continued evolution of type III cements has resulted in systems with

pump lives measured in days rather than minutes, which allow
pumping distances of several kilometres, with useful
strengths being achieved at a w:s ratio as high as 2.5:1.
Such systems are more cost effective and require less material to be transported underground than the type II cement.

Alternative products were developed based on different
chemical reactions. Long et al. (1987) described a cement
based on the formation of ettringite from $C_4A_3\bar{S}$ via reaction
(1), called here type IV cement, which is currently used in
some British mines. Its composition is compared wiyh type
II and III cements in Table 3 and it has a close relation to
type K expansive cement. Mills (1984) described an attempt
to avoid the use of the relatively expensive HAC by
returning to an accelerated OPC-based system involving a
slurry of OPC mixed with a second slurry containing PFA and
aluminium sulphate. An alternative solution based on mixing
a special clinker, rich in $C_{12}A_7$ and C_2S, with an OPC
contained in a first slurry, with anhydrite, sodium
carbonate (accelerator) and bentonite in a second slurry has
been described by Longman and Drew (1985). Ettringite is
formed by the reaction:

$$C_{12}A_7 + 21\ C\bar{S} + 9\ CH + 207\ H = 7\ C_6A\bar{S}_3H_{32} \qquad \ldots\ldots(4)$$

2.3 Chemistry of type III mine cements

Eight laboratory methods of synthesising ettringite were
described by Berman and Newman (1960). One of these was the
reaction between calcium sulphate solution and a solution of
calcium aluminate obtained by shaking HAC with water. Since
ettringite dissolves incongruently in pure water, Berman and
Newman (1960) emphasised the need to have excess lime and
also excess sulphate ions in solution at the completion of
the reaction.

This work and earlier Russian work provide the origins of
the type III mine cements. The ratio of HAC:total $C\bar{S}$ is not
far removed from that of 1.43 used by Mikhailov (1960,
1973), while the need to consider carefully the initial lime
concentration was discussed by Budinkov (1960). Beale and
Viles (1982) emphasised that care should be taken to
optimise the quantity of lime used to obtain sufficiently
rapid and sufficiently high early strength development,
together with high ultimate strength development (meaning at
least 4 MPa after 7 days in these systems). Too little lime
resulted in inadequate strength development, but too much
could lead to the formation of a gel which did not form a
hardened cement.

Much of the claimed advantage of type III mine cement
over type II related to the removal of OPC, yet the role of
this in type II cement was in part to act as a source of
lime. Added lime is an essential component of type III
cements to satisfy equation (2).

Although the ratio of CA:$C\bar{S}$ required by equation (2) is

1:3, this ratio is not used in practice, since the reaction proceeds via a through solution mechanism and involves the nucleation of growth of ettringite crystals from a super-saturated solution (Taylor, 1964; Lea, 1970). The rate of dissolution of the anhydrous phases and the concentration of ions in solution control the rate of precipitation and the morphology of the ettringite formed and hence the strength development of the systems. This point is discussed further in section 2.5.

The dissolution of calcium hydroxide is strongly influenced by the pH of the solution, which was controlled in type I cement by the addition of sodium carbonate. Too much calcium sulphate in solution suppresses the solubility of the lime and it is important to obtain a satisfactory balance between β-anhydrite and more soluble sulphate salts. Minor components, especially alkalis, have an important role, e.g. increasing the area of stability of ettringite within the $CaO-Al_2O_3-SO_3-H_2O$ system (Taylor, 1964).

A possible alternative reaction between calcium monoaluminate and calcium sulphate requires a 1:1 ratio:

$$3 \text{ CA} + 3 \text{ C}\overline{\text{S}} + 38 \text{ H} = C_6A\overline{S}_3H_{32} + 2 \text{ AH}_3 \quad \ldots \ldots (5)$$

This reaction should be avoided in rapid setting cements due to the less efficient formation of ettringite per mole of CA compared with reaction (2), and in practice packing systems tend to utilise HAC:$C\overline{S}$ ratios in the range 1.25-1.4:1. The commercial HACs used in type III mine cements contain appreciable amounts of iron and silicon as well as calcium and aluminium. Although CA is the principal crystalline phase present, other phases can be detected by X-ray diffraction and amorphous material is present. It is difficult, therefore, to specify the optimum ratio of HAC:$C\overline{S}$, as opposed to that of CA:$C\overline{S}$, because the amount of CA in the HAC is variable and the extent to which $C_{12}A_7$, ferrite and amorphous material contribute to such systems is undetermined. A further complication is that strength development is dependent upon the chemical and physical characteristics, including the surface area, of the HAC used.

Mehta (1973) demonstrated the potential for rapid formation of ettringite in a system based on CA and pointed out the importance of investigating the effect of lime additions to mixtures of calcium aluminates and calcium sulphate. In particular, he showed that the presence of lime enhances the formation of ettringite from CA, whereas it retards the formation from C_3A. Mehta (1973) suggested that the ettringite formed in the presence of calcium hydroxide is colloidal and can absorb large quantities of water.

The choice of the source of the calcium sulphate used in ettringite based high water solids systems is very important, the best results being obtained by using β-calcium sulphate. A convenient, cost-effective and consistent

Table 4. The early strength of compositions related
to type III cement

HAC/β-C$\overline{S}$	HAC/total C$\overline{S}$	C$\overline{S}$H$_2$(%)	CaO(%)	2-hour strength(MPa)
∞	1.30	77.2	5.8	0.04
2.60	1.30	38.5	5.0	0.14
1.72	1.30	19.0	5.0	0.24
1.50	1.28	11.5	5.0	0.50
1.39	1.30	5.0	3.5	0.37

source is synthetic anhydrite produced as a by-product of
the manufacture of hydrofluoric acid. In addition a more
soluble sulphate salt is required at much lower concentrations to initiate and assist ettringite production. Too much
or too little "soluble" sulphate present at the start of the
reaction is detrimental to strength development as illustrated by the data of Beale and Viles (1982) shown in Table 4.
The early strength increased with decreasing "soluble"
calcium sulphate reaching a maximum at 11.5% C$\overline{S}$H$_2$. The
optimisation of lime, sulphate, accelerator and retarder
levels with the correct choice of raw materials can lead to
2 hour strengths in excess of 2 MPa at a w:s ratio 2.5:1.

2.4 Water content of mine cements
Many claims have been made about the remarkably high water
contents of ettringite-based mine cements. Thus Nixon and
Mills (1981) stated that type II cement contains 85-90%
water by volume, Beale and Viles (1982) that type III cement
contains 91%, and Long et al. (1987) that type IV contains
up to 92%. It is important to understand the basis of these
claims, as they are critical to comparisons of cost and
strength development. In general, for a given system, the
higher the water content, the lower the cost and the lower
the strength.
    Consider a type III cement, which is made from two
slurries each containing 182 kg of solids at a w:s ratio of
2.5 to produce a pack of 1 m^3 in volume (Beale and Viles,
1982). The water content (910 kg) is determined by the w:s
ratio and its percentage is

$$\frac{910 \times 100}{1274} = 71.4\% \text{ by weight}$$

The stoichiometric reaction based on calcium hydroxide and
anhydrite, rather than calcium oxide and gypsum, is given by
equation (2a):

$$\underset{158.04}{CA} + \underset{408.42}{3\ C\overline{S}} + \underset{148.19}{2\ CH} + \underset{540.48}{30\ H} = \underset{1255.13}{C_6A\overline{S}_3H_{32}} \quad .....(2a)$$

and requires a w:s ratio of only 0.756. At a w:s ratio of
2.5, the amount of water added to 714.65 kg of solids is
1786.63 kg leading to a product containing 1255.13 kg of
ettringite and 1246.15 kg of water. The volume of the product is then (1255.13/1730) + (1246.15/1000) = 1.97166 m^3,
hence % water (by volume) = $\dfrac{1786.63 \times 100}{1000 \times 1.97166}$ = 90.62%

This is not the water content of the product, but the water
consumed in forming the product. The claims that type III
and IV cements can accommodate over 90% water by volume
suggest that they follow closely the ideal stoichiometries
(given by equation (2a) for a type III cement).

2.5 Experimental results

A modern type III cement consists of two slurries that will
remain fluid at mine temperatures for up to three days.
The initial strength development of such a system after mixing the slurries, as measured with a penetrometer, is shown
in Fig. 2 and it can be seen that the deshuttering strength
is reached after only 9 minutes. The compressive strength
reaches 2 MPa within 2h and 4 MPa after 24h (Fig. 3). The
influence of the lime content on the compressive strength
development of a type III cement is shown in Fig. 4. Incorporation of lime leads to an increase in strength up to a
maximum value, after which further addition results in a
rapid decrease in early strength. The optimum range is
relatively narrow and varies with the chemical and physical
characteristics of the raw materials and their relative
proportions, especially the ratio of HAC:C$\overline{S}$.

X-ray diffraction (XRD) patterns of a type III cement
hydrated for 7 days showed very strong peaks due to the
presence of ettringite. On drying overnight there was
little change in the pattern due to ettringite, but peaks
due to gypsum appeared, as the excess gypsum crystallised
from solution. There was no evidence for the presence of

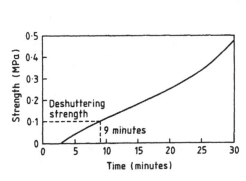

Fig. 2. Early strength development of a type
III cement

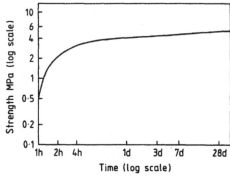

Fig. 3. Strength development
of a type III cement
up to 28 days

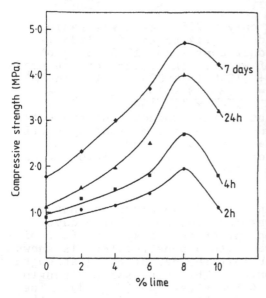

Fig. 4. Effect of lime on the compressive strength development of a type III cement

Fig. 5. SEM of ettringite needles in a hydrated type III cement

calcium hydroxide, but a few, weak, unidentified peaks were observed. Samples hydrated for 28 days were very similar to those hydrated for 7 days.

The XRD pattern of a type II cement hydrated for 7 days also had peaks due to ettringite, but these were weaker than those from the type III cements. The gypsum peaks were so weak that its presence could not be confirmed with certainty. Peaks attributable to calcium hydroxide were present, as also were weak peaks due to calcium carbonate and some form of AFm, perhaps the carbonated form of C_4AH_{13}. Little change was observed after hydration for 28 days.

Differential thermal analysis (DTA) curves were obtained on these samples and the results correlated well with those from XRD. Strong endotherms attributed to ettringite were observed at 140°C from the type III cements, whereas the type II cement had a weaker endotherm at 128°C. The XRD and DTA results clearly indicate that type II cement forms less ettringite than type III cement. Type II cement also gave endothermic peaks at 185°C and 472°C, attributed to the presence of an AFm phase and calcium carbonate, respectively. Neither of these peaks was observed in the DTA curves of type III cements, but a small endotherm at 270°C was observed which may indicate the presence of some AH_3. No other peaks were observed, not even any evidence for the presence of gypsum as a shoulder(s) on the high temperature side of the ettringite peak.

Scanning electron microscopy (SEM) studies were carried

5 μm

Fig. 6.   SEM of a type III cement with relatively low
         strength (mix A)

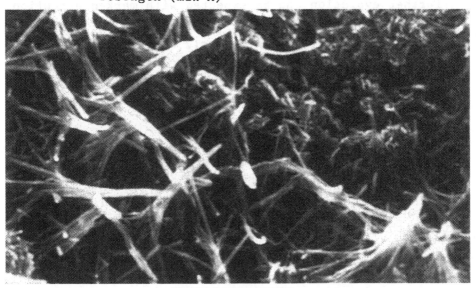

5 μm

Fig. 7.   SEM of a type III cement with the optimum
         composition for strength development (mix B)

out at the University of Oxford on two type III cement mixes exhibiting a wide difference in compressive strength. Mix A had a maximum strength about half that of mix B, but they gave closely similar XRD and DTA patterns. The striking hexagonal morphology of the ettringite needles found in mix A can be seen in Fig. 5 and 6, whereas the needles formed in mix B were much shorter, as shown in Fig. 7, taken at the same magnification as that in Fig. 6. The observed difference in strength between the two samples must be associated with this difference in morphology; the smaller needles giving the higher strength product. Mehta and Lesnikoff (1973) reported that the dimensions of ettringite crystals formed in a medium highly saturated with lime were very small, whereas those formed when lime was excluded were thicker and six times longer.

3. Conclusions

There has been a gradual evolution in the development of mine cements based on ettringite formation, leading to the high performance products available today which are extensively used in British coal mines. The use of coal fines and mine waste as aggregate has been replaced by the incorporation of large amounts of water (about 71% by weight and about 90% by volume); in effect water has become the aggregate material.

Further evolution can be expected with marked attention being given to the ability of the product reliably to maintain a long pump life, coupled with rapid setting and development of relatively high early strength for the very high w:s ratio used. In addition, commercial considerations will continue to be a major factor with regard to raw material costs and yield.

4. Acknowledgments

We are grateful to Mr I. Richardson of the University of Oxford for taking the scanning electron micrographs, to Mr D.R. Payne and Mr J.D. Cox of the University of Sheffield for help with the XRD and DTA study, and for commenting on the manuscript, respectively, and to Dr M.J. Blades of Pozament Ltd. for helpful discussions and permission to publish this paper.

5. References

Beale, J. and Viles, R.F. (1982) **U K patent**, 2,123,808 B.
Berman, H.A. and Newman, E.S. (1960) **Proc. 4th. Intl. Conf. Chem. Cem., Washington**, 247-257.
Bexon, R. (1985) **Paper presented to the Inst. of Mining**

Engineers on 15 May, 1985.

Budnikoff, P.P. (1960) **Proc. 4th. Intl. Conf. Chem. Cem.**, Washington, 469-477.

Galer, R.E. and Webb, P.C. (1983) **UK patent, 2,150,130 A.**

Kalousek, G.L. (1973) **Klein Symp. on Expansive Cement Concretes,** A.C.I. publicn. SP-38, 1-19.

Kellet, W.H. and Mills, P.S. (1980 a) **UK patent, 2,033,367 A.**

Kellet, W.H. and Mills, P.S. (1980 b) **UK patent, 2,058,037 A.**

Kesler, C.E. et al. (1970) **A.C.I. Journal,** 583-610.

Lea, F.M. (1970) **The Chemistry of Cement and Concrete,** 3rd. ed., Edward Arnold, London.

Long, G.R., Longman, P.A. and Gartshore, G.C. (1987) **Proc. 9th Intl. Conf. Cement Microscopy,** Reno, Nevada, USA, 236-246.

Longman, P.A. and Drew, N.H. (1985) **U K patent, 2,159,512 A**

Mehta, P.K. (1973) **J. Amer. Ceram. Soc.,** 56, 315-319.

Mehta, P.K. and Lesnikoff, G. (1973) **Klein Symp. on Expansive Cement Concretes,** A.C.I. publicn. SP-38, 89-105.

Midgley, H.G. and Pettifer, K. (1971) **Cem. Concr. Res.,** 1, 101-104.

Mikhailov, V.V. (1960) **Proc. 4th. Intl. Conf. Chem. Cem.,** Washington, 927-955.

Mikhailov, V.V. (1973) **Klein Symp. on Expansive Cement Concretes,** A.C.I. publicn., SP-38, 415-482.

Mills, P.S. (1984) **Colliery Guardian,** 232, 308-310.

Mills, P.S. (1988) **Paper presented to the Inst. of Mining and Mechanical Engineers** on 8 Nov. 1988.

Nixon, D.W. and Mills, P.S. (1981) **Mining Engineer,** 234, 645-652.

Robson, T.D. (1962) **High Alumina Cements and Concretes,** Contractors Record, London, 126-127.

Taylor, H.F.W. (1964) **The Chemistry of Cements,** Academic Press, London.

Whittaker, B.N., Singh, R.N. and Whittaker, P.E. (1980) **Univ. of Nottingham Mining Dept. Mag.,** 32, 1-15.

Whittaker, B.N. and Titley, H. (1971) **Colliery Guardian,** 470-476.

PART SEVEN
MISCELLANEOUS

25 EFFECTS OF AUTOCLAVING ON THE STRENGTH OF HARDENED CALCIUM ALUMINATE CEMENTS

A. SARANDILY
Jubb & Partners, Consulting Civil & Structural Engineers,
London, UK
R. BAGGOTT
Department of Civil Engineering, University of Salford,
Salford, UK

Abstract
The flexural strength of autoclaved calcium aluminate cements has been compared with room temperature cured cements. It has been shown that, prior to any loss of strength in room temperature cured cements due to conversion, significant microcracking can occur during fracture. The microcracking can be eliminated by elevated temperature hydration at 180°C, ie. autoclaving, the resulting material having significantly greater strength than room temperature cured material. Both room temperature cured and autoclaved cements were shown to behave as Griffith solids, with the latter exhibiting flexural strengths of up to 70 N/mm^2.
Keywords: Calcium Aluminates, Autoclaving, Flexural Strength, Griffith Cracks, Very High Strength.

1 Introduction

A major disadvantage of the ambient temperature hydration of high alumina cements is the metastable nature of their reaction products (Neville et al 1963, Lea 1970) since their subsequent conversion can lead to unacceptable strength losses. Slightly elevated temperature hydration increases the rate of conversion but can also lead to significant strength reductions (Robson 1962, Midgley 1967, French et al 1971, Neville 1976 and Collins and Gutt 1988). Addition of calcium carbonate is one method of overcoming the problems associated with conversion (Piasta et al 1989) as are combinations of HAC and pozzolanic materials, particularly ground granulated blastfurnace slag which is being developed commercially (Brecam). An alternative method of avoiding the problems of conversion is to hydrate at elevated temperatures and saturated steam pressures ie. autoclave (Sarandily 1986).

Very little work has been reported on the effects of higher temperature hydration in the 100-200°C range (Roy et al 1978) or the benefits that are likely to arise. Studies of the effect of pressure and humidity on the conversion reactions (Midgley et al 1975) concluded that an increase in pressure and humidity increases the rate of conversion. Reference to the literature on the lime-alumina-water system indicates that stable reaction products are formed in the temperature range 150-215°C (Peppler et al 1954) and 100-1000°C with under water pressures of 3000 atmospheres (Majumdar et al 1956), but provides no information on their mechanical properties. This contrasts with the wealth of information on the elevated temperature hydation of the lime-silica-water system (Menzel 1934, Gundlach and Ohnemuller 1967, Mindess 1970, Crennan et al 1972).

This paper presents the results of a preliminary investigation to identify whether any mechanical properties related phenomena occurred as a result of autoclaving that were worthy of more detailed study. The particular objectives were to determine whether any improvements to flexural strength resulted from autoclaving high alumina cements, to what extent very high strengths could be achieved and whether the materials behaved as Griffith solids (Griffith 1920).

2   Experimental procedure

Three types of calcium aluminate were used, HAC, Secar 51 and Secar 71. Uniform pastes were prepared in a planetary mixer following a standard procedure. The pastes were either cast directly into steel moulds, and subjected to vibration under vacuum or transferred to a cylindrical steel die and pressed to a lower water-cement ratio. The initial water-cement ratio was 0.30 and the final pressed water-cement ratio achieved was 0.14. The material was allowed to harden for a 24h period, the demoulded cast specimens were 120 x 20 x 10mm and similar sized samples were cut from the pressed discs. Half of each set of specimens were subjected to an autoclave cycle comprising two hours build up to 180°C, eight hours dwell time in saturated steam at this temperature and eight hours blow-down, the remaining specimens were cured under water at 20°C for 21 days.

A final light dressing with a diamond grinding wheel ensured parallel sided test surfaces and freedom from surface defects.

The flexural strength was determined in centre point bending across a 100mm span. The maximum dimension of the largest defect in the tensile part of the fracture surface was measured with a vernier micrometer eyepiece. In addition the fracture surfaces were examined by stereoscan microscopy.

An exploratory DTA analysis was carried out on the Secar 51 material.

3   Results

3.1 Flexural strength
Figure 1 illustrates the maximum flexural strength of the various types of material, comparing autoclaved values with room temperature cured values. It indicates that in all cases the autoclaved material is significantly stronger than the material cured at room temperature.

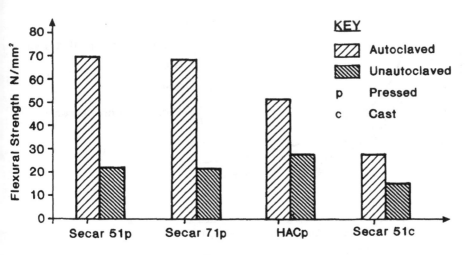

Figure 1   Ultimate flexural strength for different curing
conditions (N/mm²)

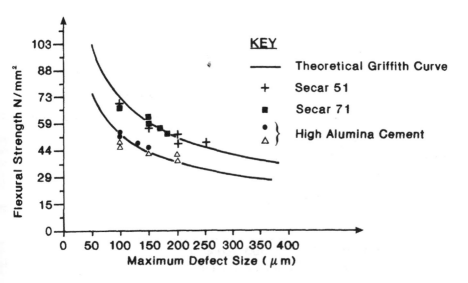

Figure 2   The relationship between flexural strength
and maximum defect size of pressed
autoclaved high alumina cement paste.

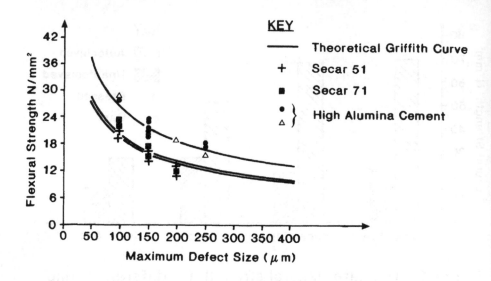

Figure 3   The relationship between flexural strength
and maximum defect size of pressed room
temperature cured high alumina cement.

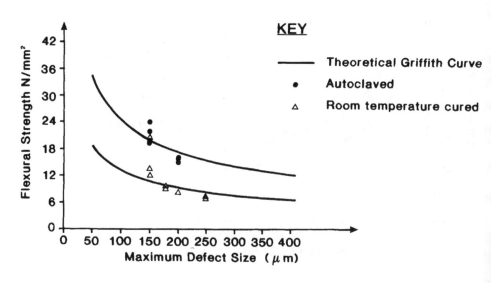

Figure 4   The relationship between flexural strength and
maximum defect size of cast secar 51 at water-
cement ratio of 0.30 for different curing conditions.

Figure 2 shows the flexural strengths of the three types of pressed aluminous cement beams after autoclaving, plotted against the nominal diameter of the maximum pore or defect in the outermost third of the tensile fracture surface and Figure 3 shows the same relationships for room temperature cured material.

If a comparison is made at the 150 micron defect size, autoclaved HAC is 100% stronger than 21 day water cured material and the two Secar mixes are 370% stronger. It is interesting to note that the relative strengths of HAC and the Secar mixes are reversed as a result of autoclaving.

The results of tests on 0.3 w/c ratio cast Secar 51, autoclaved and room temperature cured, are shown on Figure 4. The same trends are shown as with pressed material although at lower overall strength levels. In this figure the displacement of the two sets of points provides a qualitative measure of the improvement caused by autoclaving throughout the defect size range.

3.2 Microstructure

Figures 5-8 illustrate typical microstructures of both autoclaved and room temperature cured materials. The characteristic feature of the room temperature cured materials is the network cracking over the whole fracture surface observed in specimens cured for 1 day and 21 days. This was the case for both high strength and low strength materials. In the case of the autoclaved material the majority of the fracture surfaces were completely free from cracks, occasional specimens had separate cracks but in no instance was the network crack distribution of the room temperature material observed.

3.3 Differential thermal analysis

Differential thermal analyses were made on the Secar 51 material. The results, Figure 9, indicate that the effect of autoclaving is to produce significant amounts of C_3AH_6 compared to the composition after the first 24h. The autoclaved composition is also compared with the 21 day room temperature cured composition and there appears to be evidence that the predominant component at 21 days is not C_3AH_6 ie. the peaks are at 300°C and 325°C for room temperature cured and autoclaved materials respectively. The implication here is that autoclaving produces a higher density form of calcium aluminate hydrate.

4 Discussion

The first objective of the investigation was to determine whether autoclaving improved the flexural strength of calcium aluminate cements and this has been clearly shown. Very considerable improvements were observed and some microscopic evidence obtained as to a possible contributory mechanism for the strengthening. This is thought to be related to the network cracking which occurred as part of the fracture process in room temperature cured material. The network cracks are considered to be indicative of internal 'planes' of weakness in the original material which act as sources for both

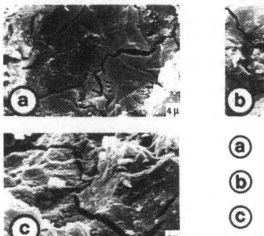

(a) = Pressed HAC:
w/c = 0.14

(b) = Pressed Secar 51:
w/c = 0.14

(c) = Pressed Secar 71:
w/c = 0.14

Figure 5   Fracture surfaces of room temperature cured
material – 1 day

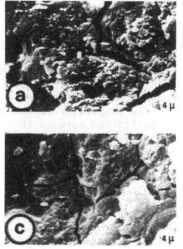

(a) = Pressed HAC:
w/c = 0.14

(b) = Pressed Secar 51:
w/c = 0.14

(c) = Pressed Secar 71:
w/c = 0.14

Figure 6   Fracture surfaces of room temperature cured
material – 24 days.

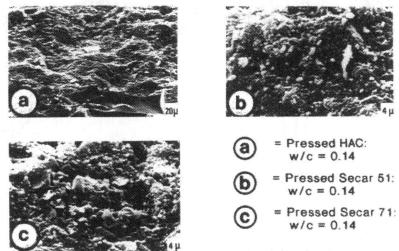

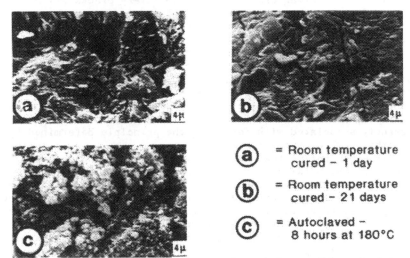

(a)	= Pressed HAC: w/c = 0.14
(b)	= Pressed Secar 51: w/c = 0.14
(c)	= Pressed Secar 71: w/c = 0.14

Figure 7   Fracture surfaces of autoclaved material
– 8 hours at 180°C.

(a)	= Room temperature cured – 1 day
(b)	= Room temperature cured – 21 days
(c)	= Autoclaved – 8 hours at 180°C

Figure 8   Fracture surfaces of cast Secar 51 material
– w/c = 0.3.

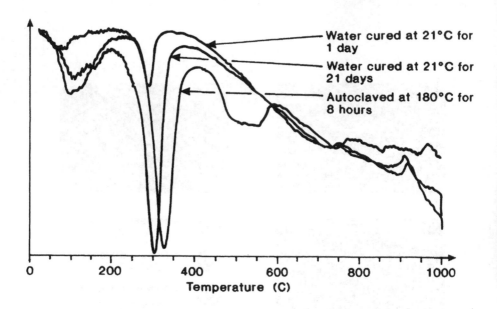

Figure 9   DTA curves for Secar 51 after different curing regimes.

the primary separating crack and the subsidiary cracks which occur as the main fracture propagates through the specimen. In the case of autoclaved material the areas of weakness are eliminated by the increased amount of reacted material and greater stresses are required to initiate crack propagation. It is also possible that a stronger phase is formed during autoclaving, the preliminary DTA results indicate the preponderance of C_3AH_6, but further work is necessary to evaluate this and other chemical factors.

The second and third objectives of the work were closely linked in so far as the achievement of very high strengths in flexure is frequently associated with following the principle determined by Griffiths.

It is worth noting that relatively high levels of flexural strength can be obtained with the 'cracked' room temperature material. The highest values obtained for Secar 51, ie. 25 N/mm², which although modest compared with polymer modified macro-defect free cements, (Birchall et al) are greater than many fibre reinforced cements. Two effects contribute to the increase in strength from that of the conventional cast material, both of which are associated with the pressing process.

Firstly the effect of reducing the water-cement ratio from 0.30 to 0.14 was to increase strength by 50% producing values of 14N/mm². In Griffith's terms this would be as a result of increasing the modulus of elasticity of the paste and/or increasing its surface energy. The second effect, ie. reducing the size of the entrapped air voids, which are the major source of maximum defect in cast material, further increased the strength to 25 N/mm². The inherent strength limiting defect remaining being the areas of weakness already referred to.

The effect of autoclaving, ie. eliminating or at least reducing the size of the areas of weakness, is then superimposed upon the same two phenomena that occur at room temperature, leading, in the case of Secar 51, to a further doubling of strength. In the case of HAC this combination of processes results in flexural strengths of up to 70 N/mm², thus producing hydrated calcium aluminates of substantially greater strength.

In order to illustrate the extent to which Griffith's theory is applicable, curves based on the theory are plotted through the experimental data as follows:

The Griffith model defines strength (σ) in terms of only three parameters ie. the size of the largest crack or equivalent feature (2c), the surface energy (γ) and the elastic modulus (E):

$$\sigma = \left[\frac{E\gamma}{2\pi c} \right]^{\frac{1}{2}}$$

For a constant surface energy and elastic modulus, stress should be proportional to the inverse square root of the equivalent defect diameter, Figures 2-4 show that reasonable agreement was obtained.

5 Conclusions

a) Autoclaving HAC, Secar 51 and Secar 71 cement pastes produces a number of significant effects compared with room temperature curing. Specifically flexural strength is increased, microcracking is suppressed and the composition of the calcium aluminate influences the magnitude of the strength increase via the formation of different distributions of phases in the final material.
b) Room temperature cured calcium aluminate and autoclaved calcium aluminate cements behave as Griffith solids.
c) Very high strength versions of the hydrated cements can be produced, with flexural strength values up to 70 N/mm², the basis of which can be explained by Griffith's theory in terms of reduced maximum defect size, increased elastic modulus and increased surface energy.

6 References

Brecam BRE. England.
Birchall, J.D., Howard, A.J. and Kendall, K (1981) Flexural strength and porosity of cements. Nature, 289, 388-390.
Collins, R.J. and Gutt, W., (1988) Research on long-term properties of high alumina cement concrete. Magazine of Concrete Research, 40 195-208.

Crennan, J.M., Dyczek, J.R.L. and Taylor, H.F.W. (1972) Quantitative phase compositions of autoclaved cement-quartz cubes. Cement and Concrete Research, 2, 277-289.

French, P.J., Montgomery, R.G.J. and Robson, T.D., (1971) High concrete strength within the hour. Concrete, 5, 253-258.

Griffith, A.A., (1920) The phenomena of rupture and flow in solids. Phil. Trans. R. Soc. Lond., A221, 163-198.

Gundlach, V.H. and Ohnemuller, W., (1967) Investigations on the influence of compressive strength on autoclaved calcium silicate products. Tonindustrie-Zeitung und Keramische Rundschau, 91, 315-330.

Lea, F.M., (1970) The Chemistry of Cement and Concrete. Arnold, London.

Majumdar, A. and Roy, R. (1956) The System CaO - Al_2O_3 - H_2O. Journal of the American Ceramic Society, 39, 434-442.

Menzel, C.A., (1936) Strength and volume change of steam cured portland cement, mortar and concrete, J. of American Concrete Institute, 1, 125-148.

Midgley, H.G., (1967) The mineralogy of set high alumina cement. Transactions of the British Ceramic Society, 66, 161-187.

Midgley, H.G. and Midgley, A., (1975) The conversion of high alumina cement. Magazine of Concrete Research, 27, 59-77.

Mindess, S., (1970) Relation between the compressive strength and porosity of autoclaved calcium silicate hydrates. J. of American Ceramic Society, 53, 621-624.

Neville, A.M., (1963) A study of deterioration of structural concrete made with high alumina cement. Proceedings of the Institution of Civil Engineers, 25, 287-324.

Neville, A.M., Wainwright, P. and Halsted P., (1976) HAC Concrete. Wiley.

Peppler, R.B. and Wells, L.S., (1954) The system of lime, alumina and water from 50° to 250°C. Journal of Research of the National Bureau of Standards, 52, 75-92.

Piasta, J., Sawicz, Z. and Piasta, W.G., (1989) Durability of high alumina cement pastes with mineral additions in water sulphate environment. Cement and Concrete Research, 19, 103-113.

Robson, T.D., (1962) High alumina cements and concretes. London, Contractors Record Ltd., London.

Roy, D.M. et al, (1978) Hydrated calcium aluminosilicate cements for hydrothermal bonding. Cement and Concrete Research, 8, 509-512.

Sarandily, A., (1986) Ph.D. Thesis, Department of Civil Engineering, University of Salford, Salford, England.

26 EFFECT OF TEMPERATURE RISE ON PROPERTIES OF HIGH ALUMINA CEMENT GROUT

S.A. JEFFERIS
Queen Mary and Westfield College, University of London, UK
R.J. MANGABHAI
TH Technology Ltd, R&D Centre, Rickmansworth, UK

Abstract
High alumina cement grouts of 0.4 water/cement ratio were cast into moulds made from steel and pvc and of various dimensions. The temperature rise of the grouts in the moulds was monitored. For moulds initially at room temperature and without cooling peak temperatures of over 100°C were observed. For steel moulds immersed in a water bath at 8°C a temperature rise of 23°C was observed. However, for pvc moulds similarly immersed in water the peak temperature was over 135°C demonstrating the substantial insulating properties offered by even a thin wall of pvc.
The results of strength measurements on samples cast in steel moulds and immersed in water at 8°C showed the expected rapid strength development. In contrast uncooled samples showed considerable reduction in strength.
These results show that for HAC grouts which necessarily have a high cement content as there is no dilution by aggregate the hydration exotherm can be very severe. Thus for test specimens and also when using HAC grouts the exotherm of the grout must be carefully considered. In general, cooling will be necessary unless the section to be grouted is very thin. Insulation provided by plastics etc. may be sufficient to prevent satisfactory temperature control.
Keywords: High alumina cement, Temperature, Peak temperature, Moulds, Strength, DTA.

1 Introduction

High alumina cement grout has been used successfully on North Sea Oil Platforms, typically for grouting the annular space between anchor piles and sleeves on structures. This form of structural connection has been used since 1969 for situations where high early strength is required (Ciment Special, 1980).

However, there has been some concern about the temperature rise of the grout during hydration. This temperature rise may be sufficient to cause loss of strength in addition to conversion.

This paper reports the findings of a short study of the temperature rise of grout specimens when subjected to two different external temperature regimes. These regimes were: (i) grout moulds immersed in water immediately after filling, water maintained at 8°C (ii) grout moulds left in free air, room temperature 15-20°C.

2 Experimental

High alumina cement (HAC) Fondu complying to BS 915 obtained from Lafarge Special Cement Co Ltd (1989) was used in the investigation. The chemical composition and surface area are given in Table 1.

Table 1. Chemical composition and surface area of HAC.

% Oxide composition

CaO :38.53, SiO_2 :4.14, Al_2O_3 :38.67, Fe_2O_3 :10.43, FeO :5.63,

SO_3 :0.183, MgO :0.02, Na_2O :0.06, K_2O :0.036, TiO_2 :1.98, CO_2 :0.42

Specific surface area : 250 m^2/kg

The grouts for the first series of tests were prepared in a Colcrete SD4 high shear mixer. The procedure was as follows: 30 litres of water were poured into the mixer, 75 kg of the cement were then added over a period of two minutes. Mixing was continued for a further 3 minutes making a total mixing time of 5 minutes.

The grout was then pumped from the mixer and poured into moulds. Three types of mould were used, 150 mm diameter by 300 mm high steel moulds, 100 mm steel cube moulds and 125 mm diameter by 460 mm high pvc moulds with a wall thickness of the order of 5 mm. After filling, a thermocouple was inserted at the centre of each mould and the moulds either placed in water at 8°C or left in free air. The output from each thermocouple was logged using a computer controlled data logger at 5 minute intervals for a period of 72 hours.

Type T thermocouples obtained from RS Components were used for these temperature measurements.

For strength measurements 100 mm cubes (without thermocouples) were cast in steel moulds and initially either immersed in water at 8°C or left in free air. The cubes were demoulded at 24 hours and cured under water at either 8°C or room temperature as appropriate.

3 Results and Discussion

3.1 Peak temperatures

The temperature-time profiles for the grouts cured at 8°C and at room temperature are shown in Figures 1a, b, 2a, b, and 3a, b. For all the samples cured at room temperature there was an initial period of cooling (the grout after mixing was always significantly above room temperature) followed by a plateau of relatively constant temperature extending to about 4 hours. This was followed by a rapid temperature rise peaking at about 5 hours and reaching 100°C for the 150 mm

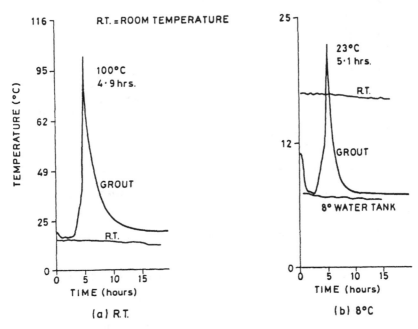

Figure 1. Temperature rise in HAC grout (150 mm dia cylinder).

cylinder, 129°C for the 100 mm cube and 134°C for the 125 mm pvc cylinder. The temperature then fell almost equally rapidly to return to room temperature at around 10 hours from mixing. The temperature then remained constant and logging was stopped at 72 hours. For clarity only the first 20 hours of temperature data are plotted in the Figures.

The results for the samples immersed in water at 8°C showed similar profiles as regards time but the peak temperatures were much reduced for the two specimens in steel moulds (showing peak temperatures of 23°C and 11°C for the 150 and 100 mm moulds respectively). In contrast the pvc tube specimen showed a peak temperature of 107°C. Although the tube had a wall thickness of the order of only 5 mm the pvc clearly represented a substantial resistance to thermal conduction (the thermal conductivity of most plastics is very substantially lower than that of most metals). Also the thin tube will add little to the thermal capacity of the system.

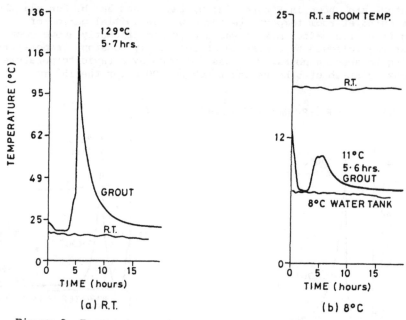

Figure 2. Temperature rise in HAC grout (100 mm cube).

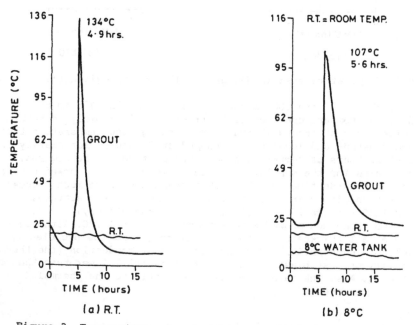

Figure 3. Temperature rise in HAC grout (125 mm pvc tube).

It is interesting to note that for the mould immersed in water the 150mm cylinder showed a somewhat higher temperature rise (peak temperature 23°C) than the 100 mm cube (peak temperature 11°C) as would be expected from consideration of the average radius of the specimen and thus the rate of heat dissipation. In contrast for the specimens cured in air the 100 mm cube shows a higher peak temperature (129°C) than 150 mm cylinder (100°C). The reasons for this are not obvious but may relate to the relative thermal capacities and thermal resistances in the two systems.

In a second series of tests to confirm the influence of the pvc four 100 mm cubes were prepared. Two of these cubes were cast in standard steel moulds and the other two in specially made pvc moulds. One pair of cubes (one cube cast in steel and one cast in pvc) was kept in air and the other was immersed in water which was initially at room temperature. The grout for these tests was prepared in a Hobart AE 125 mixer. The procedure was as follows: 2.4 litres of water were placed in the bowl of the mixer and the beater switched on at speed 1, 6 kg of HAC were then added over a period of 1 minute and mixing continued at speed 2 for a further 4 minutes. The Hobart mixer produces much more limited shearing action than the Colcrete SD4 mixer and thus the resulting grouts could be expected to show a slightly slower hydration reaction.

Figures 4a and b show the results of these tests. For the steel mould in air the peak temperature was 110°C.

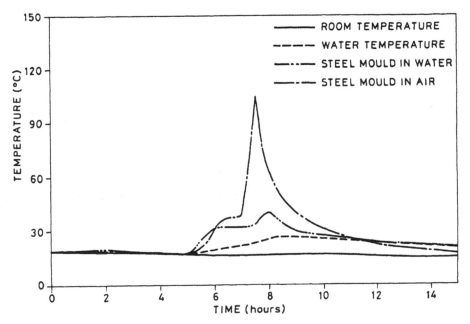

Figure 4a. Temperature rise in HAC grout (100 mm cube) steel moulds.

The plastic mould in air showed a peak temperature in air of 120°C thus demonstrating that major thermal resistance was from the mould wall to the surrounding air and the resistance of the pvc (or steel) itself was of relatively more limited significance. In contrast for the water immersed moulds the peak temperatures were 45°C and 137°C for the steel and pvc moulds respectively. Thus confirming the relatively great thermal resistance of the pvc in comparison with that of the mould wall/water interface (as distinct from the wall/air interface of the air tests). It is not obvious why the pvc mould immersed in water showed a higher peak temperature than that held in air. However, as both steel and pvc moulds were immersed in the same bath it is possible that the heat released by the steel mould slightly influenced the grout in the pvc mould. The water temperature was monitored and showed a peak value of 30°C.

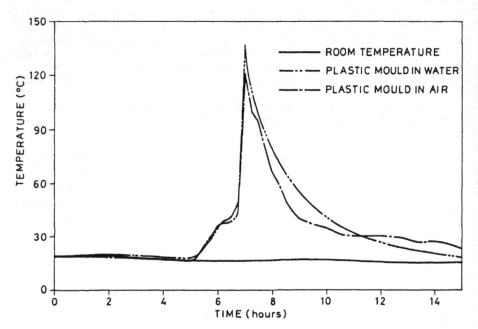

Figure 4b. Temperature rise in HAC grout (100 mm cube) plastic moulds.

3.2 Strength development

Figure 5 shows the results of the cube strength tests. For the cubes immersed in water at 8°C the results show rapid strength development which reaches 80 N/mm² at 30 days. In contrast the samples kept in air during the hydration exotherm show very low strength and no significant improvement with age. This low strength is due to the excessive exotherm developed in the specimens and due to conversion. The samples kept in air showed the brown colouration typical of conversion whereas the 8°C samples showed no colour change.

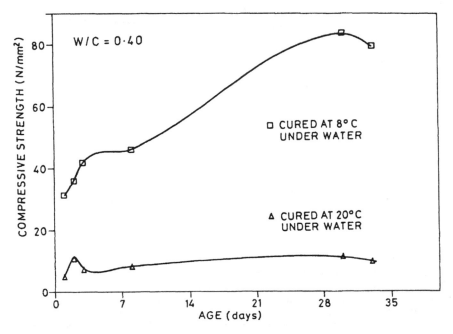

Figure 5. Strength development of HAC grout.

3.3 Differential Thermal Analysis

Exploratory Differential thermal analyses were carried out using a Stanton Redcroft DTA 673-4 with a heating rate of 10°C/min and flowing nitrogen. Figure 6 shows that a considerable amount of gel but very little AH_3 were present after 24 hours for the 8°C immersed sample. In contrast the sample kept in air shows little gel but substantial peaks for C_3AH_6 and AH_3. The results confirm that conversion had occurred in the specimen kept in air but not in the water immersed specimen. In both cases there is no evidence of the presence of C_2AH_8 and CAH_{10} from the DTA results. SEM and XRD analysis were not carried out on these samples and should be investigated.

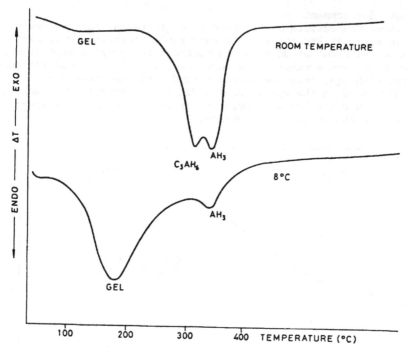

Figure 6. DTA of HAC grout after 24 hours of hydration.

4 Conclusions

The results of these simple tests show that the hydration exotherm of HAC grouts may lead to a peak temperature of over 100°C if the grout samples are kept in air. Excessive exotherms also may occur in water cooled grouts if plastic or other low thermal conductivity moulds or formwork are used. This temperature rise is sufficient to damage the grout so that the strength is drastically reduced. The damage is possibly mainly mechanical due to steam generation and thermal contraction on cooling but conversion also occurs.

Samples cast in steel moulds and immersed in water at 8°C showed much reduced peak temperatures and satisfactory strength development.

It should be noted that external water cooling can be effective only in relatively thin grout sections. In thick sections the insulating effect of the grout itself will be such that an excessive exotherm must occur at the centre. Thus there will be a maximum useable thickness of grout which will depend on the water/cement ratio and the initial temperature of the grout. Ordinary Portland cement grouts will also have a maximum useable thickness but this will be rather greater than for HAC. The effect of section thickness is being further investigated by computer simulation.

From the work it would seem that temperature matched curing may have a role for HAC grouts as well as Portland cement concretes. However, for HAC the role of temperature matched curing would be to ensure that the hydration exotherm was not so great as to damage the properties of the grout.

Finally it would seem that lightweight plastics moulds should not be used to prepare HAC grout cubes as they may overly insulate the grout and thus lead to excessive exotherms in the test cubes.

5 Acknowledgement

The authors gratefully acknowledge the financial support for this work provided by the Marine Technology Directorate Limited. The work forms part of a larger programme of work on the influence of early age thermal history on the properties of cementitious grouts.

The authors would also like to thank Dr. Charles Fentiman, Technical Manager of Lafarge Aluminous Cement Co. Ltd., for supplying the HAC (Fondu).

6 References

BS 915, Part 2 (1972). High alumina cement, British Standard Institution, London.

Ciment Special (1980). Lafarge Aluminous Cement Co. Ltd., Grays, Essex, U.K.

Lafarge Aluminous Cement Co. Ltd., (1989). England.

27 ACTIVATION OF HYDRAULIC PROPERTIES OF THE COMPOUND CaO.2Al$_2$O$_3$

T.W. SONG, S.H. CHOI, K.S. HAN
Inha University, Inchon, Korea

Abstract
Activated CaO.2Al$_2$O$_3$ was prepared by successive calcination, hydration and clinkering. Material prepared by this hydration-burning method has been found to be of superior hydration characteristic to that of CaO.Al$_2$O$_3$ synthesized by a conventional heating technique. Activated hydraulic properties of CA$_2$ depend largely on the hydration condition of the calcined material and the sintering temperature of the hydrates. CA$_2$-based clinker prepared by the hydration-burning technique has a large surface area and its microstructure and hydration characteristics have been compared with CaO.Al$_2$O$_3$ by microcalorimetry, X-ray diffraction, derivative thermogravimetric analysis and scanning electon microscopic observation. Heat evolution of CA$_2$ is much faster than that of CaO.Al$_2$O$_3$ at 30°C, and it is remarkable that large amounts of AH$_3$ are produced from the beginning of hydration even at ambient temperature.
Keywords: CaO.2Al$_2$O$_3$, Hydraulic properties, hydration-burning method, CaO.Al$_2$O$_3$, Surface area, BET, Heat of hydration, Hydration products, AH$_3$.

1 Introduction

The compound CaO.2Al$_2$O$_3$ contains the most amount of Al$_2$O$_3$ among the hydraulic calcium aluminates, but its cementing property is too weak to be applied for refractory use (Hallstedt 1990, Parker and Sharp, 1982). Adding small amounts of rapid setting materials such as C$_{12}$A$_7$ or CA and increasing the surface area of CA$_2$ by finer grinding were proposed for accelerating the slow hydraulicity of CA$_2$ sintered at high temperature by conventional method, however they were not recognized as a desirable method (Tseung and Carruthers (1963), Bobrov et al (1974), Buttler and Taylor (1959), Mehta (1964)).

Activated CA$_2$ compound with microstructure prepared in this study has rather superior hydraulicity to that of CA, which is the major cementitious component of conventional aluminous cements. The mixture of CaCO$_3$ and Al$_2$O$_3$ in the ratio of CA$_2$ composition was calcined to form a primary clinker at the temperature at which no free CaO can remain, and then hydrated hardened paste of the primary clinker was sintered to produce activated CA$_2$-based clinker. The hydration temperature of primary clinker and the sintering temperature of the final clinker exerted a great influence on the hydration characteristics of activated CA_2.

2 Experimental

The primary clinker of CA_2 was prepared from the mixture of analytical-reagent grade $CaCO_3$ and Al_2O_3 in the ratio of 1:2. The mixture was decarbonated, fired in pelletized form at 1300°C for 1 hour and ground to pass through a 74μm screen. The ground primary clinker was hydrated to form hardened paste with a water/clinker ratio of 0.7 at 35°C for 24 hours, then it was crushed to around 1 cm dia. mass and reburnt to 1250°C for activated CA_2-based clinker. Hence, it is called hydration-burning method. Analysis by X-ray diffraction showed that CA_2-based clinker consisted mainly of CA_2 compound, small amounts of CA and unreacted alumina. For comparison with experiment of new CA_2, CA was prepared from an equimolar mixture of $CaCO_3$ and Al_2O_3 by a conventional heating method (at 1450°C, three times).

Prepared CA_2-based clinker was ground to pass through a 47μm screen and the surface area was measured by the BET method and the heat evolution by microcalorimeter. The hydration products of CA_2 and CA hydrated with a w/c ratio of 0.5 at different temperature and time were detected by X-ray diffraction and derivative thermogravimetric analysis. The fracture surface of the hydrated hardened paste were observed by scanning electron microscope.

3 Results and discussion

Figure 1 shows the relationship between BET surface area and heating temperature for hydrates of the primary clinker which was hydrated with w/c = 0.7 at 10 and 40°C for 24 hours. The specific surface area of unheated and sintered hydrates at 1450°C were about 30,000 and 6000 cm²/g respectively without regard to hydration temperature, but at the temperature of intermediate stages, the specific surface area of hydrated sample at 40°C was much larger than that of the sample hydrated at 10°C. The specific surface area in the temperature range of 1250-1300°C, over which sufficient CA_2 was produced, was more than 40,000 cm²/g, but at the temperature above 1350°C, it was rapidly lowered and represented lower activation of the CA_2.

The large difference in the surface area between two samples hydrated at elevated and lower temperature is assumed to be due to the fact that the kinds and composition of the hydration products, sizes of their decomposed pores and reaction behaviour of the hydrated products were different in each case.

The rate of heat evolution from the hydration of activated CA_2-based clinker and CA are shown in Figure 2. A sharp peak at 2.5 hours is observed for activated CA_2-based clinker with a w/c = 0.5 at 30°C, whilst for CA this peak is broader and occurs after 13 hours. These results show that the hydraulicity of CA_2-based is highly activated.

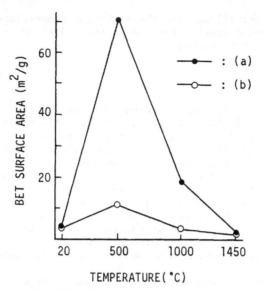

Figure 1. Relationship between BET surface area and heating
temperature for hydrates of primary clinker hydrated with
w/c = 0.7 for 24 hours at (a) 40°C and (b) 10°C.

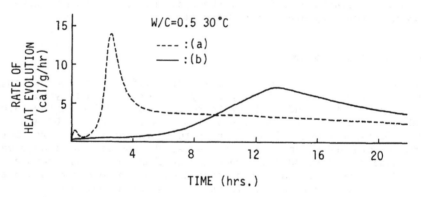

Figure 2. Rate of heat evolution vs time for (a) Activated CA_2-based
clinker and (b) CA.

Hydration products of the activated CA_2-based clinker and CA
hydrated with w/c = 0.5 at various temperatures were analysed by
X-ray diffraction and derivative thermogravimetry. Below 20°C alumina
gel, CAH_{10} and C_2AH_8 were favoured and above 30°C C_3AH_6 products on
hydration of both compounds. AH_3 was mainly produced from the very
beginning of the hydration of the activated CA_2-based clinker even
below 30°C, without being accompanied by C_3AH_6, but in the case of CA
hydration, considerable amount of C_3AH_6 was produced after 4 hours
hydration. Figures 3 and 4 show typical XRD and DTG for both
compounds hydrated at 30°C. AH_3 production is increased with

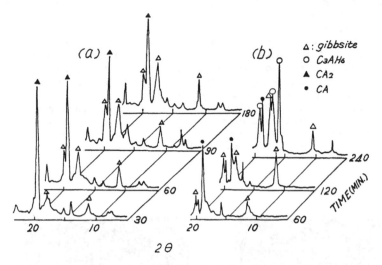

Figure 3. XRD patterns of (a) activated CA$_2$-based clinker and (b) CA hydrated for various periods at 30°C w/c = 0.5.

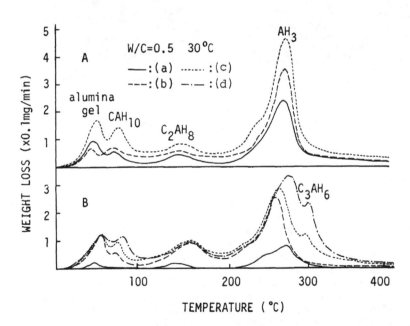

Figure 4. DTG curves of (A) activated CA$_2$-based clinker and (B) CA hydrated for various periods of time at 30°C, w/c = 0.5. (a) 1 hrs, (b) 2 hrs, (c) 6 hrs and (c) 12 hrs.

hydration temperature and the amount of AH_3 from activated CA_2 at 30°C after 6 hours of hydration measured by a semi-isothermogravimetric technique (El-Jazairi, 1977), was around 70% of the total products. The activated hydraulic characteristics of CA_2 prepared by the new method is mainly presumed to be due to the fact that AH_3 is produced from the beginning of the hydration.

Figure 5 shows scanning electron micrographs of the fracture surface of the hardened pastes of activated CA_2 and CA hydrated with w/c = 0.5 at 30°C for 120 minutes. Gel-like ambiguous crystals were observed for CA hydrate and gibbsite-like laminar crystals crystallized well in the hydrate of activated CA_2. It was recognized that the hydration of activated CA_2 is much developed than that of CA.

Figure 5. SEM micrographs of fracture surface of (A) activated
CA_2-basedd clinker and (B) CA hydrated with w/c = 0.50 at
30°C for 120 minutes

4 Conclusion

Activated CA_2-based clinker sintered at 1350°C by a hydration-burning method and ground to pass through a 74μm screen has more than 40,000 cm^2/g of specific surface area measured by the BET method.

At 30°C hydration, a sharp calorimeter peak occurred after 2.5 hours on the activated CA_2-based clinker, whilst for the CA a broader peak occurs after 13 hours.

Below 30°C hydration of CA, considerable amount of C_3AH_6 was produced after 4 hours of hydration, whilst for the activated CA_2-based clinker, AH_3 was mainly produced from the beginning of hydration without being accompanied by C_3AH_6.

It is recognized that the hardened paste hydrated from activated CA_2-based clinker for 120 minutes was well crystallized as gibbsite-like laminar crystal than the case of CA.

5 References

Bobrov, B.S., Zaldat, G.I., Zalizovski, E.V., (1974). **Izvakad. Nauk. SSSR. Neorg. Mater.**, 10, 2187.

Buttler, F.G., and Taylor, H.F.W. (1959). **J. Appl. Chem.** 20, 616.

El-Jazairi, B. (1977). **Thermochimica Acta**, 21, 381-389.

Hallstedt, Bengt (1990). **J. Am. Ceram. Soc.** 73 (1), 15-23.

Mehta, P.K. (1964). **Miner. Process.** 2, 16.

Parker, K.M. and Sharp, J.H. (1982). **Trans. Brit. Ceram. Soc.** 81 (2), 35-42.

Tseung, A.C.C. and Carruthers, K.G. (1963). **Trans. Brit. Ceramic. Soc.** 62, 305.

INDEX

This index has been compiled using the keywords supplied by the authors in conjunction with keywords provided by the Concrete Infortion Service of the British Cement Association. The assistance of Association in this is gratefully acknowledged. The numbers refer the opening page of the papers.

Accelerators 113, 127, 142, 167
Acids 230
Admixtures 113, 127, 142
Age 113
Air-entraining agents 142
Aluminium compounds 52, 81, 155, 259
Aluminium oxides 27
Ambient temperature 353, 363
Autoclaving 52, 353

Beams 353
Bending strengths 241, 353
Blastfurnace slag 259, 272, 294
Buildings 1
Burning temperature 27, 181

Calcium aluminates 17, 27, 41, 259, 335, 372
Calcium compounds 52, 127, 259
Calcium sulphate 320
Calorimetry 96, 155, 167, 259, 282, 363, 372
Carbonation 208
Cement clinker 27
Cement paste 41, 230, 241, 272, 282, 292, 353
Chemical analysis 65
Chemical kinetics 65, 155, 320
Chemical properties 1, 181, 230
Chlorides 127, 167
Citric acid 127
Codes of practice 1
Composition 17, 27, 41, 113, 127, 222, 241, 272, 282, 294, 320
Compressive strength 17, 113, 181, 208, 259, 272, 320, 335, 363, 3
Concrete 27, 241
Conductivity 65, 363
Construction 1
Conversion 1, 113, 181, 208, 272, 294
Corrosion 230
Cracking 353
Curing 96, 241, 294, 353

Differential scanning calorimetry 282
Differential thermal analysis 127, 167, 241, 272, 335, 353, 363, 3
Durability 181, 208, 222

Electrochemical effects 65
Electron microprobe analysis 282
Electron microscopy 335
Ettringite 208, 320, 335
Expansion properties 27, 208, 222, 241, 320, 335

Failure 1
Fillers 241
Flexural strength 241, 353
Fracture properties 353

Granulated slag 259, 272, 282
Groundwater 208
Grouts 363
Gypsum 208, 320

Hardening 96
Heat of hydration 52, 65, 167, 259, 372
High pressure steam curing 52, 96, 353
Hydration 17, 27, 41, 52, 65, 81, 96, 113, 127, 155, 167, 181,
          259, 272, 282, 294, 320, 335, 372
Hydroxides 127, 155

Image analysis 41
Ions 155
In situ concrete 208

Kinetics 65, 142, 320

Laboratory testing 208, 363
Lime 320
Limestone 241
Liquids 65
Lithium compounds 127, 155
Low cement castables 52

Magnesium compounds 222
Manufacturing 27, 181
Marine structures 142, 363
Metals 155
Microhardness 222
Microscopy 17, 41
Microstructure 17, 41, 127, 222, 282, 335, 353, 372
Mines 222, 335
Mixing water 142
Mixtures 96, 282
Mortars 113, 142, 241
Moulds 363

Nickel compounds 155
NMR 52, 81

OPC 208

Particle size 259
Phases   27, 41, 65, 96, 113, 222, 282, 294
Pores, pore size distribution   52, 96, 181
Porosity 230
Precast concrete 208
Prestressed concrete 1, 181

Quality control 181
Quartz 241

Reactions 52, 155, 208, 222, 230, 259, 272
Refractories 27, 52, 127
Resistance 1, 208, 241
Retarders 127, 167

Salts 230
Seawater 142, 167
Segregation 142
Setting 17, 65, 127, 142, 320
Silica fume 96, 294
Site testing 208
Slag 259, 272, 282
Sodium compounds 222
Soils 208
Solubility 230
Solutions 155, 208, 222, 241
Spectrum analysis 81
Stability 272
Structural properties 1
Sulphates and sulphate resistance 167, 208, 241, 335
Superplasticizers 113, 142
Surface area 52, 372

Temperature 41, 52, 65, 113, 142, 167, 230, 241, 259, 272, 294,
              353, 363
Test specimens 208, 222, 363
Thermal analysis 155, 259
Time 52, 222, 272, 294

Ultrasonic pulse velocity 113
Use 1

Washout 142
Waterproofing 142
Weight loss 230
Workability 113, 142

XRD 27, 52, 127, 208, 222, 241, 272, 282, 294, 320, 335, 363, 372